Molluscan Shellfish Depuration

W.S. Otwell
G.E. Rodrick
R.E. Martin

 CRC Press
Taylor & Francis Group
Boca Raton London New York

CRC Press is an imprint of the
Taylor & Francis Group, an informa business

First published 1991 by CRC Press
Taylor & Francis Group
6000 Broken Sound Parkway NW, Suite 300
Boca Raton, FL 33487-2742

Reissued 2018 by CRC Press

Library of Congress Cataloging-in-Publication Data

Molluscan shellfish depuration / editors, W. Steven Otwell, Gary E.
    Rodrick, Roy E. Martin.
        p.   cm.
    "Based, in part, on the proceedings of the First International
Molluscan Shellfish Depuration Conference, held November 5-8, 1989
in Orlando, Florida"—Foreword.
    Includes bibliographical references and index.
    ISBN 0-8493-4295-3
    1. Shellfish fisheries—Sanitation—Congresses. 2. Shellfish
fisheries—Sanitation—Standards—United States—Congresses.
    I. Otwell, W. Steven. II. Rodrick, Gary Eugene, 1943-
    III. Martin, Roy E. IV. International Molluscan Shellfish
    Depuration Conference (1st : 1989 : Orlando, Fla.)
RA602.S6M65 1991
664'.94—dc20   91-11909

A Library of Congress record exists under LC control number: 91011909

ISBN 13: 978-1-315-89571-0 (hbk)
ISBN 13: 978-1-351-07481-0 (ebk)

Visit the Taylor & Francis Web site at http://www.taylorandfrancis.com and the
CRC Press Web site at http://www.crcpress.com

FOREWORD

This text serves as a status report on shellfish depuration technology and the associated commercial and regulatory concerns evidenced as we approach the 21st century. The chapters are based, in part, on the proceedings of the First International Molluscan Shellfish Depuration Conference held November 5-8, 1989 in Orlando, Florida, USA. The intent of the Conference was to bring the concerns and advances to the attention of the industry and regulators. One principle objective was to review prior research and practice from which additional developments could be planned and judged. The justification and timing for this information realized increasing interstate and global shellfish commerce, emerging aquaculture in amounts and diversity unlike any prior decade, and increasing scrutiny for seafood product safety. Domestic shellfish handling and processing will come under greater food safety concern with anticipated seafood inspection legislation, issues noted in the recently released National Academy of Science study on Seafood Safety, recommendations of the National Committee on the Microbiology of Foods, recent decisions of the Interstate Shellfish Sanitation Conference, and from recommendations contained in the NMFS/HACCP reports on molluscan shellfish. Similar pressures will evolve in international commerce, particularly evident in attempts to harmonize concerns through the EEC (European Economic Community).

It is therefore appropriate and timely that the technology and potential of depuration be addressed in formal and permanent means. It is our opinion that the United States is lagging behind the rest of the world in support of this technology. It was for that reason that we invited other country participation in the conference to discuss technology development and transfer.

These published proceedings are the first of their kind, and we trust that future advances will build on this base.

W. Steven Otwell
Gary E. Rodrick
Roy E. Martin

ACKNOWLEDGEMENT

This text and supporting Conference were conceived jointly by the National Fisheries Institute and the University of Florida's Sea Grant Marine Advisory Program and Food Science and Human Nutrition Department. NFI secured the major portion of support via a federal Saltonstall-Kennedy Grant through the U.S. Department of Commerce, National Marine Fisheries Service. Further funds and assistance were provided by the National Coastal Resource Research and Development Institute (Newport, OR), the Gulf and South Atlantic Fisheries Development Foundation, Inc. (Tampa, FL), the Virginia Sea Grant College Program, and the Southeastern Fisheries Association (Tallahassee, FL). These supporters represent industry, government and consumer interest throughout the United States.

EDITORS

W. Steven Otwell, Ph.D., is a professor in the Food Science and Human Nutrition Department of the University of Florida, Institute of Food and Agricultural Sciences in Gainesville, FL, where he has served as the Extension Seafood Specialist since 1979. His formal training culminated in a Ph.D. in Food Science from North Carolina State University following an M.S. degree in Marine Sciences from the University of Virginia (VA Inst. Marine Science) and B.S. degree in Biology from the Virginia Military Institute. His chairmanship and lead editorial work with the First International Molluscan Shellfish Depuration Conference proceedings exemplifies his approach to addressing seafood product quality and safety issues. Through numerous applied research projects, demonstrations, manuals for commercial practice, workshops and lectures he has addressed a large variety of product quality and safety issues for different fish ad shellfish in production through processing and retail settings. He has just completed a committee assignment for the National Academy of Science's, Institute of Medicine, Food and Nutrition Board to draft a 1991 report on Seafood Safety. Since 1982 has served as the secretary for the Institute of Food Technology Seafood Product Technology Group; whereas he became chairman-elect in 1990. He has served as the Executive Director and annual TSFT conference proceedings Editor for the Tropical and Subtropical Fisheries Technological Society of the America's since 1986. He is on the editorial board for the new Journal for Aquatic Product Technology.

Gary E. Rodrick, Ph.D., is an associate professor in the Department of Food Science and Human Nutrition at the University of Florida. He received both his B.S. and M.S. degrees at Kansas State College in 1966 and 1967, respectively. In 1971, he received his Ph.D. in Zoology at the University of Oklahoma. He was awarded a N.I.H. post doctoral fellowship at the University of Massachusetts from 1971 to 1973. In 1973, he was a lecturer in biology and associate director of the Institute for Pathology at Lehigh University. In 1978, he joined the Department of Comprehensive Medicine as an associate professor and associate chairman in the College of Medicine at the University of South Florida. In 1987, he joined the Department of Food Science and Human Nutrition at the University of Florida where he has initiated research in shellfish depuration.

Dr. Rodrick is presently a member of the editorial committee for both the Journal of Food Protection and the Journal of Invertebrate Pathology. He has served on numerous national committees and has extensive international experience. He has published more than 100 research articles and has been awarded numerous research grants dealing with shellfish depuration. His current research interest relates to Vibrio changes in shellfish during UV and ozone assisted depuration.

Roy E. Martin, is Vice President for Science and Technology at the National Fisheries Institute, Arlington, VA. He is also the Executive Director of the Shellfish Institute of America and the National Blue Crab Industry Association. He received his training at both Valpariso and Northwestern Universities. Prior to assuming his present position he spent 17 years with Swift and Co., Chicago, Illinois, leaving there in 1972 as general manager of the Edible Protein Division. Prior to that position Mr. Martin worked on basic physical chemistry of proteins, amino acids and enzyme systems. Mr. Martin holds memberships in the American Chemical Society, American Oil Chemists Society, Institute of Food Technologists, American Fisheries Society, Association of Food and Drug Officials of the U.S., American Public Health Association, the Marine Technology Society, Atlantic Fisheries Technologists, Tropical and Subtropical Fisheries Technological Society of the Americas, Pacific Fisheries Technologists, American Association of Cereal Chemists, Washington Academy of Sciences and the New York Academy of Science, Fellow - American Institute of Chemists. In addition, he is an advisor to several Codex Aliminetarius Committee, editorial advisory board member of "Sea Grant Abstracts", "Journal of Aquatic Food Product Technology" and the "Journal of Muscle Foods". In 1980 Mr. Martin received the Earl P. McFee Award for excellence in the field of fishery technology development. Mr. Martin has published more than 50 papers and co-authored 4 books. His present interest in the subject of this book relates to his involvement with the Interstate Shellfish Sanitation Conference.

CONTRIBUTORS

Chuck M. Adams, Ph.D.
Food and Resource Economics Dept.
University of Florida
Gainesville, FL 32611

Atilla Alpbaz, Ph.D.
Fisheries College
Ege University
Bornova. Izmir, TURKEY

Robert E. Arnold
Marriott Corporation
International Headquarters
Marriott Drive
Washington, DC 20058

Peter A. Ayres, Ph.D.
97 Burns Rd.
Springwood, NSW 2777
Australia

Murat O. Balaban, Ph.D.
Food Science & Human Nutrition
  Department
University of Florida
Gainesville, FL 32611

Mariá José Prol Bao
Depuradora de Mariscos de Lorbe S.A.
(DERMARLOSA)
P.O.B. 2, Sada, 15160
La Coruña, Spain

Leon E. Beghian, Ph.D.
Biological Sciences
Dept. of Biology Radiation Laboratory
Olson Bldg.
University of Lowell
Lowell, MA 01854

Walter J. Blogoslawski, Ph.D.
U.S. Department of Commerce/NOAA
Northeast Fisheries Center
National Marine Fisheries Service
2l2 Rogers Avenue
Milford, CT 06460

Walter J. Canzonier, Ph.D.
Rutgers Shellfish Research Laboratory
and
Maurice River Oyster Culture Foudation,
  Inc.
Port Norris, NJ 08349

John D. Castell, Ph.D.
Department of Fisheries and Oceans
P.O. Box 550
Halifax, Nova Scotia, Canada,
  B3J 2S7

Allan D. Cembella, Ph.D.
Maurice Lamontagne Institute
Dept. of Fisheries and Oceans
850 route de la Mer
Mont-Joli, Quebec, Canada
  G5H 3Z4

Dave Doncaster
Area Inspection Chief
Department of Fisheries and Oceans
P.O. Box 270
Black's Harbour, New Brunswick
Canada E0G 1H0

David M. Dressel, Ph.D.
Shellfish Sanitation Branch
FDA - Center for Food Safety and
  Applied Nutrition
Washington, D.C. 20204

Grete Ellemann (retired)
Fish. Inspection Service
Fiskeriministeriets Industritilsyn
Droningens Tvaergade 21
Postboks 9050 1022 Kobenhavn k
Denmark

Graham H. Fleet, Ph.D.
Department of Food Science &
  Technology
University of New South Wales,
PO Box 1, Kensington, NSW, Australia
2033

Grahm C. Fletcher, Ph.D.
Seafood Research Laboratory
CROP, Dept. of Scientific and Industrial
  Research
120 Mt. Albert Rd.
Private Bag, Auckland,
  New Zealand

Santo A. Furfari, Deputy Chief
Shellfish Sanitation Branch
Northeast Technical Service Unit
U.S. Food and Drug Administration
Davisville, RI 02852

Michael W. Gilgan, Ph.D.
Department of Fisheries and Oceans
P.O. Box 550
Halifax, N.S., Canada, B3J2S7

D. Jay Grimes, Ph.D.
Office of Enery Research
U.S. Department of Energy
Ecological Research Division
Washington, DC 20585

B.E. Hay
AquaBio Consultants, Ltd.
C.P.O. Box 560
Auckland 1, New Zealand

B.J. Hayden, Ph.D.
MAFFish Fisheries Research Centre
P.O. Box 297
Wellington, New Zealand.

Baldomero Puerta Henche
Depuradora de Mariscos de Lorbe S.A.
(DERMARLOSA)
P.O.B. 2, Sada, 15160
La Coruña, Spain

Thomas L. Herrington
Southeastern Regional Shellfish Branch
U.S. Food and Drug Administration
Atlanta, GA 30309

Thomas L. Howell, Vice President
Spinney Creek Oyster Co.
1 Howell Lane
Elliot, ME 03903

Lee-Ann Jaykus, student
Department of Environmental Sciences
  and Engineering
School of Public Health
University of North Carolina
CB# 7400, Rosenau Hall
Chapel Hill, NC 27599

Stephen H. Jones, Ph.D.
Department of Natural Resources and
Jackson Estuarine Laboratory
University of New Hampshire
Durham, NH 03824

Kenneth L. Kasweck, Ph.D.
Shellfish Testing Service
Box 541
Grant, FL 32949

Karl C. Klontz, M.D., M.P.H.
Clinical Nutrition Branch
Center for Food Safety and Applied
  Nutrition
Food and Drug Administration
200 C St., S.W.
Washington, DC 20204

Gilles Lamoureux, Ph.D.
Centre de Recherche en Immunologie
Institute Armand-Frappier
531 Boulevard de Prairies
Laval, Quebec, Canada H7N 4Z9

Bruno Langlais, Ph.D.
Anjou Recherche
Centre de Recherche de la
Compagnie Générale des Eaux
Chemin de la Digue
B.P. 76 - 78600 Maisons, Laffitte
France

J. Le Pauloue
Compagnie Générale Des Eaux
52 rue d'Anjou
Paris, France

Ronald F. Malone, Ph.D.
Department of Civil Engineering
Louisiana State University
Baton Rouge, LA 70803

John C. Mallet, M.D.
Dept. of Biology
Radiation Laboratory
Olson Bldg., Rm 611
University of Lowell
Lowell, MA 01854

Stephen H. McNamara
Hyman, Phelps & McNamara, P.C.
1120 G Street, NW, Suite 1040
Washington, DC

Theodore G. Metcalf, Ph.D.
Dept. of Virology
Baylor College of Medicine
Houston, TX 77030

John J. Miescer
Shellfish Sanitation Branch
U.S. Food & Drug Administration
Washington, D.C. 20204

Fernando Lopez Monroy
Director Tecnico
Bao DEMARLOSA
P.O.B. 2, Sada, 15160
La Coruna, Spain

Ken B. Moore, Chairman
Interstate Shellfish Sanitation
  Conference
and South Carolina Dept. Health and
  Environmental Control
2600 Bull Street
Columbia, SC 29201

Kathleen O'Neill, Ph.D.
Research Assistant
Department of Microbiology
University of New Hampshire
Durham, NH 03824

Robert B. Poggi, Ph.D.
Institut Francais de Recherche
pour l' Exploitation de la Mer
  (IFREMER)
Centre de Nantes
rue de l 'lle d'Yeu
B.P. 1049, 44037
NANTES CEDEX 01
France

J.Y. Perrot
Trailigaz
Compagnue Générale de l'Ozone
29-31 BD de la Muette
95140 Garges-les-Gonesse
France

Roger Pocklington, Ph.D.
Department of Fisheries and Oceans
P.O. Box 550
Halifax, Nova Scotia, Canada, B3J2S7

Scott R. Rippey, Ph.D.
Northeast Technical Services Unit
Food and Drug Administration
CBC Building S-26
North Kingston, RI 02852
Raymond J. Rhodes
Aquafood Business Associates
P.O. Box 16190
Charleston, SC 29412

Kenneth J. Roberts, Ph.D.
Louisiana Cooperative Extension
  Service
Knapp Hall
Louisiana State University
Baton Rouge, LA 70803-1900

James E. Robin, Ph.D.
Department of Civil Engineering
Louisiana State University
Baton Rouge, LA 70803

Gary E. Rodrick, Ph.D.
Food Science and Human Nutrition
  Dept.
University of Florida
Gainesville, FL 32611

Kelly A. Rusch, Ph.D.
Department of Civil Engineering
Louisiana State University
Baton Rouge, LA 70803

Mark D. Sobsey, Ph.D.
Department of Environmental Sciences
  and Engineering
School of Public Health
University of North Carolina
CB# 7400, Rosenau Hall
Chapel Hill, NC 27599

David J. Scarratt, Ph.D.
Department of Fisheries and Oceans
P.O. Box 550
Halifax, Nova Scotia, Canada, B3J2S7

Keith Schneider, student
Department of Food Science and
    Human Nutrition
University of Florida
Gainesville, FL   32611

P.D. Scotti
DSIR Plant Protection
Private Bag, Auckland, New Zealand

Mary I. Snyder, Ph.D.
Division of Regulatory Guidance
FDA - Center for Food Safety and
  Applied Nutrition
Washington, DC 20204

Ira J. Somerset
U.S. Food and Drug Administration
One Montvale Avenue-4th Floor
Stoneham, MA 02180

Richard A. Souness , Ph.D.
Department of Food Science & Technol-
    ogy
University of New South Wales
P.O. Box 1
Kensington, NSW, Australia 2033

John E. Supan
Louisiana Cooperative Extension
    Service
Louisiana Sea Grant College Program
P.O. Box 2440
Covinton, LA 70434

Paul A. West
North West Water Limited
Great Sankey
Warrington WA5 3LW
United Kingdom

CONFERENCE STEERING COMMITTEE

The following list of expertise assisted in planning the first International Molluscan Shellfish Depuration Conference which was the impetus for this text.

Dr. James R. Brooker
National Marine Fisheries Service
Washington, DC

Dr. James C. Cato, Director
Florida Sea Grant Program
University of Florida
Gainesville, FL

Dr. David Dressel, Chief
FDA's Shellfish Sanitation Branch
Washington, DC

Dr. Robert Kifer, Director
National Marine Fisheries Service Lab
Charleston, SC

Mr. Roy Martin, Vice President
Science and Technology
National Fisheries Institute
Arlington, VA

Mr. Ken B. Moore, Chairman
Interstate Shellfish Sanitation Conf.
Columbia, SC

Dr. Thomas J. Murray, Director
Gulf & South Atlantic Fisheries
  Development, Inc.
Tampa, FL

Dr. W. Steven Otwell, Professor
University of Florida
Gainesville, FL

Dr. William Rickards, Director
Virginia Sea Grant College Program
University of Virginia
Charlottesville, VA

Dr. Gary E. Rodrick, Associate
  Professor
University of Florida
Gainesville, FL

Mr. John Schneider, Bureau Chief
  (previous)
Shellfish Branch
Florida Department of Natural Re-
  sources
Tallahassee, FL

MOLLUSCAN
SHELLFISH DEPURATION

CONTENTS

PERSPECTIVES

AN INDUSTRY PERSPECTIVE ON DEPURATION 3
  Robert E. Arnold

HISTORICAL PERSPECTIVE ON COMMERCIAL DEPURATION OF SHELLFISH 7
  Walter J. Canzonier

UNITED STATES REGULATIONS AND OPERATIONS

DEPURATION - THE REGULATORY PERSPECTIVE 19
  David Dressel and Mary I. Snyder

CURRENT U.S. COMMERCIAL SHELLFISH DEPURATION 25
  Ira J. Sommerset

STATE MONITORING AND CERTIFICATION FOR CONTROLLED PURIFICATION 31
  Ken B. Moore

REGULATORY REQUIREMENTS FOR SUBSTANCES ADDED TO DEPURATION WATER:
POSSIBLE "FOOD ADDITIVE" STATUS 35
  Stephen H. McNamara

MICROBIAL CONCERNS

EPIDEMIOLOGY OF MOLLUSCAN-BORNE ILLNESSES 47
  Karl C. Klontz and Scott R. Rippy

BACTERIAL AGENTS IN SHELLFISH DEPURATION 59
  Richard A. Souness and Graham H. Fleet

HUMAN ENTERIC VIRUSES AND DEPURATION OF BIVALVE MOLLUSKS 71
  Mark D. Sobsey and Lee-Ann Jaykus

VIBRIOS IN DEPURATION 115
  Gary E. Rodrick and Keith R. Schneider

TECHNICAL AND ECONOMICS CONSIDERATIONS

DESIGN OF DEPURATION SYSTEMS 129
  Santo A. Furfari

USE OF ULTRAVIOLET LIGHT IN DEPURATION..137
    Thomas L. Herrington

ENHANCING SHELLFISH DEPURATION..145
    Walter J. Blogoslawski

ROUTINE TESTS TO MONITOR DEPURATION..151
    John J. Miescier

ECONOMIC CONSIDERATIONS FOR CLAM DEPURATION..159
    Raymond J. Rhodes and Kenneth L. Kasweck

ECONOMIC CONSIDERATIONS FOR OYSTER DEPURATION..163
    Kenneth J. Roberts, John E. Supan and Charles Adams

RELATED ISSUES AND DEVELOPMENTS

BACTERIAL EVALUATION OF A COMMERCIAL CONTROLLED
PURIFICATION PLANT IN MAINE..181
    Stephen H. Jones, Thomas L. Howell and Kathleen O'Neill

OCCURRENCE OF VIBRIO VULNIFICUS IN WATER AND SHELLFISH
FROM MAINE AND NEW HAMPSHIRE..189
    Kathleen O'Neill, Stephen H. Jones, Thomas L. Howell and D. Jay Grimes

DESIGN OF A BENCH SCALE AUTOMATED RECIRCULATING SYSTEM
FOR USE IN THE DEVELOPMENT OF PURGING/TASTE ENHANCEMENT
CRITERIA FOR THE RANGIA CLAM (RANGIA CUNEATA)..195
    Kelly A. Rusch, James E. Robin and Ronald F. Malone

CONTAINER-RELAYING OF OYSTERS: AN ALTERNATIVE TO
DEPURATION AND RELAYING..205
    John Supan

MONITORING THE ACCUMULATION AND DEPURATION OF PARALYTIC
SHELLFISH TOXINS IN MOLLUSCAN SHELLFISH BY HIGH-PERFORMANCE
LIQUID CHROMATOGRAPHY AND IMMUNOLOGICAL METHODS..217
    Allan D. Cembella and Gilles Lamoureux

THE DEPURATION OF PACIFIC OYSTERS (CRASSOTREA GIGAS)..227
    Graham C. Fletcher, P.D. Scott and B.E. Hay

DETOXIFICATION OF BIVALVE MOLLUSCS NATURALLY
CONTAMINATED WITH DOMOIC ACID..239
    David J. Scarratt, M.W. Gilgan, R. Pocklington and J.D. Castell

POTENTIAL OF IRRADIATION TECHNOLOGY FOR IMPROVED
    SHELLFISH SANITATION 247
        John C. Mallett, Leon E. Beghian and Theodore Metcalf

INTERNATIONAL SETTINGS

A PRELIMINARY REPORT OF INTERNATIONAL TRADE IN MAJOR
    MOLLUSCAN SHELLFISH...261
        Myles Raizin

CANADA DEPURATION 271
        Dave Doncaster

AN OVERVIEW OF THE BIVALVE MOLLUSCAN SHELLFISH INDUSTRY
    AND DEPURATION PRACTICES IN THE UNITED KINGDOM 275
        Paul A. West

THE STATUS OF SHELLFISH DEPURATION IN AUSTRALIA
    AND SOUTH-EAST ASIA...287
        Peter A. Ayres

OVERVIEW OF THE MOLLUSCAN SHELLFISH INDUSTRY IN NEW ZEALAND
    RELATIVE TO DEPURATION 323
        Graham C. Fletcher and B.J. Hayden

CURRENT PRODUCTION AND REGULATORY PRACTICE IN MOLLUSCAN
    DEPURATION IN SPAIN..331
        Fernando López Monroy, Baldomero Puerta Henehe
        and Mariá José Prol Bao

FRENCH SHELLFISH INDUSTRY REGULATORY STATUS AND DEPURATION
    TECHNIQUES 341
        J. Le Paulouë, Bruno Langlais, Robert B. Poggi and J.Y. Perrot

OVERVIEW OF MOLLUSK INDUSTRY AND DEPURATION IN DENMARK
    WITH SPECIAL REFERENCE TO OYSTERS...361
        Grete Ellemann

AN OVERVIEW OF MOLLUSCAN SHELLFISH INDUSTRY IN TURKEY 365
        A. Alpbaz and Murat O. Balaban

BIVALVE SHELLFISH DEPURATION IN ITALY BACKGROUND AND CURRENT
    STATUS 369
        Walter J. Canzonier

INDEX...375

Perspectives

AN INDUSTRY PERSPECTIVE ON DEPURATION

Robert E. Arnold

We truly have an exciting opportunity and challenge ahead of us as we approach the decade of the 90's and reflect on the changes we have all witnessed and shared in the 80's. We know any task we undertake will be fraught with uncertainties, but given the dynamics of today's society and the rapidity with which change occurs, we can settle for nothing less than a game plan for success and a timetable for implementation. It is absolutely essential this first International Conference on Molluscan Shellfish Depuration produces deliverable results for a struggling shellfish industry and for a hesitant, unsure consumer. We cannot allow this opportunity to pass. We cannot allow the activities that take place here to become proceedings that are published two years later and then sit on a shelf looking pretty and intelligent, but also unread and unused.

I was delighted to be asked to provide some comments representing a user's perspective on shellfish products. I can honestly say this is the sixth complete speech I have written for this presentation It is also the most honest, almost militant, of the six. I believe there is an urgency to our efforts here today because of the changes I see on the purchasing end of our business, the sales end, and on the quality assurance aspects of shellfish

If any corporation in American has been built on the commitment to principles of hard work, quality, value, and service, it is Marriott Corporation. Since 1964, Marriott has grown at an annual compounded rate of approximately 20 percent in sales, net income, earnings per share and stock price. In 1964, Marriott had total sales of $84 million, and 9,600 employees. In 1989, sales should exceed $8.5 billion and Marriott now has over 230,000 employees, making it one of the eight largest employers in America.

Today, Marriott is a diversified lodging and food service company, with operations and franchises in all 50 states and 26 countries. Its businesses include nearly 500 hotels in all segments of the lodging market – from full service hotels and resorts to economy motels, over 2,200 institutional food and services management accounts; in-flight catering operations serving more than 150 airlines; airport terminal restaurant and merchandise operations at 51 airports; over 1,100 fast food, family and turnpike restaurants; and a growing number of retirement communities. Marriott serves over five million meals a day.

Success like this doesn't just happen, but comes about by being responsive to the changing needs of our customers and guests. Five million meals a day gives us amble opportunities to be tested. It also provides the challenge for continued change and, I believe change is what this conference is going to be all about.

Marriott utilizes as many as 40,000 bushels of clams and oysters a year primarily in two divisions, Hotels and the Host, our airport feeding contractor business. We serve oysters because our customers demand them, because they are an attractive addition to any menu, because they are profitable for us, and because they are consistent with the dining experience the customer has come to expect of Marriott. Implicit in all of these reasons, is the faith that has been built up over time that Marriott has taken all the necessary precautions to ensure that the dinning experience will also be a safe experience. We go to extraordinary efforts to maintain that "faith" with our customers. Any vendors in the audience will attest to our absolute, and resolute, commitment to quality.

These opening couple of minutes have been to set the stage for the rest of my presentation. We, Marriott specifically and the industry in general, need your help! Oyster and clam usage and consumption are going nowhere with us. They are like the statistics National Fisheries Institute (1989) compiled which indicated about 15% of the respondents of a recent survey had eaten oysters in the last 3 months at home and in restaurants. About two-thirds of the respondents were eating oysters at about the same rate as always. The rest indicated on a near equal basis that they were eating more or eating less. This translates to indicate no growth in shellfish consumption.

Another recent survey from the research department at Restaurant Business, Inc. (Anon , 1989) indicated that "oysters, served in 26 percent of restaurants in 1986, were served in 19 percent of restaurants in 1988". The population surveyed included restaurant chains and franchises, multiunit headquarters, and operators in leisure-service segments.

These aren't good numbers! For a variety of reasons customers are turning away from shellfish, both those who have eaten shellfish products for years, and even more importantly, those people who could be potential consumers if they were not scared away. As a businessman, I would not want to be facing the future of a product with statistics like these. We are dealing with two issues, consumer perception, and a very scary reality.

Perception! Perception! Perception! The stories we hear, and I'm sure you do too, are seldom based on fact. It seems as though each day brings a new story about molluscan shellfish products being the primary suspect in making a person or persons ill. Often times this is just not true but we never hear that aspect of the story. The consumer is totally confused as to any aspect of shellfish safety. Typical comments might include:

"I only eat them in winter months"
"I hear they are bad for you"
"They are too high in cholesterol"
"They aren't safe"
"I hear too many people die from eating them"
"Will you guarantee I won't die?"

These are serious comments and need to be addressed in conjunction with any efforts that this Conference may undertake. Consumer demand for items thought to be healthy is a reality. Everyday, in our four test kitchens, we see menu development experts responding to this demand, and more importantly, turning away from foods that have a "perception" problem. Oysters and clams certainly fall in this latter category. Moving away from perceptions, the reality of molluscan shellfish problems is why you are here. Assuring a healthful supply of shellfish in the face of persistent challenges is an industry and public responsibility. When there is adverse publicity anywhere in the country, it affects shellfish consumption everywhere for us. Within a week of an occurrence, I may have as many as 20 copies of news releases sitting on my desk and every article published about the issue.

Again I say – we need help! And here in no specific order are the areas I suggest. Your customers need assurance that the molluscan shellfish products we purchase are safe. A tag system is nice but it's only real value is after the fact. We have not, but we certainly do not want to have a customer die, and then be able to find out why. There is very little value in that scenario for any of us.

The time for depuration is here! We are not neutral on the issue. Marriott must be able to buy an oyster with confidence and sell it with confidence. You give us an assured safe oyster or clam and we will promote it all day long! Then watch consumption. We realize we will pay more but we can "sell the sizzle" of a safe product – and that is what the customer wants to hear. You give the customer what they want and everyone will profit.

It's not enough anymore to only monitor for coliform bacteria. We must move onto eliminating dangerous viruses and vibrio bacteria. Questions such as the utilization of ozone need to be reexamined. We need assurances in these areas. Also helpful would be quick monitoring procedures for us to do our own quality assurance tests on site in our properties.

We need phase-in plans that take effect now so that bed certification means something. We need communication and cooperation among the various regulatory bodies so that local, state, and federal efforts are united to solve the issue. We need new research and development efforts to advance available technology and future technology. We need immediate action to blend the accomplishments of this conference into a call for better food safety and the development of a HACCP (Hazard Analysis Critical Control Point) system for the industry We need to make frank and honest evaluations of where we are and where we should be. We can not accept previous answers or conclusions. We need a newer more advanced sophistication in our methodologies. The industry needs to unite behind its most effective trade or lobbying association and fight for stronger water regulations.

Federal programs like Saltonstall-Kennedy grants must continue to fund projects like the New England Fisheries Development Association. Residential, industrial, and agricultural pollution must be battled on any field available. The Chesapeake Bay needs help! The Gulf of Mexico needs help! The promising West coast production can not be allowed to diminish for the scenario of events affecting the other sources.

The industry must be proactive. Educational efforts must be significant and extensive. There must be cohesive teamwork to address any crisis that may come up. Training and informational materials that can be used in our properties to educate our people about molluscan shellfish will help us sustain your industry efforts. The media should be used as the powerful force it is by restoring public confidence with a series of profiles telling the consumer what a safe oyster or clam is and that they are available. Product safety must be communicated.

I can hear the grumblings in the audience of "It's fine for him to stand up there and say this is what we need, but how do we pay for it? We are not making any money now!" My answer to that is if you don't do these things in five years you will not be knocking on our door with anything to sell. You must value-add your product and that value-add must be safety. If you do not give us safety for our customers, we will not serve your products – its as simple as that We are running out of time. This gathering must be the corner we turn for the future of your industry.

For change to occur we need three things: information, education, and incentives. Information is data. Education is getting that data to the right people, at the right time, in an understandable and timely format. Information and education can change attitudes, but incentive will change behavior. Economics and public health are two of the strongest incentives we have.

Your efforts here in Orlando can certainly be, in fact must be, the information, the education, and the incentive. Customers of the industry such as Marriott need these three areas fulfilled and in return we can become your strongest advocate. Give us an assured supply of safe oysters and clams and we can provide you with a return on investment unlike you have ever seen. It starts today, this week. For the good of the ultimate customer and the continued viability of your industry, everyone of you must commit yourself to make truly significant changes Our customers will reward your diligence and success in this effort.

REFERENCES

Anonymous. 1989. What's really on the menu. Institutional Distribution magazine, Restaurant Business, Inc., Vol. 10, p. 109.

National Fisheries Institute. 1989. Personal Communications. NFI, Arlington, VA.

HISTORICAL PERSPECTIVE ON COMMERCIAL DEPURATION OF SHELLFISH
A review of the history, philosophy, research and
commercial application of the process in North America.

Walter J. Canzonier

By working definition, depuration is the transfer of contaminated shellfish from their polluted growing waters to a cleaner aqueous environment that permits them to open and function in a quasi-normal physiological mode, thus favoring the unloading, or elimination of sequestered contaminants of public health concern. As typically applied, the definition is somewhat restrictive, implying a rather tight control over and manipulation of the aqueous environment to which the mollusks have been transferred. This more restrictive definition is the one that will be inferred in this text.

The concept of cleansing contaminated bivalves is certainly not new, and its origin is lost in antiquity. The custom of transferring shellfish to cleaner waters prior to consumption seems to have been well established in Mediterranean countries long before the germ concept of disease. The "modern" concept of depuration seems to have bloomed in the latter part of the last century in several European countries (Canzonier, 1988). Indeed, by the end of the first decade of this century several attempts at rudimentary depuration were already in progress The problems of typhoid fever associated with consumption of raw shellfish were widely recognized by health authorities in North America, and the concept of depuration was soon introduced as a potential remedy for reducing this source of infection. As early as 1911, Phelps conducted some limited relay experiments with contaminated oysters in Narrangansett Bay, RI. He found good elimination of the indicator bacteria after only two days sojourn in clean waters. Kelly (1966) implies that this was the first recognition of the need for depuration on this continent. Some early experimental systems had been tried at New Haven, Connecticut at about the same. This work, most probably relaying, was noted by Wells (1929), but further details and dates have not been found in the literature.

A major proponent of the depuration process in the early part of this century was a rather controversial and outspoken individual by the name of William Firth Wells. This man, an engineer by training, was involved in a variety of research and development projects that ranged from oyster hatchery and rearing procedures to sewage treatment plant design to shellfish sanitation (Wells, 1926). He was apparently influenced by the work of Phelps (1911) in Rhode Island and Johnstone (1914) in the United Kingdom. His extensive work in the state of New York paralleled that of Dodgson (1928) in Wales. From about 1914 to 1923 he was employed by the U S Public Health Service (USPHS) and was involved with some of the earliest ventures into experimental depuration. He conducted demonstration projects for the agency, first at Fisherman's Island in Virginia and later in the New York-New Jersey area.

The original work done by Wells, under the direction of Hugh S. Cumming of the USPHS, involved suspending polluted oysters in baskets in some area "at the mouth of Chesapeake Bay". He later did tank experiments at the USPHS facility on Fisherman's Island on the southern end of the Delmarva peninsula, using a 10% calcium hypochlorite "suspension" to disinfect the seawater. Use of chlorine at this level did not appear to influence the activity of the oysters. He obtained approximately a 99.6% reduction in coliforms within 24 hours. The results were promising and were reported by Wells (1916).

About 1916, the USPHS conducted a demonstration of oyster purification at a site on Jamaica Bay. This was followed by a second demonstration of depuration using chlorinated water in large (50 ft x 9 ft) scows that could be flooded and drained during the process. The site chosen was Great Kills Harbor on Staten Island. Wells was the mastermind of both projects . The results of these demonstration projects stimulated considerable interest and were reported by the USPHS (Carmelia, 1921).

The first attempt at depuration on a commercial scale was initiated about 1921 at Inwood, NY. This operation was documented by Wells in 1922 in a report to the New York State Conservation Commission (Krumweide et al. 1926a.) and lead to formalization of an oyster depuration protocol using chlorinated water. This plant operated until about 1925. The oysters depurated were presumably from Jamaica Bay.

The second half of the 1920's saw considerable agitation for the improvement of shellfish sanitary quality. The motive was a series of typhoid fever outbreaks associated with the consumption of eastern oysters. Indeed, these outbreaks were probably the major stimulus for corrective action that lead to the establishment of the National Shellfish Sanitation Program (NSSP), currently based in the U.S. Food and Drug Administration. The topic of health problems related to shellfish consumption created sufficient interest to warrant a special session during the American Public Health Association meeting in 1926. At this session a number of papers were presented that dealt with the process of depuration. Krumweide et al. (1926a) reported on work which suggested caution in using coliforms alone in evaluating the efficacy of the process, or of shellfish sanitary quality in general. He gives details on some experimental work done using "B. typhosus" [Salmonella typhosa] as an indicator, and notes need for prudence in attempting to depurate shellfish from grossly polluted growing waters. His presentation, and the paper that followed (Krumweide et al. 1926b.) generated considerable discussion. It is interesting to note some of the comments concerning the process made by this author based on his experience with experimental systems both in the lab and in the field:

"...a high B. coli [probably fecal coliforms by assay used] score is an indication of danger, while a low B. coli score is not necessarily an index of safety."

"...the only safe oyster is one which has been protected from any contamination with fecal pathogens for at least some months prior to harvesting."

These quotes clearly ruffled a few feathers, but then the audience could not have been aware of future problems with infectious hepatitis and other concerns.

At this same meeting, Tarbett (1926) pointed out that merely using the level of indicator in the final product, without considering other data such as the sanitary survey of the growing waters, was " . . . misleading and open to criticism." The basic problem, in his mind, was " one of water pollution", and "it should be borne in mind that such treatment [depuration] is not a substitute for sanitary controls over growing areas . . . . but simply an additional safeguard". This paper elicited a very critical response by W.F. Wells based on his findings at the Fisherman's Island facility in 1914 and 1915. Wells claimed that his conclusions regarding the efficacy of the depuration process are supported by unpublished reports to the USPHS, the US Bureau of Chemistry, the state health departments of New York and New Jersey, and the city health departments of New York and Newark. He pleaded for the use of a sanitary standard based on product bacteriological titre, rather than a rigid control of growing water quality.

In the same issue, as a sequel to the joint session, an editorial by Ravenel (1926) points out that the "apparent benefits of chlorine treatment [depuration] will vary with the efficiency of the methods of isolation employed," and that "the only safe procedure is to protect oyster beds from pollution." However, he conceded that, "although chlorination cannot be depended upon to render oysters sterile, as long as they continue to be used as food it provides an added safeguard particularly for reducing gross pollution should this occur in spite of the strictest sanitary precautions." This editor felt that "wherever it is applied, chlorination cannot be relied upon to render safe any oyster supply known to be or suspected of being polluted." Clearly, some of these observations are still worthy of serious consideration even in the high-tech world in which we operate today.

Wells mentions seven plants on Long Island that were operating in the 1920s, four of which were built in 1926 (Maier, 1941). Three of these were for oysters and clams and one for clams only. Cost of operation was 4 to 10 cents per bushel, and total annual capacity was about 250,000 bushels. The largest of these was capable of depurating up to 1000 bushels per day in twelve tanks with a total flow of treated sea water of 150 gpm. The water flow was not continuous; rather, there were several changes during the day. This large plant was presumably Bluepoints Co. Inc. at West Sayville, though this is not directly stated in the report. Large numbers of samples collected at various times after initiation of the process (zero – 120 hours) indicated a rather good reduction of coliforms.

A sequel to the presentation of Wells (1929) is a discussion section in which Fuller (1929) makes some rather interesting remarks:

> "Steps in this direction [i.e. depuration of oysters] must be made cautiously however, because the safeguards depend not only upon soundness of the method, but upon the kind of plant equipment employed and the adequate character and performance of the operating personnel "

> "Improved methods of safeguarding shellfish are no more an excuse for slackening sewage treatment than is the construction of a modern filteration plant an excuse for substantial pollution of the raw water supply. Neither does pasteurization give a license to unsanitary handling of milk."

In the same period, at least three plants were constructed in Massachusetts for soft clams Mya arenaria (Wright, 1931). The first of these, an experimental unit using hypochlorite and batch changes of water at 12 hour intervals, was used by a Clark in 1927 (Wright, 1931) to develop operating parameters for a pilot scale municipal plant that was built in 1928. This plant treated 1500 bushels of soft clams in one year. In 1930 the town of Newburyport, MA constructed a commercial scale plant that continued to operate until rebuilt in recent years to incorporate ultraviolet lighting for treatment of the process water. The original plant, with batch use of process water, utilized hypochlorite with a required residual of 0.5 ppm for at least 15 min. after addition. Dissolved oxygen was maintained at a minimal value of 30% saturation. The quantities processed ranged from 19,000 to 59,000 bushels in the early years of operation, at a cost of about 27 cents per bushel (Wright, 1931). Also in 1930 a private plant (Pioneer Fisheries Co.) was built at Plymouth; another private plant was built at Scituate (Scituate Certified Clam Co.); and a third was under construction at Winthrop in 1931. The peak of production for soft clam depuration seems to be about 1934 (annual production, 10,000 bushels) (Kelly, 1966). In recent years, utilizing a rebuilt plant at the Newburyport site, production in excess of 50000 bushels per year has been common (Roach, 1989).

Wright (1931) mentions some work on soft clam depuration in New Jersey in the late 1920s, and on the basis of this a soft clam depuration plant was constructed on the Shrewsbury River (Highlands, NJ) that began operation in early 1930s. It had a capacity of 50 to 60 bushels per cycle and continued to process clams until the late 1940s. Cost of the operation was about 6 cents per bushel. This facility, though privately operated (Kohlenbusch, 1989) was constructed and monitored by the New Jersey Department of Health, in collaboration with the USPHS. Though originally intended for desanding of soft clams the results so impressed the health officials that they authorized its use for purification of clams from closed areas (Fisher, 1941). The dosing of hypochlorite was accomplished with a sophisticated chemical feed pump. There were provisions for heating and recirculating the process water, and daily laboratory services were provided by a state operated field laboratory located nearby. There is mention in a report by Fisher (1941) that the State of New Jersey on the basis of results obtained in Monmouth county, built three seawater chlorination systems in other areas of the State to provide clean water for other depuration or conditioning tanks to be operated by private individuals. One of these plants was presumably operated by J. Richards Nelson of the Rutgers Department of Oyster Culture in Bivalue, NJ (Nelson, 1934). There does not appear to be any further information on the other two plants, or even any indication of their utilization.

During the 1930s, some experimental work was conducted by Maj. Richard E. Messer in Virginia, on Willoughby Spit near Norfolk, as a collaborative study between the state health authorities and the USPHS. There are a few other references to shellfish purification activities in the 1930s and 1940s (Fisher, 1941; Maier, 1941; Medcof, 1942; Renn, 1947; Tarbett, 1934), but no significant effort seems to have been devoted to the topic in this period. Interest was probably diverted by preoccupations with more pressing problems and by the rapid demise of typhoid fever as a serious public health threat.

In the mid-1950s some initial studies of soft-clam depuration, on a laboratory scale, were carried out near Boothbay Harbor, ME (Goggins and Hurst, 1955; Kelly, 1966). A soft-clam depuration plant was established in Maine in 1962 and was utilized for many years to depurate on a commercial scale as well as to develop physiological and microbiological parameters related to the depuration of soft-shell clams. The operational experiences and plant design criteria, reported by Goggins et. al. (1964) and Sterl et al. (1964) are still valuable sources of basic information on this species. Simultaneously, the operation of the plant in Newburyport, MA served to accumulate important observational data on the same species.

There was revitalized activity in federal labs at Woods Hole, MA; Pensacola, FL; and Purdy, WA during the 1950s. Arsciz and Kelly (1955) and others, using newer methods to treat the process water (e.g. ultraviolet lighting), conducted a variety of depuration trials on a lab and pilot plant scale, reflecting a surge of interest in Europe (France, England and Spain). These studies, and commercial operations of the period, were primarily addressing the reduction of risk from bacterial enteric infection associated with the consumption of shellfish from marginally acceptable waters. Reflecting this renewed interest, Furfari (1966) compiled a construction and operation manual for depuration plants. In this publication, produced by the USPHS, he draws together much of the pertinent work in the fields of bivalve physiology and microbiology and practical depuration experiences, and attempts to develop design criteria and operating protocols for the commercial application of the process.

In the late 1950's and early 1960's, a number of European investigators began to examine depuration as a system for reducing enterovirus contamination in bivalves. A similar interest was developing in North America. This new approach saw a definite burst of activity due to the occurrence of hepatitis-A cases associated with the consumption of raw shellfish in the early 1960s (Beck and Hoff, 1966). The USPHS accentuated its role in sponsoring depuration studies and establishing shellfish sanitation research centers at Narangansett, RI and Dauphine Island, Alabama.

Starting in the early 1960s, with the aid of both federal and State funding, several groups attacked some of the basic questions of molluscan physiology, microbiology and epidemiology as they pertain to shellfish sanitation and depuration. From that point until present, government, university and private research groups have generated an impressive volume of data that addresses various aspects of the process. A good example would be the highly detailed, in depth study of hard clam depuration done on the south shore of Long Island in the 1960s. In 1964 the State of New York, with funding assistance from the US Fish and Wildlife Service, Bureau of Commercial Fisheries initiated a comprehensive pilot scale operation at West Sayville. A wealth of basic information on techniques, microbiology and clam physiology, as well as plant design and operation was reported by Bennett (1969). It would be prudent for those contemplating future work on this species to examine this report thoroughly before designing experiments or building plants.

In the same period there was considerable interest in examining the feasibility of the depuration process in Canada. Most of the work that has been reported involved pilot scale plants operated in both the Atlantic and Pacific provinces (Devlin 1973; Devlin and Neufeld 1971; Erdman and Tennant, 1956; Neufeld et al. 1975; Rowell et al. 1976; Swansburg, 1957).

Within the constraints of this presentation, it would be difficult to do justice to the numerous research projects relevant to the topic that have been reported since the early 1960s. Furfari (1966) provided a brief synopsis of technology aspects. The recent review of Richards (1988) would, however, be a good starting point for those interested in a more detailed coverage of depuration literature for this period.

At this point it might be opportune to digress a bit and mention the role of the USPHS in stimulating and supporting activity in the field of bivalve depuration Older workers in the field will certainly recall C. B. Kelly, and newcomers will recognize the name from the literature. This member of the USPHS played a major part in fostering depuration as the panacea for rapidly worsening pollution problems confronting the shellfish producers. Such statements at an early 1960s workshop that "within five years all raw shellfish sold in the US will be depurated" helped to set the theme for the various groups studying the process. Indeed, the universal use of depuration seemed to be deeply entrenched in the philosophy of some members of the public health community, as was expressed by R. A. Prindle in a keynote address before an NSSP Workshop in 1968

> "As you review the national situation, we of the Public Health Service hope that you will give serious consideration to the provision of another barrier to possible disease transmission by shellfish. We believe that it would be provided by addition of controlled processing such as depuration. Because of the increase in variables in pollution sources, including the explosive growth of watercraft that ply the same shellfish waters, we hope that the National Shellfish Sanitation Program will take this additional step by preparing a timetable now for the adoption of depuration for all raw products."

This is practically a mandate to make depuration an obligatory component of marketing, whether the benefits were fully understood, and regardless of the quality of the waters from which the shellfish were harvested.

Even as early as 1931 the USPHS seemed convinced of the utility of the process (Fisher, 1931). But, as Fisher points out, though he was pleased with the results of the studies with soft clams that had been funded by his agency, and would consider requiring obligatory depuration for all clams, there were still technical problems associated with the process that required further study and he was quick to recognize the risk of potential "short-circuiting". In this same report he observes that even when clams were placed in contaminated water they were able to purge themselves and even decontaminate the water, correctly attributing this phenomenon to the sequestering of the bacteria in feces and pseudofeces that settled to the bottom of the tank

The concept of holding of shellfish in chlorinated water had previously been suggested by USPHS officials (Tarbett, 1926) as a possible method for reducing the risk of typhoid infections. Tarbett reiterated this concept as a way to circumvent some sanitation problems that were confronting the Delaware Bay oyster industry (Tarbett, 1928).

Again in 1934, Tarbett, a high-ranking USPHS representative, noted the need for rigid control of the process by state health authorities, and cautions the responsible agencies regarding the inadequacy of the currently available microbiological techniques used to monitor the efficacy of the purging operation. This agency spokesman solicited more research on the techniques used to evaluate shellfish sanitary quality.

It is interesting to note that such reservations regarding the appropriateness of the microbiological assay techniques with regard to depuration monitoring have been reiterated more recently (Canzonier, 1971; Furfari, 1973). The latter author quite clearly points out that though there is a probability of reducing the public health risk by the application of depuration, this reduction is not quantifiable, since even the risk factor associated with the sacrosanct "less than 70 coliforms per 100 ml" has not been defined. This shortcoming in the shellfish sanitation program has now gained broad recognition and there is a comprehensive National Indicator Study in the course of development [1989-90] that is designed to address the problem.

During September and October 1989, shellfish sanitation and natural resource officials of all the coastal states and provinces in the US and Canada were polled to determine the status of depuration operations in their jurisdictions. As of 31 October 1989, there were 10 depuration plants active in North America; three of these are in Canada; one plant is under process of certification on Prince Edward Island and one planned for Newfoundland; a large hard clam plant is planned for NJ; an authorized plant is inactive in Florida, as is another in New Jersey.

Less than ten years ago there were twenty plants operating in the US alone. The reduction in the number of plants operating cannot be attributed to a single factor. Most often mentioned is the added cost, which makes the application of the process non–competitive. Certain industry members, salesmen of depuration hardware, and an occasional public health official, would like to see this competitive constraint removed, by making depuration obligatory for all shellfish, regardless of the quality of the waters in which they are grown Some plants in Florida and New Jersey are inactive due to reduction in the volume of resource that is available for depuration Others based in NJ have been closed due to the illicit operations of the owners. Still others have probably been abandoned due to technical problems.

Furfari, in a status report in 1973, provides a synopsis of research priorities in the field of depuration at that time. Today, many of these priorities are yet to be addressed to the satisfaction of all concerned. In conclusion, he mentions the alternative approach (philosophy), that of some European countries, to build the plants and operate them to gather the necessary epidemiological evidence to evaluate the process, or in his words, "let the chips fall where they may". This reiterates the comment of Oscar Liu, at the 1968 NSSP Workshop, concerning the need for a real–world evaluation of the efficacy of the process, to determine if the risk of viral infection is truly reduced to acceptable levels.

CONCLUSIONS

I would hazard that few would object to the conclusion that the depuration process is not just a simple exercise in hydraulic engineering, such as building swimming pools or designing a bathroom. The objects of the process are living, active and complex organisms, and they must be treated as such. A recent conversation with Victor Cabelli, a microbiologist with long–term interest and productivity in the field of shellfish sanitation, addressed the enigmas of depuration by stating, "The water is not the problem, the animals are the problem."

A careful perusal of the literature would also indicate that there are limits to the efficacy of the process for reducing the contaminant load in the bivalves. Enter here the factors of type of sequestered contaminant (bacterial/viral/chemical), and the titre of the contaminant load in the animals at the beginning of the process. There are also serious problems related to monitoring of the efficacy of the process; i.e., is the elimination of the indicator of choice truly reflecting the elimination of the contaminant of concern.

Finally, as is often discovered too late by practitioners of the art and science of shellfish depuration, there is no 'best' process. The system and protocol chosen for a particular plant is a function of the species of bivalve, its origin and contamination load, and the site at which the plant is to operate. Perhaps it is time for a refinement and restatement of the concept, and the principles of application of the depuration process, which seriously considers, without bias, the information available from past experiments and commercial experiences, judiciously integrating the newer information.

ACKNOWLEDGEMENTS

It would be impractical to list by name and affiliation all those persons that have so graciously supplied information used in the preparation of this paper. They range from long-time colleagues in the research field to total strangers in commercial operations and governmental agencies. Many have devoted considerable effort in tracking down esoteric and long forgotten reports. Commercial plant owners have afforded a point of view that is lacking in the purely academic sphere.

REFERENCES

Arcisz, W. and Kelly, C.B. 1955. Self purification of the soft clam Mya arenaria. Public Health Reports, 70(6):605–614.

Beck, W.J. and Hoff, J.C.. 1966. Northwest Shellfish Sanitation Research Conference, 1966, NW Shellfish Sanit. Lab., Gig Harbor, WA. 90 pp.

Bennet, Q.R. 1969. Operation of a Depuration Plant for Hard Clams (Mercenaria mercenaria). Final Report, Bur. Mar. Fish., Shellfish Sanitation & Engin. Serv., Ronkonkoma, NY, 146 pp.

Canzonier, W.J. 1971. Accumulation and elimination of coliphage S-13 by the hard clam, Mercenaria mercenaria. Appl. Microbiol. 21:1024–1031.

Canzonier, W.J. 1988. Public health component of bivalve shellfish production and marketing. J. Shellfish Res. 7(2)261–266.

Carmelia, F.A 1921. Hypochlorite process of oyster purification. U.S. Public Health Serv., Public Health Report, No. 652, l0pp.

Dodgson, R.W. 1928. Report on Mussel Purification. Min. Agri. & Fisheries, Fishery Invest. Series II, Vol. X, No. 1, pp. 498 + 15 plates.

Devlin, I.H. 1973. Operation Report: Oyster Depuration Plant, Ladysmith, B.C.; Tech. Report, Fisheries and Marine Sciences, Ottawa, Canada; 108 pp.

Devlin, I.H. and Neufeld, N. 1971. Oyster Depuration Plant, Ladysmith, B.C. Tech. Report No. 43, Fisheries and Marine Services, Ottawa, Canada; 29 pp + 5 appendices.

Erdman, I.E. and Tennant, A.D. 1956. The self-cleansing of soft-shell clams: Bacteriological and public health aspects. Can. J. Publ. Hlth. 47:196–202.

Fisher, L.M. 1931. Soft clam purification studies. Natl. Shellfish. Assoc., 1931 Convention Addresses, pp. 175-183 mimeo.

Fisher, L.M. 1941. Progress in shellfish sanitation. Natl. Shellfish. Assoc., 1941 Convention Addresses, pp. 4 mimeo.

Fuller, G.W. 1929. Discussion of presentation by Wells at Am. Public Health Assoc. Meeting, 1929 Amer. J. Public Health 19(1)79.

Furfari, S.A. 1966. Depuration Plant Design. US Public Health Serv. Publ. 999–FP-7, pp l09.

Furfari, S A 1973. Current status of depuration. Shellfish Sanitation Research Planning Conference, Washington, DC, 3-5 December 1973; U.S. Public Health Service, Food and Drug Adiminstration mimeo report, 14 pp.

Furfari, S.A. l976. Shellfish purification: A review of current technology. FAO Tech Confr Aquaculture, Kyoto, May–June 1976. FAO Tech. Publ. Series FIR:AQ/Conf/76/R.11.

Goggins, P.L. and Hurst, J.W. 1955. Progress Report on Clam Cleansing Tech Report (s/n) to Maine Dept. Sea & Shore Fisheries, Boothbay Harbor. 37 pp. mimeo.

Goggins, P.L., Hurst, J.W. and Mooney, P.B. 1964. In: Soft Clam Depuration Studies: II. Laboratory studies on shellfish purification. Maine Dept. Sea & Shore Fish. pp. 19–36.

Johnstone, J. 1914. The methods of cleansing living mussels from sewage bacteria. Lancashire Sea–Fisheries Lab , Report No. 23, pp. 57–108.

Kelly, C.B. 1966. Status of depuration in the United States. National Conference on Depuration, Kingston, RI, 19-22 July 1966; U.S.P.H. Service; mimeo proceedings, pp 19-22.

Kohlenbusch, C. 1989. Personal communication. Operator of former depuration plant in Highland, NJ.

Krumweide, C , Park, W.H., Cooper, G., Grund, M., Tyler, C. and Rosenstein, C. 1926a. The chlorine treatment contaminated oysters. J. Am. Public Health Assoc. 16:142–152.

Krumweide, C., Park, W.H., Cooper, G., Grund, M., Tyler, C.H and Rosenstein, C. 1926b. Effect of storage and changing seawater on contaminated oysters J Am Public Health Assoc. 16:263–268.

Maier, F. J. 1941. Present status of shellfish cleansing and conditioning processes. Natl. Shellfish. Assoc., 1944, Convention Addresses, pp. 4 mimeo.

Medcof, J.C. 1942. Self-purification of sewage–contaminated oysters Fish Res. Bd. Can. ms report, No. 331, (section lOb), pp. 104 (ref. letter by J.C. Medcof; not currently available).

Nelson, J.R. 1934. Controlled water storage and conditioning of oysters for market. N.J. Agr. Expt. Station, Circular 322, pp. 8.

Neufeld, N., Tremblett, A. and Jackson, K. 1975. Clam Depuration Project. Tech. Report No. 83, Fisheries and Marine Service, Ottawa, Canada, 35 pp.

Phelps, E.J. 1911. Some experiments upon the removal of oysters from polluted to unpolluted waters. Amer. J. Publ. Hlth. 1(5)305–308.

Prindle, R.A. 1968. Partnership and Responsibility. Proc. 6th National Shellfish Sanitation Workshop, Wash. DC,; G. Morrison ed., pp 6-7.

Ravenel, M.P. 1926 The purification of contaminated oysters Editorial in J. Am. Publ Hlth Assoc. 16: 292-293.

Renn, C. E. 1947. Report of the Special Commission Relative to Increasing the Supply of Shellfish and the Propogation of Soft Clams and Relative to Shellfish Chlorinating Plants . Commonwealth of Massachusetts, H.R. 1850, Dec. 1947. (not available, cited by Arcisz and Kelly, 1955).

Richards, G.P. 1988. Microbial purification of shellfish: A review of depuration and relaying. J. Food Protec. 51(3)218–251.

Roach, D. 1989.   Personal communication. Massachusetts Div. Marine Fisheries, Shellfish Purification Plant, Newburyport, MA 01950.

Rowell, T.W., Robert, G., Swansburg, K.B., and Davis, R. 1976. Soft–Shell Clam Depuration, Digby, Nova Scotia. Fish. & Mar. Serv., Ottawa, Canada. Tech Report No. 687, pp. 121

Sterl, B., DeRocher, P.J., and Hurst, J.W. 1964. Soft Clam Depuration Studies. I. Design and operation of a cleansing plant. Maine Dept. Sea & Shore Fisheries, Proj. Report, pp. 1–18.

Swansburg, K.B. 1957. Review of quahaug (Venus mercenaria L.) self–cleansing program, Prince Edward Island, 1957. Fish. Res. Bd. Can., Fish. Inspection Lab., P.E I., Report No. 58. (not available, cited by Rowell et al. 1976).

Tarbett, R E 1926. The engineering aspects of oyster pollution. J. Am Publ. Hlth Assoc. l6:5-12

Tarbett, R. E. 1928. Conference Relative to Shellfish Sanitation at Maurice River. Philadelphia, 13 Dec. 1928, typed min. pp. 23 (available Rutgers Shellfish Lab. Port Norris, NJ).

Tarbett, R. E. l934. Some factors to consider in the development of the practice of conditioning shellfish for market purposes. National Shellfish. Assoc., 1934 Convention Addresses, pp. 6 mimeo.

Wells, W.F. 1916. Artificial purification of oysters. USPHS, Washington DC. Public Health Reports 31:1848–1852.

Wells, W. F. 1926. Early oyster culture investigations by the New York State Conservation Commission, (1920–1926). Ann. Report to the Legislature, reprinted l969, NY Cons Dept., Div. Mar. & Coastal Res. pp. 56–59.

Wells, W.F. 1929. Chlorination as a factor of safety in shellfish production. Amer. J. Public Health 19(1)72–79.

Wright, E. 1931. Shellfish purification by chlorine in Massachusetts. National Shellfish. Assoc., 1931 Convention Addresses, pp. 7 mimeo .

United States Regulations and Operations

DEPURATION - THE REGULATORY PERSPECTIVE

David M. Dressel and Mary I. Snyder

INTRODUCTION

The regulation of molluscan shellfish products in interstate commerce receives special attention in the United States. Molluscan shellfish pose unique public health problems. They are filter feeders capable of concentrating pathogenic bacteria, viruses, and contaminants during feeding; they are commonly consumed whole, including the gastrointestinal tract; and they are commonly eaten raw. Fresh and frozen oysters, clams, and mussels must meet general Federal regulations governing the manufacturing, holding, and distribution of foods. In addition they must comply with detailed state regulations developed under the National Shellfish Sanitation Program (NSSP) and other programs. Frequently the jurisdictional division of authority between Federal and State regulations is not immediately apparent.

FEDERAL REGULATIONS

The Food and Drug Administration (FDA) is the federal agency responsible for enforcing the Federal Food, Drug, and Cosmetic (FDC) Act, Title 21 United States Code (U.S.C.) Section (§) 301 et. seq., and certain portions of the U.S. Public Health Service (PHS) Act, including 42 U.S.C., § 241, 242, 243, and 254. These laws require that foods shipped in interstate commerce, including shellfish, be prepared, packed, and held at all times under sanitary conditions; that they be honestly and informatively labeled; and that the food itself be safe, clean, and sanitary. The major law is the Federal FDC Act. This law and the others are the major consumer protection laws in the country They give FDA the authority to conduct inspections, to collect samples in order to monitor the products under our jurisdiction for compliance, and to remove from interstate commerce those products which do not comply. These laws apply to both domestically produced and imported foods.

The shellfish industry must comply with the Good Manufacturing Practices Regulations (GMPRs), which require that all foods be prepared, packaged, and held under sanitary conditions. The GMPRs found in Title 21, Code of Federal Regulations, part 110 (21 CFR 110) are general regulations that are applied to all food processing and warehousing operations. They are based on common sense practices to promote food safety and they never go out of date. Topics covered include plant construction and grounds, sanitary facilities and controls, sanitary operations, personal hygiene, equipmental and processing controls.

Shellfish, like other foods shipped in interstate commerce, must comply with FDA's food labeling laws and regulations. These laws, the Federal FDC Act and the Fair Packaging and Labeling Act and regulations, 21 CFR 100-101, require that foods in package form bear certain specific mandatory information. That information includes a statement of identity, a statement of the quantity of contents, a list of ingredients, and the name and place of business of the manufacturer, packer, or distributor. For imported foods which have not been substantially transformed, the U.S. Customs Service requires that the country of origin be declared. The NSSP requires that molluscan shellfish meet additional identification and certification requirements.

Unshucked oysters, clams, and mussels must be accompanied by a tag which provides specific information necessary to ascertain the harvesting site and date of harvest. In addition, depurated shellfish require an increased level of control compared to shellfish from approved areas because of the increased potential for contamination. These controls must include packaging and labeling that will help to identify the depuration cycle of each harvest lot and deter illegal commingling with nondepurated shellfish. Such controls include prohibition against commingling of harvest lots during packing, tags that identify the shellfish as being depurated, and a prohibition against repackaging after the shellfish leave the depuration plant. Tamper-evident seals on the packages are recommended as a further deterrent.

In addition to the GMPRs and the food labeling regulations, food additive regulations ensure that no substance that becomes a component of food will render it injurious to health. Food additives are deemed to be unsafe unless their safety is demonstrated under Generally Recognized as Safe (GRAS) criteria or a specific regulation permitting its use. These regulations are found in 21 CFR 170-199.

In addition to its own resources, FDA is authorized to accept assistance from state and local authorities, in the enforcement of laws to prevent and suppress the spread of communicable disease, 42 U.S.C., § 243 and 21 U.S.P.H.S., § 372. This latter authority gave rise to the NSSP, which was initiated in 1925 and continues to date as a voluntary FDA/State/Industry program.

STATE REGULATIONS/FEDERAL GUIDELINES

The NSSP is a voluntary cooperative effort among 23 shellfish producing states, nine foreign governments, the shellfish industry, the FDA (formerly the Public Health Service), and the National Marine Fisheries Service (formerly the Bureau of Commercial Fisheries). FDA administers the NSSP.

The NSSP (1990 a, b) Manual of Operations, last published in 1965 and updated annually since 1987, sets forth the standards and procedures that govern the NSSP. States have codified the recommendations in the NSSP to form a broad-based regulatory system to promote shellfish sanitation. One key provision of this network is that states will not accept molluscan shellfish in interstate commerce unless they are from a certified source of origin.

Certified source of origin is an important concept in assuring shellfish safety and the free flow of molluscan shellfish in interstate commerce State laws require that molluscan shellfish dealers be certified by their respective shellfish producing states or foreign governments. Dealers are certified to indicate that their products were harvested, processed, and transported in accordance with NSSP criteria. States assure that these conditions are met when they submit dealer certifications to FDA for listing in the Interstate Certified Shellfish Shippers List. This publication is printed monthly by FDA.

In addition to the laws of 23 shellfish producing states, 40 of the 50 states have adopted similar source of origin requirements under state laws developed pursuant to provisions of the Food Service Sanitation Manual (FDA, 1976) and the Retail Food Store Sanitation Code (AFDO and FDA,1982). Certified source of origin provisions are also part of the newly proposed Food Protection Unicode (FDA, 1987), now under review by the Conference for Food Protection.

In short, dealers in fresh and frozen molluscan shellfish must comply with specific sanitary requirements of the FDC Act and provisions of the NSSP as adopted under state laws.

THE INTERSTATE SHELLFISH SANITATION CONFERENCE

Recently the NSSP guidelines for the sanitary production of molluscan shellfish products have been significantly strengthened with the formation of the Interstate Shellfish Sanitation Conference (ISSC). The ISSC was formed on September 21, 1982 by ratification of a constitution and bylaws. The purpose of the ISSC is to provide a formal structure wherein the state regulatory authorities can establish updated guidelines for the shellfish industry, and procedures for their uniform application. The ISSC is a voluntary organization that is open to all persons interested in fostering controls that will assure sources of safe and sanitary shellfish.

In accordance with FDA authority to accept assistance from state and local authorities in the enforcement of laws to prevent and suppress the communicable disease, FDA entered into a memorandum of understanding (MOU) with the ISSC to foster and improve shellfish sanitation and quality. The ISSC membership is a tripartite organization which expanded the original NSSP structure established in 1925. Membership was expanded to include shellfish receiving states and the Environmental Protection Agency. The ISSC is the active body whose annual conference provides a structured forum to advise FDA and recommend changes in the NSSP Manuals of Operation. The candid discussion of shellfish sanitation issues, research advancements, and peer pressure to assure that states meet NSSP certification criteria has added a new dimension to improving shellfish sanitation.

DEPURATION

The NSSP (1990 b) Manual of Operations, Part II defines controlled purification (depuration) as "the process of using a controlled aquatic environment to reduce the level of bacteria and viruses that may be present in [live] shellfish harvested from moderately polluted [restricted] waters to such levels that the shellfish will be acceptable for human consumption without further processing." The process is not intended for shellfish from heavily polluted (prohibited) waters. The controlled purification process whereby mildly contaminated products are harvested, processed, and released for consumption requires stringent controls to protect public health. The ISSC and FDA readily recognize the threats that depuration can pose if all parties do not meet their responsibilities in assuring safe products. In 1988, the ISSC endorsed FDA's position that new program areas, including depuration and wet storage, must meet FDA approval prior to allowing industry to market products. Both the State and industry share responsibilities to assure safe products. Some of these responsibilities are as follows.

STATE RESPONSIBILITIES

States must demonstrate that they have the technical skills, supporting laws and regulations, and sufficient management and supporting staff to oversee and monitor the depuration process. Specific responsibilities include:

1)   proper classification and continual monitoring of harvesting areas;

2)   sufficient patrol to oversee harvesting and assure that all potentially contaminated products are subjected to depuration;

3)   a laboratory certification officer to certify that private or state laboratories that analyze plant processing waters and products do so in accordance with NSSP accepted methods;

4)   technical staff to review and approve depuration plant design and operational characteristics prior to beginning operation; and

5)   plant inspection and monitoring capabilities to assure that only products meeting end point criteria are released for marketing and that quality assurance programs are in place and are always followed.

These requirements pose a significant burden on state shellfish programs and require agreements among the agencies that share responsibility for growing area classifications, patrol, lab evaluations, and depuration plant inspections. Only if all parties effectively meet their responsibilities can depuration operations function properly without compromising public health.

INDUSTRY RESPONSIBILITIES

Prior to planning investments in depuration facilities, the industry should be aware of its responsibilities. In addition, prudence would indicate that industry work closely with the state shellfish sanitation control personnel to ascertain if the state can and will make the requisite commitments to obtain FDA approval for its depuration program. Only when there is a state commitment to support the necessary infrastructure can industry responsibilities be placed in perspective. Two major points the industry must address are l) securing certified laboratory services to analyze water and shellfish meats, and 2) designing a plant and daily operation plan that will meet NSSP and FDA approved certification criteria.

LABORATORY SUPPORT

The NSSP requires that all microbiological analyses be conducted by certified laboratories. The FDA does not certify private laboratories. Therefore, depuration facilities must contract with a certified state or private laboratory to run its analyses or establish its own laboratory facilities. If the latter course is chosen, FDA must approve a State laboratory certification officer who can then certify the private laboratory. The costs of establishing a certified on-site laboratory with trained personnel and backup support can be very costly and contribute significantly to the depuration facility's daily operating expenses.

APPROVED PROCESSES

Currently, the only FDA accepted depuration processes use ultraviolet (UV) light to assure bacteria-free processing waters. The use of ozone, chlorine, peroxides, or other oxidizing agents that may come into contact with shellfish are not recognized by FDA as being safe for these uses. When water treatment processes subject shellfish to potentially toxic chemicals, these chemicals become food additives. Food additives require GRAS status before they will be sanctioned in depuration facilities. If the industry is contemplating using treatments other than UV light to purify depuration processing waters, it is industry's responsibility to prove that any compounds or their derivatives formed do not contact shellfish. Otherwise, GRAS approval must be received prior to operation. It is also industry's responsibility to prove that these compounds are safe – either by historical use or detailed toxicological studies. Obtaining GRAS affirmation is usually a slow, very technical procedure.

CONCLUSION

Although depuration is a recognized process that allows mildly contaminated products to be cleansed under controlled conditions and then marketed, the practice requires strict supervision and controls. Both the states and industry share responsibility for producing safe products certified under approved practices and quality assurance programs.

The FDA will not certify depuration plants unless it is assured that the requisite criteria are met to assure product safety. Depuration is a costly, management intensive program that involves an industry and state partnership prior to becoming operational. The NSSP Manuals of Operation specify criteria which must be rigidly followed. FDA will sanction depuration operations only when the states have demonstrated that all certification criteria are met.

REFERENCES

AFDO and FDA. 1982. Retail Food Sanitation Code. Association of Food and Drug Officials, P.O. Box 3425, York, PA 17402-3425 and U.S. Department of Health, Education and Welfare, Public Health Service, Food and Drug Administration, Division of Retail Food Protection, Washington, D.C.

FDA. 1976. Food Service Sanitation Manual. U.S. Department of Health, Education and Welfare, Public Health Service, Food and Drug Administration, Division of Retail Food Protection, Washington, D.C.

FDA. 1987. Development of Model Food Protection Unicode. Federal Register 52(70): 11885-11886, April 13, Washington, DC.

National Shellfish Sanitation Program (NSSP). 1990a. Manual of Operations, Part I. Sanitation of Shellfish Growing Areas. Public Health Service, U.S. Food and Drug Administration, Washington, DC.

National Shellfish Sanitation Program (NSSP). 1990b. Manual of Operations, Part II: Sanitation of the Harvesting, Processing and Distribution of Shellfish, Public Health Service, U.S. Food and Drug Administration, Washington, DC.

CURRENT U.S. COMMERCIAL SHELLFISH DEPURATION

Ira J. Somerset

OVERVIEW

A unique feature of the shellfish program is that we deal with animals that are frequently eaten alive, raw, and noneviscerated. To ensure that the product reaching the consumer is safe and wholesome, public health professionals have endorsed the National Shellfish Sanitation Program (NSSP, 1990a, b) requirements for the harvesting, handling, processing, and shipping of shellfish. This program is also concerned with one type of shellfish processing - depuration.

Depuration, also called controlled purification, is "the process of using a controlled aquatic environment to reduce the level of bacteria and viruses in live shellfish" (NSSP, 1990b). A depuration processor is "a person who receives shellstock from approved or restricted growing areas and submits such shellstock to an approved controlled purification process" (NSSP, 1990b).

This unique food product, when taken from restricted areas, is unsuitable for food use in its harvested state. To produce a safe, wholesome depurated product, specific growing area classification, process approval, and process controls are required. The growing areas must be surveyed and classified in accordance with the certification requirements of the NSSP. Areas are classified as approved, restricted or prohibited on the basis of shoreline, hydrographic and water quality surveys. The approved and restricted classifications may be further refined into conditionally approved and conditionally restricted. For conditional designation, the conditions which govern acceptability must be definable (e.g., seasonality, sewage treatment plant discharge, river stage, marina use, rainfall runoff), and manageable in that regulators can restrict harvesting when conditions would make the shellfish unsuitable for their intended use. Complex management plans are required for conditional area classification.

Shellfish from approved and conditionally approved areas may be harvested for marketing without purification when the approved area criteria are met. Shellfish from restricted and open conditionally restricted areas may be harvested for relay or depuration, provided that the required controls are in place and enforced. Required controls, discussed in several sections of the NSSP Manual of Operations include proper classification of the harvest area, licenses or permits, restrictions on harvest location, enforcement oversight, and oversight of plant operations.

Depuration is sometimes confused with wet storage, which is defined by the NSSP as "the temporary storage of shellfish from approved sources, intended for marketing, in containers or floats in natural bodies of water or in tanks containing natural or synthetic seawater (NSSP, 1990b)." The key difference in the processes is that the shellfish from restricted areas are not suitable for consumption without treatment. Depuration is one of the ways in bacteria may be reduced to make the shellfish safe for consumption. In wet storage, one is simply moving marketable shellfish to a more convenient storage location. Sometimes salts are added to the storage water to increase the salinity and improve product flavor. Wet storage also allows the shellfish to reduce their gut contents and eliminate grit or sand within the mantle cavity, providing some cleansing and a longer shelf life. The primary objective of depuration is to reclaim a product which is not safe for consumption. In both treatments, increased quality and wholesomeness are achieved as by-products of the process. It is important to note that both processes require moving water because stagnant water in the process tanks is not acceptable under the current NSSP Manual of Operations.

DEPURATION IN THE UNITED STATES

Currently, there is active depuration in three states – Maine (three plants), Massachusetts (one state-run plant), and Florida (two plants). Historically, the industry has passed a phase of rapid growth to a total of 21 plants, gone through a decline, and reached the current level of six approved plants nationwide. This decline has occurred despite the increase in the number of closed areas. The Florida plants depurate <u>Mercenaria</u> (hard-shell clams, quahogs); the Massachusetts plant depurates <u>Mya</u> (soft-shell clams); and in Maine two plants depurate <u>Mya</u> and one depurates <u>Crassostrea</u> (eastern oysters).

MASSACHUSETTS

The oldest operating depuration plant in the U.S. (since 1928) is located in Newburyport, Massachusetts, and is owned and operated by the Commonwealth of Massachusetts, Division of Marine Fisheries (DMR). This plant has nine tanks, each of which hold 60 bushels of softshell clams. The plant obtains its water from a salt water well and treats it for low dissolved oxygen by aeration. Acceptable bacteriological quality is achieved by treatment with commercial ultraviolet (UV) light treatment units. After 24 hours, the tanks are drained, cleaned, and refilled with new water. The sampling regimen consists of a O-hour sample from each area dug. Two samples each of treated clams (after 24 and 48 hours) every other day, making a total of six samples of each per week. If the initial (O-hour) simple exceeds the allowable maximum for the plant (1,000 fecal coliform/100g), duplicate samples are taken at 24 hours and duplicate or triplicate samples are collected and run of shellstock processed for 48 hours. This multiple sample policy has been instituted recently. All laboratory tests are conducted in the plant laboratory, which is certified in accordance with the NSSP procedures. Analyses of shellfish for fecal coliform bacteria are conducted using a pour plate method at elevated temperature, which has been tentatively approved pending completion of a collaborative study of this methodology (Miescier, 1989). The protocol for the plant process verification study was developed in cooperation with the U.S. Food and Drug Administration and is undergoing final review. Plant operating procedures have been developed and are used on a daily basis at the plant.

MAINE

In Maine, new regulations which took effect in mid-July were consistent with the NSSP requirements at that time. The oyster depuration plant can process up to 50 bushels every 24 hours. Water is obtained from an approved area, treated with a diatomaceous earth filter and a commercial UV treatment unit. The water is replaced after each lot is treated. The plant has its own laboratory and runs initial, 24-hour and finished product samples of each lot.

The Maine softshell clam plants likewise have a capacity of approximately 100 bushels in 48 hours and have flow-through water supplies treated with UV units. One firm has located its intake in an approved area, while the other uses shallow wellpoints in a closed area. Each plant has either established its own laboratory or contracted with a private laboratory to meet the new monitoring requirements. The Maine Department of Marine Resources (DMR) collects weekly water and shellfish samples at each plant. Operating procedures have been developed and are in use at each plant.

FLORIDA

In Florida, where the plant operations have begun within the last five years, the requirements for certification are more stringent. Currently there are two operating depuration plants in Florida, both processing hardshell clams. Although their capacities differ (350 bushels versus 850 bushels), the procedures and requirements are similar. They both operate sporadically, averaging one run per week.

Both plants obtain their water from the restricted area adjacent to the plant and then either filter it though a sand filter or settle it in a spare tank if excessive turbidity is present (greater than 20 turbidity units). The water is treated by UV before it enters the process batch tank. These are recirculating systems which are drained, cleaned, and sanitized between batches. Each plant is required to collect one O-hour sample from each different harvesting area, three 24-hour and three 48-hour samples, which are analyzed in certified laboratories approved by the state. The state conducts a complete laboratory and administrative evaluation annually at each laboratory. The process verification studies were conducted by the state five years ago, and are being updated to meet the revised NSSP requirements.

In summary, it is evident that the three states have different requirements and are at different states of development. However, as new firms begin operating and the opportunity occurs to modify existing regulations, they will become more uniform and more consistent with the NSSP requirements.

WET STORAGE

Wet storage, which is primarily a short-term storage of market quality shellfish, also serves to remove sand and grit. Naturally-occurring marine organisms, such as the Vibrios, may also be reduced or removed during depuration and wet storage. The amount of reduction will vary and may not be significant in a particular system. If reduction is desirable, tests should be conducted to determine that it takes place consistently. Salt may be added to closed systems to increase the salinity and improve product flavor. Wet storage, in one form or another, is practiced in at least eight states. The possible configuration options are shown in Figure 1. All are currently in use. Oysters, hardshell clams, ocean quahogs, and softshell clams are frequently wet-stored. The water supply for a tank system may be from an approved area, restricted area, or artificial seawater; flow-through or recirculated; filtered through sand, charcoal, or fiberglass; and untreated (flow-through system only) or treated with UV light. Tracing the flow chart of Figure 2 will clarify the operations.

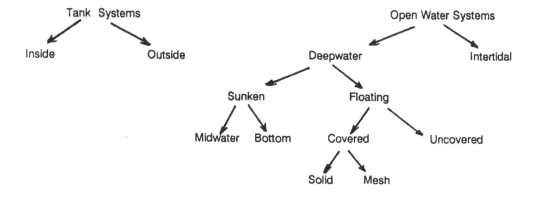

FIGURE 1. Types of Wet Storage Systems

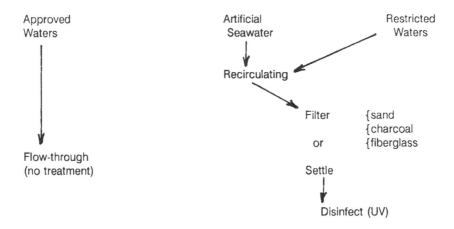

FIGURE 2. Water Sources for Wet Storage Systems

Wet storage has the potential either to improve or contaminate a good, safe product. Which of these alternatives occur depends upon the design, construction, operation, and maintenance of the facility. The requirements of the NSSP are contained in the Manual of Operations, Part II, Section C (NSSP, 1990b). Shellfish placed in wet storage are ready for direct marketing from a public health point of view.

Because one is holding them in a location of our choice, one must control as many of the variables as possible to ensure that they are protected from contamination. This makes proper operation of a wet storage facility more difficult since it requires extra vigilance to ensure that nothing happens to degrade the shellfish. Lapses in maintenance and sanitation can allow, or even encourage, the growth of bacteria (e.g., Vibrio), or adversely affect the shellfish, decreasing shelf-life and quality.

SUMMARY

In summary, we have fewer depuration plants in the U.S. now than we had a few years ago, and half of them are operational only part-time. The construction, operation, maintenance, and testing requirements for a depuration plant have become more stringent and, consequently, more expensive. This translates into a higher cost and a greater price differential between the depurated and the non-depurated shellfish. To be economically feasible, a depuration plant requires a long-term continuous supply of shellfish. Depuration allows us to utilize the shellfish that are marginal, i.e., that are harvested from areas that almost meet the approved area criteria. However, because the shellfish are contaminated and not suitable for use as food without cleansing, special precautions and requirements are necessary to ensure that cleansing is effective.

Wet storage, practiced in at least eight states, has several functions and provides a reservoir of shellfish that will help to even out supply and demand peaks and valleys, cleanse grit from the mantle cavity, and allow the salinization of the stored shellfish. Care must be exercised to ensure that the market quality product placed in wet storage is not contaminated during storage.

REFERENCES

Miescier, J. 1989. Personal communication. U.S. Food and Drug Administration, Washington, DC.

National Shellfish Sanitation Program (NSSP). 1990a. Manual of Operations, Part I: Sanitation of Shellfish Growing Areas. Public Health Service, U.S. Food and Drug Administration, Washington, DC.

National Shellfish Sanitation Program (NSSP). 1990b. Manual of Operations, Part II: Sanitation of the Harvesting, Processing and Distribution of Shellfish, Public Health Service, U.S. Food and Drug Administration, Washington, DC.

STATE MONITORING AND CERTIFICATION
FOR CONTROLLED PURIFICATION

Ken B. Moore

The development of controlled purification in recent years has been economically beneficial to the shellfish industry in the United States. Controlled purification allows the harvest and marketing of shellfish for human consumption which otherwise could not be utilized for food purposes. Increases in controlled purification facilities during the last twenty years have resulted from declines in the productivity of approved shellfish growing areas as well as increases in the acreage of shellfish growing areas restricted to harvesting for direct marketing purposes. Although controlled purification is financially beneficial for some, that is not the case for state shellfish control agencies with responsibility for monitoring this activity.

To protect public health, states must establish a program to monitor closely all activities involving controlled purification. The classification of the harvesting areas used to supply shellfish for controlled purification is a critical point. There is a definite potential human health danger if shellfish are allowed to reach markets without purification, therefore, strict supervision is essential. States have both administrative and supervisory responsibilities which must be addressed in monitoring the operations. The administrative requirements include.

1. FDA approval of the state program;
2. adequate laws and necessary resources;
3. plant design and process review;
4. establishment of a certification process;
5. a separate licensing process;
6. a plant inspection program;
7 record keeping; and
8. Memorandums of Understanding with all involved shellfish control agencies.

If controlled purification facilities plan to ship shellfish interstate, the state shellfish control agency must apply to the U.S. Food and Drug Administration for evaluation and approval of the state program. State program evaluations are necessary to ensure that states have essential public health protection elements in place to ensure that shellfish involved in controlled purification do not create a health hazard.

Promulgating adequate laws and obtaining necessary resources for monitoring and enforcement are the most difficult of the administrative responsibilities states must assume. Drafting laws and acquiring state legislative approval is a cumbersome and very time consuming task. States allowing controlled purification must adopt administrative guidelines or rules and regulations which empower the state to govern this activity. In developing shellfish rules and regulations, states are encouraged to follow National Shellfish Sanitation Program (NSSP) guidelines. All shellfish producing states participate in the NSSP and are involved in the Interstate Shellfish Sanitation Conference (ISSC). The ISSC is a national organization which, with the concurrence of FDA, establishes NSSP guidelines for shellfish sanitation.

Acquiring adequate resources is also a very difficult problem for states considering control purification. Increased budgetary problems in state government have affected controlled purification. Increased acreage of restricted waters have resulted from inadequate state resources dedicated to classification of shellfish growing areas and pollution control and abatement programs.

As the acreage classified restricted to harvesting increase as a result of funding shortages, a growth in controlled purification would be expected; however, states without adequate funding for classification and pollution control and abatement usually also lack the necessary resources to manage an adequate controlled purification program.

Other administrative requirements of an effective controlled purification program also demand considerable time and effort. A plant design and process review of proposed facilities must be conducted by the state shellfish control agency prior to their construction. State shellfish control agencies must also establish a certification process for controlled purification and then issue certification numbers to approved facilities. Once certified by a state with a FDA approved program, controlled purification facilities are included in the Interstate Certified Shellfish Shippers List which is published monthly by FDA.

A separate licensing process must be established for harvesters involved in controlled purification harvesting. A controlled purification inspection program must be established for a periodic sanitary evaluation of the facility. All sanitary items necessary to assure a safe product must be included in this inspection program. Memorandums of Understanding must also be established with all state shellfish agencies which have responsibility for controlled purification activities.

States must keep detailed records of the state controlled purification plan, sanitary survey reports, harvest area water quality data, process verification data, periodic process analyses data, and routine sanitary inspection reports.

The supervisory responsibilities states must assume are also very demanding. These responsibilities include:
1. classification of harvesting waters;
2. control of harvesting;
3. control of transportation;
4. process verification;
5. routine inspection of plants and associated follow-up; and
6. bacteriological sampling of shellfish meats and processing waters.

Each of these supervisory responsibilities are demanding and many state shellfish control agencies have found these responsibilities difficult to assume.

Shellfish intended for controlled purification must be harvested from growing areas meeting water quality requirements for approved or restricted areas (NSSP, 1990a, b). States with controlled purification programs must incorporate upper limit bacteriological criteria in establishing harvest areas for controlled purification. Classification efforts for the purpose of controlled purification are more demanding of resources than classification for controlled relaying purposes, which do not have upper limit bacteriological criteria.

Harvest control is the most difficult of all supervisory responsibilities for states. States have experimented with many techniques including bonding of harvesters, designation of reputable harvesters, and hiring of off-duty law enforcement personnel to fulfill this labor intensive responsibility, but most rely on their own shellfish patrol agency (Heil, 1988; Hickey, 1988; Wolf, 1988). Depending upon the number of harvesters involved, the man-hours consumed in harvest monitoring can be extensive. In an effort to reduce this impact, South Carolina limits harvesting for controlled purification purposes by establishing a controlled purification harvest season. The season extends from November 15 to March 15. The effort required for monitoring does not conclude with harvesting; it continues during transportation. Harvesters must be monitored from the beginning of harvest until the shellfish reach the purification facility to ensure that all shellfish harvested from restricted waters reach the proper destination.

Following classification and control of harvesting and transportation, state monitoring shifts to the processing facility. After administrative approval of plant design, field inspections are necessary to ensure proper construction, a prerequisite if equipment is to operate effectively. Once the facility is constructed and the equipment is properly installed, the state shellfish control agency must conduct verification tests to ensure that the process is capable of cleansing the shellfish to acceptable levels. The verification, which must be conducted prior to certification, includes bacteriological sampling of harvesting waters, process water, and shellfish meats both before and after purification. This sampling effort, which includes several process runs, is necessary to ensure process effectiveness, and reliability.

Monthly inspections of facilities are another necessary element of an adequate state controlled purification program. When inspections reveal deficiencies, follow-up inspections must be conducted to ensure corrective action. The inspection must include a review of all sanitation items necessary to ensure safe shellfish for human consumption. In Massachusetts, the State owns and operates the only controlled purification facility. While State ownership improves quality control, it does not relieve the financial burden. The State of Massachusetts has been unable to consistently operate the facility at a profit (Hickey, 1988).

An adequate routine sampling program must also be developed and maintained to ensure that plant efficacy continues in accordance with the established process. The sampling program must also ensure that each lot of shellfish meets established end product criteria. The sampling requirements are immense and some state shellfish control agencies, in an attempt to alleviate this resource burden, require the purification facility to acquire the services of a state certified laboratory to conduct the majority of process sampling. Although this approach can eliminate much of the sampling responsibilities of the state, verification sampling still must be conducted by the state.

The establishment and management of an FDA approved controlled purification program is not a simple task for states. Controlled purification is the most labor intensive element which exists in shellfish sanitation. States presently involved in controlled purification programs have experienced the effects of this labor intensive and resource demanding activity. These states generally feel controlled purification is being subsidized by an area of state government which simply cannot afford it. Recent declines in state funds allocated for shellfish sanitation have created fiscal problems for many state shellfish programs. This fiscal problem has occurred during a period of increased shellfish sanitation requirements and states are finding it difficult to meet NSSP guidelines in other more critical program areas. Budgetary problems are evident in many recent state evaluations conducted by FDA. The general consensus of involved states is that the resources presently being dedicated to controlled purification could best be utilized in other shellfish program areas, most notably classification (Heil, 1988; Hickey, 1988; Wolf, 1988). However, if states choose to implement a program of controlled purification, they should strive to establish funding sources capable of handling the cost of controlled purification without reducing funding for other existing shellfish sanitation program elements.

REFERENCES

Heil, D. 1988. Personal communication. Florida Departmetn of Natural Resources, Tallahassee.

Hickey, M. 1988. Personal communication. Massachusetts Division of Marine Fisheries, Sandwich.

National Shellfish Sanitation Program (NSSP). 1990a. Manual of Operations, Part I: Sanitation of Shellfish Growing Areas. Public Health Service, U.S. Food and Drug Administration, Washington, DC.

National Shellfish Sanitation Program (NSSP). 1990b. Manual of Operations, Part II: Sanitation of the Harvesting, Processing and Distribution of Shellfish, Public Health Service, U.S. Food and Drug Administration, Washington, DC.

Wolf, G. 1988. Personal communication. New Jersey Department of Health, Trenton.

REGULATORY REQUIREMENTS FOR SUBSTANCES ADDED TO DEPURATION WATER: POSSIBLE "FOOD ADDITIVE" STATUS

Stephen H. McNamara

INTRODUCTION

The U.S. Food and Drug Administration (FDA) has recently stated that ozone added to depuration water may be subject to regulation as a "food additive," and accordingly this paper will address this concern relative to (1) the definition of a food additive, (2) the regulatory consequences if a substance is deemed to be a food additive, and (3) possible options for a company that is considering the addition of a chemical substance to depuration water.

FDA REGULATION OF SHELLFISH PRODUCTION

Shellfish are subject to regulation as "food" under the Federal Food, Drug, and Cosmetic (FDC) Act Title 21, United States Code (U.S.C.) Section (§) 301, which is enforced by the FDA The FDC Act provides for seizure of "adulterated" or "misbranded" foods, and for initiation of injunction actions or criminal prosecutions against persons responsible for adulteration or misbranding of foods, 21 U.S.C. § 321 (f), 331-334, 342-343. Some of FDA's earliest recorded enforcement actions concerned shellfish. For example, the United States v. 408 Bushels of Oysters, S.D.N.Y. 1916, reported in U.S. Department of Agriculture, Decisions of Courts in Cases Under the Federal Food and Drugs Act, pp. 718-722 (U.S. Government Printing Office, l934) was a civil seizure action for condemnation of adulterated "oysters in the shell .. for the reason that [they] consisted in .. [part] of a partially filthy, decomposed, and putrid animal product, to wit, polluted oysters".

More recently, the Interstate Shellfish Sanitation Conference (ISSC) evolved to better facilitate federal and state cooperation in addressing shellfish sanitation (49 Fed. Reg. 12751; March 30, 1984). The ISSC consists of agencies from shellfish producing and receiving States, FDA, the shellfish industry, and the National Marine Fisheries Service The ISSC is recognized by FDA as "the primary voluntary national organization of State shellfish regulatory officials that will provide guidance and counsel on matters for the sanitary control of shellfish FDA has executed a formal memorandum of understanding with the ISSC "to improve the sanitation and quality of shellfish". Additional structure is noted in 50 Fed. Reg. 7797; February 26, 1985

In conjunction with the ISSC, FDA coordinates, and administers the National Shellfish Sanitation Program (NSSP). A principal objective of the NSSP is to provide a mechanism for certifying that shellfish shipped in interstate commerce meet agreed upon, specific sanitation and quality criteria (FDA Shellfish Sanitation Branch, The National Shellfish Sanitation Program -- A Guide for State Participation (undated, but in distribution in 1989). The NSSP certification system requires fresh and "fresh frozen" oysters, clams, and mussels in interstate commerce to be "tagged" by a certified dealer. The dealer must also maintain a
file identifying the source of each lot of shellfish shipped in interstate commerce. Minimum plant sanitation requirements are described in the NSSP Manual of Operations (1990a,b). Only those shellfish firms that meet the guidelines in the Manual are eligible for certification and listing in FDA's monthly publication, the Interstate Certified Shellfish Shippers List (ICSSL).

As a practical matter, FDA generally relies upon the ISSC/NSSP for regulation of fresh and "fresh frozen" shellfish in interstate commerce. However, one should remember that although FDA has agreed to "accept assistance" (49 Fed. Reg. 12751) from the ISSC, and to work together with the ISSC in implementing the NSSP, FDA has not relinquished its continuing authority independently to enforce requirements of the FDC Act whenever the agency believes such action is needed.

FDA REGULATION OF SHELLFISH DEPURATION

Controlled purification of living shellfish, also known as depuration, is a process intended to reduce the number of pathogenic organisms that may be present in shellfish harvested from moderately polluted (restricted) waters to such levels that the shellfish will be acceptable for human consumption without further processing. The process is not intended for shellfish from heavily polluted (prohibited) waters, nor to reduce the levels of poisonous or deleterious substances that the shellfish may have accumulated from their environment Procedures for regulation of depuration appear in the NSSP Manual (1990 b), Part II, Sanitation of the Harvesting, Processing and Distribution of Shellfish, section I, "Controlled Purification," pages I-I through I-22.

SUBSTANCES ADDED TO DEPURATION WATER

The NSSP Manual (1990 b) provides for use of a "water treatment system ... when necessary, to provide an adequate quantity and quality of water for operating the controlled purification process." (Part II, section I, subsection 5, "Equipment Construction and Facility Design," pages I-8 through I-II). "Currently, all plants in the United States use ultraviolet light for disinfection of process water.... Numerous studies have shown UV treatment to be highly effective for inactivating bacteria and viruses provided the units are properly maintained." (Same sections, page I-10).

However, the same section of the Manual also states, "Ozone has been used effectively for many years in Europe for treating depuration process water. Care must be taken in using other chemicals for disinfection which may react with organic components of the water supply and form compounds which adversely affect physiological activity." (same NSSP manual, 1990 b, sections, I-11).

Recently, FDA has issued a letter to the Louisiana Department of Health and Hospitals stating, inter alia, "The use of free ozone in shellfish contact waters ... [used for depuration] ... will constitute a food additive situation requiring FDA approval before use" Addendum A; FDA letter dated August 18, 1989 to Mr. Charles Conrad, Administrator, Seafood Sanitation Unit, Louisiana Department of Health and Hospitals, from Mr. L. Robert Lake, Director, Office of Compliance, FDA Center for Food Safety and Applied Nutrition. Given the significance of this FDA letter, not only for the use of ozone but also by implication, for the addition of any other substances to depuration water, it is important for persons who engage in depuration to understand the law concerning food additives, including the definition of a food additive, the consequences of food additive status, and possible options for a company that may want to add substances to depuration water.

FDA REGULATION OF FOOD ADDITIVES AND GRAS SUBSTANCES

Substances that are intentionally added to food or that are used in such a way that they may reasonably be expected to become a component of food are subject to regulation by FDA pursuant to various provisions of the FDC Act. The Act establishes two classifications that are especially pertinent to our present discussion: (1) substances that are "generally recognized as safe" (GRAS) for their intended use; and (2) substances that are not generally recognized as safe (not GRAS) for their intended use, which are termed "food additives." These classifications, which are mutually exclusive (by definition, a GRAS substance is not a food additive), are discussed below;

FOOD ADDITIVES

A "food additive" is defined by the FDC Act as follows:

> ... any substance the intended use of which results or may
> reasonably be expected to result, directly or indirectly, in its
> becoming a component or otherwise affecting the characteristics
> of any food (including any substance intended for use in
> producing, manufacturing, packing, processing, preparing,
> treating, packaging, transporting, or holding food; and including
> any source of radiation intended for any such use), if such
> substance is not generally recognized, among experts qualified
> by scientific training and experience to evaluate its safety, as
> having been adequately shown through scientific procedures (or,
> in the case of a substance used in food prior to January 1, 1958,
> through either scientific procedures or experience based on
> common use in food) to be safe under the conditions of its
> intended use.... 21 U.S.C. §321(s).

Note that this definition is truly a term of art. For example, in the common parlance, a chemical that is added to food as a preservative or to enhance flavor would be regarded as a food additive. Under the FDC Act's definition, however, such a chemical would not be deemed a food additive if it is generally recognized as safe, by experts, for its intended use. Thus, for example, monosodium glutamate is deemed by FDA to be a GRAS substance and therefore is not a food additive, 21 C.F.R. §182.1(a).

The FDC Act provides that a food additive shall be "deemed to be unsafe" if it is used without an approving food additive regulation (or an exemption), 21 U.S.C. §348(a) The Act provides further that a food shall be "deemed to be adulterated" if it "is, or it bears or contains" an unapproved food additive, 21 U.S.C. §342(a) (2) (C). Thus, the FDC Act effectively prohibits commercial use of a food additive until FDA first has published a food additive regulation that authorizes the intended use of the substance. (There are a number of exceptional situations, not discussed here, e.g., 21 U S.C. §321(s)(l)-(5). One exemption, for example, provides for investigational use of an unapproved food additive, 21 U.S.C. §348(a)(1), (i).)

Obtaining the issuance of a food additive regulation can be a lengthy and difficult undertaking. To initiate the process, an interested person must file a food additive petition, which is required to provide extensive information concerning the identity of the substance and other relevant matters, including information about intended use and, especially, "[f]ull reports of investigations made with respect to the safety of the food additive," 21 C.F.R. §171.1(c).

After the petition is filed, FDA publishes a notice of filing in the Federal Register, providing the name of the petitioner and a brief description of the proposal in general terms, 21 C.F.R. §171 1(i). The FDC Act and FDA regulations provide that, within 90 days after the filing of the petition (or within 180 days if the time is extended by the agency, as authorized by the Act), FDA either will publish in the Federal Register a food additive regulation prescribing the conditions under which the additive may be safely used, or will deny the petition. 21 U.S.C. §348(c) (1), (2); 21 C.F.R. §171.100(a).

In practice, however, FDA action on a food additive petition routinely takes much longer than 180 days. It has been estimated that the average total time (prefiling research plus FDA review) required for approval of a direct food additive is five to seven years, 49 Fed. Reg. 50, 859; December 31, 19884.

GRAS SUBSTANCES

A GRAS substance (i e., a substance that is "generally recognized, among experts ... as having been adequately shown . . to be safe under the conditions of its intended use") is not a food additive, 21 U.S.C. §321(s), and is not required by the FDC Act to be approved by FDA prior to use. If a substance is GRAS, a food manufacturer is free to use it without ever obtaining FDA approval or even notifying the agency of its use.

FDA regulations provide the agency's interpretation of "general recognition of safety" from 21 C.F.R. § 170.30(a), (b);

> General recognition of safety requires common knowledge about the substance throughout the scientific community knowledgeable about the safety of substances directly or indirectly added to food.

> * * *

> General recognition of safety through scientific procedures shall ordinarily be based upon published studies which may be corroborated by unpublished studies and other data and information.

FDA has published numerous regulations that recognize the GRAS status of particular food substances, 21 C.F.R. Parts 182, 184, 186. These regulations are commonly known as the FDA "GRAS list." Nevertheless, FDA also has explicitly acknowledged that the food ingredients listed as GRAS in the agency's regulations -

> do not include all substances that are generally recognized as safe for their intended use in food. Because of the large number of substances the intended use of which results or may reasonably be expected to result, directly or indirectly, in their becoming a component or otherwise affecting the characteristics of food, it is impracticable to list all such substances that are GRAS, 21 C.F.R. §170.30(d) (Emphasis added.) and to same effect, 21 C.F.R. §182.1(a).

If a manufacturer chooses to use a food substance that he believes is GRAS but that FDA has not officially listed as GRAS, he assumes the risk that someday the agency may take note of his marketing and disagree with his assessment of GRAS status. If FDA should conclude that such a substance is not GRAS, and thus an unapproved food additive, the agency could undertake regulatory action to stop its marketing.

Because of the uncertainties and regulatory risks attendant upon use of a food substance without explicit prior approval by FDA, many food companies are reluctant to use a new substance that the agency has not either acknowledged to be GRAS or approved by a food additive regulation, even if they believe the substance is in fact GRAS and that it properly may be used without FDA approval.

In general, it appears that in recent years, FDA enforcement personnel have become increasingly hostile toward a company's self-determination, without consultation with FDA, that a substance is GRAS. Furthermore, when a company asks FDA about a novel use of a substance with which the agency is not familiar, FDA personnel generally will advise that a formal petition requesting either food additive approval or GRAS affirmation should be filed with the agency to resolve the matter. Moreover, FDA personnel may encourage a company to refrain from use not only of a substance for which a food additive petition is pending, but even for a substance that is the subject of a pending petition for GRAS affirmation -- until the agency reaches a final decision on the matter. See, e.g., Food Chemical News, October 30, 1989, page 24.

When asked by the press or the general public about the status of a novel substance for which there is no existing food additive or GRAS regulation, FDA personnel tend to make statements to the effect that the agency has not sanctioned the use of the substance, even if data in the published scientific literature might support the view that the substance is in fact GRAS for food use. As a practical matter, of course, such statements tend to deter use of substances that have not been officially approved by FDA.

APPLICATION TO SUBSTANCES ADDED TO DEPURATION WATER

Applying the law discussed above to substances that are intended to be added to depuration water, one can make the following observations:

(1) If a substance that is added to depuration water would reasonably be expected to come into contact with the shellfish, and if the substance is not "generally recognized," by experts, to be safe for such use, based on data available in the published scientific literature, the substance may be deemed by FDA to be a "food additive." Food additive status would require that the substance not be used until a food additive petition has been filed with FDA and the agency has published a regulation approving use. For example, FDA personnel have stated that the use of "free ozone in shellfish contact waters ... will constitute a food additive situation requiring FDA approval prior to use", (FDA letter to Louisiana Department of Health and Hospitals; Addendum A).

(2) If, however, after treatment of the depuration water a substance ceases to be present in the water and leaves no reaction products (e.g., to quote FDA in the case of ozone, if there is "no residual ozone or oxidation products" in the water), FDA would not regard the substance as a "food additive "

(3) FDA's position that ozone added to depuration water may constitute a food additive might possibly be questioned on the basis that resulting levels of ozone and/or oxidation products present in the shellfish would be so negligible as to present no public health or safety concerns. Monsanto Co. v. Kennedy, 613 F.2d 947 (D.C. Cir. 1979) is instructive here. In that case, the court held that a substance intended for use as a food contact material (i.e., in beverage containers) was not necessarily a food additive just because of the possibility of very low level of migration to the food (beverage), 613 F.2d at 954-55. The court stated specifically that:

> For the ... ["component' element of the food additive definition, 21 U.S.C. §321(s)] ... to be satisfied, Congress must have intended the Commissioner to determine with a fair degree of confidence that a substance migrates into food in more than insignificant amounts .... Thus, the Commissioner may determine based on the evidence before him that the level of migration into food of a particular chemical is so negligible as to present no public health or safety concerns.... 613 F.2d 955.

(4) Furthermore, if a substance is generally recognized, among experts, to be safe for its intended use, based upon published scientific literature (i.e., if the substance is "GRAS"), the substance is not a food additive, and FDA approval would not be required.

(5) However, even if a company believes a substance is GRAS and possesses substantial data to support that view, nevertheless, unless FDA is consulted, there would be a risk that the

agency might subsequently take note of any use of the substance and disagree with the company's assessment of GRAS status. Use of the substance without FDA concurrence would be at continuing risk of the possibility that FDA might assert that the substance is an unapproved food additive and undertake regulatory action to try to stop its use.

(6) When asked by a company whether a novel use of a substance is GRAS, FDA often asks the company to file either a petition for GRAS affirmation or a food additive petition as the means for resolving the question -- which, in effect, asks the company to obtain formal FDA approval even for a substance the company believes is GRAS.

(7) If the NSSP Manual were to identify a particular substance as appropriate for use in depuration water, this could be strong evidence that such a use is in fact GRAS, and might avoid the need for preparation of a petition for GRAS affirmation. Furthermore, if FDA were to participate favorably in such an amendment of the Manual, such action would provide reasonable assurance that the agency would not take regulatory action against the described usage. However, it is not clear whether FDA would be willing to sanction the addition of particular substances to depuration water in this manner. It should be noted that FDA issued its warning letter to the Louisiana Department of Health and Hospitals, stating that the "use of free ozone in shellfish contact waters" would "constitute a food additive situation requiring FDA approval before use," notwithstanding the fact that the NSSP Manual, Part II, section I, page I-11, states, "Ozone has been used effectively for many years in Europe for treating depuration process water." This does not sound like an agency that is willing to resolve issues concerning GRAS status of depuration water additives by amendments to the NSSP Manual.

CONCLUSION

If a substance added to shellfish depuration water is not "generally recognized" by experts, based on published scientific literature, to be safe for such use, the substance may be subject to regulation as a food additive and require FDA approval of a food additive petition prior to use. Accordingly, companies should be certain to evaluate carefully the status of any substance intended to be added to shellfish depuration water, to assure that the substance complies with any applicable FDA-enforced requirements prior to use.

REFERENCES

National Shellfish Sanitation Program (NSSP). 1990a. Manual of Operations, Part I: Sanitation of Shellfish Growing Areas. Public Health Service, U.S. Food and Drug Administration, Washington, DC.

National Shellfish Sanitation Program (NSSP). 1990b. Manual of Operations, Part II: Sanitation of the Harvesting, Processing and Distribution of Shellfish, Public Health Service, U.S. Food and Drug Administration, Washington, DC.

Addendum A.   Letter from U.S. Food and Drug Administration to the Louisiana Department of Health and Hospitals concerning use of 'ozone' to purify water used in shellfish wet storage or depuration.

 DEPARTMENT OF HEALTH & HUMAN SERVICES   Public Health Service

_____

Food and Drug Administration
Washington DC 20204

August 18, 1989

Mr. 'Charles Conrad, Administrator
Seafood Sanitation Unit
Louisiana Department of Health and Hospitals
Post Office Box 60630
New Orleans, Louisiana   70160

Dear Mr. Conrad:

This is in response to your letter of April 10, 1989, to Mr. David M. Dressel, and your subsequent telephone conversation with him, requesting that the Food and Drug Administration evaluate, for approval purposes, the use of ozone or "Photozone" to purify process water in shellfish wet storage and depuration applications.

Based on the literature you provided us, we understand "Photozone" to be ozone which is produced by exposing air to ultraviolet (UV) light radiation in the range of 185 nanometers. Ozone is a powerful germicidal agent as well as an oxidant. Our concern over the use of ozone is limited to the extent that its use might result in changes in the composition of the shellfish and that its use is effective. Requisite performance criteria to address our concerns differ between wet storage and depuration operations as follows.

Wet Storage

The wet storage process as practiced in Louisiana involves the use of aerated, ozone-treated, artificial seawater made from well water and sea salts. Prior to introduction into the wet storage tanks the process water is subjected to aeration, ozonation, sand-filtration (when necessary) and UV irradiation. The water then enters the storage tanks by means of submerged perforated distribution pipes to provide uniform flow. After exiting the wet storage tanks, the water passes through a lamella separator to remove solids and then recirculates through the system.

Based on our review of this process and the information submitted by the Shellfish Sanitation Branch, we have no objection to the use of ozone in wet storage facilities for shellfish, provided the following conditions are met:

   1. The shellfish must come from approved growing areas.

2. The turbidity of water receiving ozonation or UV disinfection shall not exceed 20 nephelometric turbidity units.

3. The ozonated water should retain a residual concentration of ozone of approximately 0.5 mg/L for at least 20 minutes. These conditions should provide adequate disinfection so that there are no detectable levels of coliform organisms/100 ml of water entering the wet storage tanks.

4. Water in contact with shellfish shall contain no free ozone. Residual ozone shall be removed by exposing the water to UV radiation at 254 nanometers prior to contact with shellfish.

5. The recirculated water should be discarded after 72 hours to prevent the buildup of potentially harmful oxidation by-products.

6. In the artificial seawater formulations, bromides should be avoided to reduce formation of secondary oxidants such as hypobromites which are potentially toxic to shellfish.

Depuration

The use of ozone, "Photozone", or the simultaneous application of ozone and UV light in shellfish depuration is not presently sanctioned by FDA for use in the United States. Ozone use under these circumstances may constitute a food additive situation requiring FDA approval before use.

If FDA could be assured that water used for depuration contains no residual ozone or oxidation products we would have no objection to this use of ozone. Satisfactory performance criteria which must be met in addition to those for wet storage include: 1) shellfish contact waters must be constantly monitored to assure there is no free ozone, 2) process waters must be discarded after each batch depuration, and 3) end product microbiological testing must confirm that safety criteria have been met before depurated shellfish products are marketed.

The use of free ozone in shellfish contact waters or conditions not covered above will constitute a food additive situation requiring FDA approval before use. Then, it will be necessary to submit either a food additive or generally recognized as safe (GRAS) affirmation petition seeking FDA approval in accordance with 21 CFR 171.1 or 21 CFR 170.35, respectively.

Affirmation of GRAS status of ozone to disinfect process water used in shellfish depuration may be based on its historical use in Europe if a literature review indicates that it has pre-1958 use without known public health problems. The procedures for this evaluation are contained in 21 CFR 170.35.

Once safety of the use of ozone for depuration has been determined, satisfactory compliance criteria will be provided under the National Shellfish Sanitation Program (NSSP) through the Manual of Operations. The issue dealing with free ozone in shellfish contact waters to remove <u>Vibrio vulnificus</u> and other contaminants from the shellfish will require significant research. Depending upon whether ozone or oxidation products directly contact the shellfish, toxicological studies may be required to verify that toxic compounds are not formed when shellfish take in active ozone.

We trust that this information will be helpful to you. If you have any other questions, please do not hesitate to contact me.

Sincerely yours,

L. Robert Lake
Director
Office of Compliance
Center for Food Safety
and Applied Nutrition

Microbial Concerns

EPIDEMIOLOGY OF MOLLUSCAN - BORNE ILLNESSES

Karl C. Klontz and Scott R. Rippey

INTRODUCTION

Illnesses associated with the consumption of molluscan shellfish (defined here as oysters, clams, cockles, and mussels) are caused by infectious agents (bacteria and viruses), chemical toxins, and allergies. This review emphasizes the epidemiology of the principal recorded illnesses resulting from infectious agents and chemical toxins.

GENERAL CONSIDERATIONS

Several factors contribute to the role of mollusks as vehicles in the transmission of illness caused by bacteria and viruses. First, mollusks are harvested primarily from estuarine and near-coastal environments, and these areas are generally contaminated, to varying degrees, with human waste. Second, as filter feeders, mollusks concentrate potential viral or bacterial pathogens present in seawater. Third, humans characteristically eat the entire mollusk, including the gastrointestinal tract where pathogens are concentrated. Finally, because shellfish are frequently eaten raw or only partially cooked, viable pathogens may he introduced to the human alimentary tract.

Symptomatic molluscan-borne infections in humans may present themselves in one or more clinical syndromes. Table 1 lists the major clinical syndromes associated with bacterial and viral molluscan-borne infections in humans.

BACTERIAL AGENTS

Bacterial agents that have been associated with molluscan-borne illness in humans include Salmonella typhi (Pennington et al. 1902; Lumsden et al. 1925; Ramsey et al. 1928; Hart, 1945) and paratyphi (Ramsey et al. 1928; Anon, 1973), Vibrio species (Morris et al. 1980; Blake et al. 1980a; Pollitzer, 1959; Klontz et al. 1987; Pavia et al. 1987; Anon, 1989; Blake et al. 1980b; Lin et al. 1986; Lowry et al. 1989; Anon, 1981-1987; Baine et al. 1974; Di Lorenzo et al. 1974; Blake et al. 1977, McIntyre et al. 1979; Salmaso et al. 1980; Morris et al. 1984; Tacket et al. 1984; Spellman et al. 1986; Morris et al. 1982, Blake et al. 1979; Klontz et al. 1988), Shigella species (Reeve et al. 1989, Black et al. 1978), and Campylobacter jejuni (Griffin et al. 1983) In addition, Plesiomonas shigelloides and Aeromonas hydrophila have been implicated as possible shellfish-borne disease agents (Holmberg et al. 1984).

At least four outbreaks of typhoid fever linked to the consumption of contaminated mollusks occurred during this century in the United States (Pennington et al. 1902; Lumsden et al 1925; Ramsey et al 1928; Hart, 1945). One of the earliest described oyster-associated outbreaks occurred in 1902 in Atlantic City, New Jersey, where 82 cases of typhoid fever were reported. The oysters had been "freshened" and "fattened" in adjacent city waters that were contaminated with human sewage (Pennington et al. 1902). The largest molluscan-borne outbreak of typhoid fever in the United States, linked to raw oysters, occurred in 1924-1925 (Lumsden et al. 1925). Outbreaks were reported in Chicago, New York City, and Washington, D.C., with 10 other cities reporting a markedly increased prevalence of typhoid fever. Nationwide, more than 1,500 cases and 150 deaths occurred above normal expectancy for the period, and nearly 80 percent of cases reported eating raw oysters in the 30-day period before becoming ill. The implicated oysters were harvested from approved New York waters. However, they were subsequently stored in waters that may have been contaminated with human sewage.

TABLE 1.   Clinical presentations of molluscan-borne bacterial and viral infections in humans.

Organism	Clinical presentation		
	Gastro-enteritis	Hepatitis	Septicemia
Bacteria			
Salmonella			
S. typhi			xx
S. paratyphi			xx
Vibrio			
V. cholerae O1	xx		
V. cholerae non-O1	xx		x
V. parahaemolyticus	xx		x
V. mimicus	xx		
V. hollisae	xx		
V. fluvialis	xx		
V. vulnificus	x		xx
Shigella	xx		
Campylobacter*	xx		
Viruses			
Hepatitis A virus		xx	
Norwalk-like viruses			
Norwalk virus	xx		
Cockle virus	xx		
Snow Mountain agent	xx		

xx = most common presentation
x = less common presentation
* = role not well defined

One important result of the 1924-1925 typhoid fever outbreak was the initiation, in the United States, of an organized effort to improve the sanitary environment of the shellfish industry (Anon, 1925). These initiatives developed into the current National Shellfish Sanitation Program and the recently created Interstate Shellfish Sanitation Conference (ISSC), a cooperative alliance between federal, state, and industry groups formed to maintain high standards in the shellfish industry.

At least five species of bacteria in the genus Vibrio have been associated with molluscan-borne illness in humans (Table 1). While the most common clinical presentation of molluscan-borne Vibrio infection is gastroenteritis, septicemia is the most common presentation in patients with V. vulnificus infection. Septicemia also occurs occasionally in persons with V parahaemolyticus and V. cholerae non-O1 infections (Morris et al. 1985; Blake et al. 1980a).

While circumstantial evidence incriminating oysters in the spread of cholera was recorded as early as 1907 (Pollitzer, 1959), only recently have epidemiologic studies focused on the specific role sf raw oysters as a vehicle for toxigenic V. cholera O1. In the United States the first case of raw oyster-associated cholera was reported in 1986 (Klontz et al. 1987); by the end of 1988, seven more cases of raw oyster-associated cholera were reported (Pavia et al. 1987; Anon, 1989; and personal communication, William Levine, Centers for Disease Control). Cases of crab or shrimp associated cholera have occurred sporadically since 1973 in persons who live or have visited states along the Gulf of Mexico (Blake et al. 1980; Lin et al. 1986; Lowry et al. 1989), suggesting that toxigenic V. cholerae O1 may be endemic in this region (Blake et al. 1980b). This area also yields nearly one half of the annual United States production of oysters (Anon, 1981-1987).

Several outbreaks of molluscan-borne cholera (O1, toxigenic) have been reported recently outside the United States. During a 1973 outbreak of cholera in Italy involving 278 cases, some of which were linked to the ingestion of mussels, significant differences were first documented between cases and controls as to whether any seafood was eaten in the five days prior to the patient's illness (Baine et al. 1974; Di Lorenzo et al. 1974). A case-control study conducted during an epidemic of cholera in Portugal in 1974, with 2,467 hospitalized cases, demonstrated cockles to be an important risk factor for illness (Blake et al. 1977). The implicated cockles were eaten raw or partially cooked after being harvested from mud flats contaminated with untreated human sewage. In the Gilbert Islands in 1977, clams obtained from a sewage-contaminated lagoon were one of several seafood vehicles shown to be significantly associated with illness in an outbreak of cholera involving 572 cases (McIntyre et al. 1979). In 1979 a cluster of 12 cases of cholera, linked to the consumption of clams, was identified in Sardinia, Italy (Salmaso et al. 1980).

In addition to their role as vehicles in the transmission of toxigenic V. cholera O1, mollusks have been implicated as vehicles in cases of gastroenteritis and/or septicemia in patients from whom other Vibrio strains or species have been recovered, including nontoxigenic V. cholerae O1 (Morris et al. 1985), non-O1 V. cholerae (Blake et al. 1980a), V. mimicus (Morris et al. 1985), V. vulnificus (Tacket et al 1984), V. fluvialis (Spellman et al. 1986); and V. hollisae (Morris et al 1982).

The role of raw oysters as vehicles of transmission in infections with V. vulnificus only recently has been described (Tacket et al. 1984; Blake et al 1979). Of serious concern with this pathogen is the high mortality rates (>45%) observed among patients developing primary septicemia (Blake et al. 1979; Klontz et al. 1988). In a series of V. vulnificus infections in humans in Florida between 1981-1987, 64.5 percent occurred in persons who ate raw oysters in the week prior to onset of illness (Klontz et al. 1988). Two-thirds of patients with primary septicemia in this series had liver disease, which has been shown to be a significant risk factor in a case-control study (Tacket et al. 1984).

Mollusks have also been implicated in at least two outbreaks of shigellosis (Reeve et al. 1989; Black et al. 1978) and one outbreak of Campylobacter (Griffin et al. 1983) gastroenteritis. A raw oyster-associated outbreak of shigellosis occurred in Texas in 1986 involving 24 cases The oysters eaten by the patients were harvested from approved beds that may have been contaminated by the feces of an infected oysterman (Reeve et al. 1989). Another shigella outbreak resulted from the consumption of raw oysters harvested from waters that were shown to be polluted (Black et al. 1978).

VIRAL AGENTS

Viral agents that have been associated with molluscan-borne illness in humans include hepatitis A virus (Roos, 1956; Dienstag et al. 1976; Dougherty et al. 1962; O'Mahoney et al. 1983; Mason et al. 1962; Rindge, 1962; Ruddy et al. 1969; Anon, 1964; Dismukes et al. 1969; Feingold, 1973; Bostock et al. 1979; Ohara et al. 1983; Mele et al. 1989; Koff et al. 1967), Norwalk virus (Murphy et al. 1979; Linco et al. 1980; Grohmann et al. 1981; Morse at al. 1986), and Norwalk-like viruses (including cockle virus and Snow Mountain agent) (Appleton et al. 1977; Truman et al. 1985).

The first reported outbreak of molluscan-borne hepatitis A occurred in 1955-1956 in Sweden, where 629 persons became ill after ingesting oysters that had been held in a sewage-contaminated harbor (Roos, 1956). Since then, mussels (Dienstag et al. 1976), clams (Dougherty et al. 1962), and cockles (O'Mahoney et al. 1983) have also been implicated as vehicles of infection in outbreaks of hepatitis A. In the United States, the first two published reports of molluscan-borne hepatitis A outbreaks occurred in 1961 (Dougherty et al. 1962; Mason et al. 1962). The first reported outbreak of clam-associated hepatitis A occurred in New Jersey, and involved 459 patients (Dougherty et al. 1962). The clams were harvested from Raritan Bay, which received a great volume of sewage from New Jersey and New York municipalities. An outbreak of raw oyster-associated hepatitis A involving 84 cases occurred in Alabama and Mississippi (Mason et al. 1962). The implicated oysters were harvested from the Pascagoula River, which was contaminated with raw sewage from the city of Pascagoula. Between 1963-1973, at least seven other molluscan-borne hepatitis A outbreaks involving over 700 cases were reported in the United States (Rindge, 1962; Ruddy et al 1969; Anon, 1964; Dismukes et al. 1969; Feingold, 1973). Published reports of outbreaks of hepatitis A linked to mollusk ingestion have been described in Australia (Dienstag et al 1976), England (Bostock et al. 1979) the Philippines (Ohara et al 1983), and Italy (Mele et al. 1989) as well

Molluscan shellfish have been shown to be an important vehicle as well in the transmission of hepatitis A during time periods when no distinct outbreaks are detectable. In a study of viral hepatitis in a group of hospitals, Koff et al. (1967) showed that in a period of declining incidence of hepatitis in Boston, ingestion of raw clams and oysters was as common a source of infection as contact with jaundiced persons.

An Australia-wide outbreak in 1978 of raw oyster-associated gastroenteritis, with at least 2,000 cases, was the first time Norwalk virus was definitively identified and implicated as the cause of molluscan-borne illness (Murphy et al. 1979). In December 1978 a second outbreak of Norwalk virus-associated gastroenteritis linked to oyster consumption occurred in Darwin, Australia, involving at least 150 persons (Linco et al. 1980). The oysters implicated in both outbreaks were harvested from beds in the Georges River. Subsequent testing of this growing area demonstrated elevated fecal coliform densities indicative of recent fecal contamination. The oysters consumed by persons in the Darwin outbreak were frozen for 15 weeks prior to being eaten, suggesting that Norwalk virus may remain viable in oysters, held under these conditions, for protracted periods (Linco et al. 1980).

Following the 1978 Australia Norwalk virus outbreaks, several programs were introduced in Australia to minimize the possibility of further outbreaks. A program of oyster depuration was commenced in late 1978, and monitoring of the effectiveness of the process was carried out using human volunteers (Grohmann et al. 1981). These depurated oysters consumed by the volunteers were below the legal limit of bacterial contamination (< 2.3 coliform organisms per gram of oyster flesh). Fifty-two volunteers who ate oysters harvested from the Georges River experienced gastroenteritis, representing an average sickness rate of 1.8% over 32 weeks of testing. Norwalk virus was identified in a fecal specimen from eight ill persons (Grohmann et al. 1981). Three other outbreaks of gastroenteritis of probable viral etiology have been reported in England among persons who consumed depurated mollusks (Gunn et al. 1969; Gill et al. 1983; Heller et al. 1986). In only one outbreak was a viral candidate for illness observed in human stool specimens (a small round structured virus) (Gill et al. 1983).

In the United States, an epidemic of clam- and oyster-associated Norwalk virus gastroenteritis occurred in 1982 in New York state, with 103 separate outbreaks involving 1,017 cases identified (Morse et al. 1986). A dose-response relation was demonstrated for the number of clams eaten and the risk of illness. All clam and oyster specimens tested had total and fecal coliform counts within acceptable standards. However, 4 of 6 shellfish specimens from two outbreaks yielded Norwalk virus antigen by radioimmunoassay (Morse et al. 1986).

Molluscan-borne outbreaks of gastroenteritis have been reported among persons from whom two other Norwalk-like viruses have been identified (Appleton et al. 1977; Truman et al. 1985). In England in 1976-1977, 797 cases of gastroenteritis were recorded in 33 outbreaks linked to the consumption of cockles that were harvested from waters where pollution by sewage was known to occur (Appleton et al. 1977). Stool samples from some ill persons yielded a 25-26 nm Norwalk-like virus, subsequently termed cockle virus. Similar but smaller outbreaks occurred in 1978, 1979, and 1980 in England (Turnbull et al. 1982). The cause of the outbreak in 1976-1977 was believed to be an underprocessing of cockles in recently installed steaming equipment (Turnbull et al. 1982). In Rochester, New York in 1983, 84 cases of gastroenteritis were identified among persons who ate raw or baked clams that had been harvested off the southeast coast of Massachusetts (Truman et al. 1985). During the week prior to gathering the clams, raw sewage was discharged into waters near the clam harvest sites. The virus recovered from stool specimens from some of the patients was identified as Snow Mountain agent, a virus similar in size and shape to the Norwalk virus.

INFECTIOUS AGENTS--ETIOLOGY UNKNOWN

Many outbreaks resulting from shellfish consumption involve gastroenteritis of unknown origin (Gunn et al. 1969; Gill et al. 1983; Heller et al. 1986; Ratzan et al. 1969). Most of these outbreaks are characterized by a similar incubation period (24-48 hours), symptomology (mild gastroenteritis), illness duration (<48 hours), and prognosis (uneventful resolution) A physician's attention is generally not required. Some of these illnesses may represent undiagnosed infections with Norwalk virus or Norwalk-like viruses, a group of viruses that requires relatively sophisticated methodologies to identify.

CHEMICAL TOXINS

At least four syndromes in humans resulting from the ingestion of toxins produced by marine dinoflagellates have been associated with molluscan shellfish consumption. These include paralytic shellfish poisoning (PSP) (Meyer, 1953; Ayres, 1975, Shimizu, 1989; Vancouver, 1789, Sommer et al. 1937; Medcof, 1985; Meyer et al. 1928; Hughes, 1979; McCollum et al. 1968), neurotoxic shellfish poisoning (NSP) (Shimizu, 1989; Sakamoto et al. 1987; Baden, 1989), diarrheic shellfish poisoning (DSP) (Baden, 1989; Underdal et al. 1985; Yasumoto et al. 1984; Krogh et al. 1985), and a newly described intoxication presently being referred to as amnesic shellfish poisoning (ASP) (Gray, 1988; Perl et al. 1988). Illness in humans generally occurs during periods of intense multiplication of the planktonic dinoflagellates, a phenomenon termed "red tides", and the subsequent accumulation of these organisms by shellfish. Paralytic shellfish poisoning (PSP) is one of the most severe forms of food poisoning caused by ingestion of seafood. The case-fatality rate for patients with PSP in two large series of reported cases was 8.5 and 9.5 percent, respectively (Meyer, 1953; Ayres, 1975). Initially, it was believed that a single toxin, saxitoxin, was the cause of the clinical manifestations of illness in humans. However, recent advances in PSP toxin research have elucidated more than a dozen additional toxins produced by several different species of marine dinoflagellates (Shimizu, 1989). The most common clinical features of PSP include circumoral paresthesia, numbness of the tongue, paresthesia of the fingers and toes, a feeling of floating, and weakness of the limbs. Death results from respiratory paralysis. There is no way to detoxify the toxins, and treatment is supportive.

Outbreaks of molluscan-borne PSP have occurred world-wide (Meyer, 1953). One of the earliest cases to be described occurred in 1793 in a seaman on a vessel exploring passages off the mainland coast of what is now British Columbia; the man died 6 hours after eating roasted mussels (Vancouver, 1789). Several studies have reported the occurrence of PSP outbreaks in a defined geographic region over multi-year periods (Meyer, 1953; Ayres, 1975; Sommer et al. 1937, Medcof, 1985). Along the Pacific coast of the United States between 1927-1936, PSP was recorded in 243 persons, 16 of whom died (Sommer et al. 1937). All cases occurred in the months between May through October. Included in this series was an outbreak of mussel-associated PSP, involving 102 cases and 6 deaths, that occurred in 1927 in San Francisco (Meyer et al. 1928). From 1971 to 1977, 10 outbreaks of PSP, totaling 63 patients and no deaths, were reported to the Centers for Disease Control (Hughes, 1979). Information was available on method of shellfish preparation for 8 outbreaks: the mussels or clams were eaten raw in 2 outbreaks, and steamed, boiled, or fried in the remaining outbreaks. The median incubation period in the 10 reported PSP outbreaks was 45 minutes, with a range of 15 minutes to 3 hours, while the median duration of illness was 72 hours, with a range of 6 to 168 hours. Twenty (32%) patients required hospitalization, and 4 (6%) required mechanical ventilatory assistance.

In Britain between 1827-1968, 10 incidents, all associated with mussels, were sufficiently documented to be reliably classified as PSP (Ayres, 1975). The largest epidemic of the series occurred in 1968 in persons who ate cooked mussels gathered from the northeast Coast of England (McCollum et al. 1968). Seventy-eight cases were identified, with no fatalities. In Canada in the Fundy and St. Lawrence regions between 1899-1970, there were 246 known cases of PSP, including 24 deaths (Medcoff 1985). Most cases were associated with clams or mussels.

Far less severe than PSP, neurotoxic shellfish poisoning (NSP) in humans results from the consumption of mollusks contaminated with toxins produced by the marine dinoflagellate Ptychodiscus brevis (formerly Gymnodinum breve). Red tides composed of P. brevis generally occur in the Gulf of Mexico, and are the cause of massive fish kills (Shimizu, 1989). In humans, the symptoms of NSP are mild. Illness characteristically begins within minutes to 3 hours after ingestion of contaminated mollusks, and subsides within 36 hours (Sakamoto et al. 1987). Symptoms include paresthesia around the mouth and fingers, vomiting and diarrhea, ataxia, and hot-cold reversal of temperature sensation (Baden, 1989). Although P. brevis has been found in sea water from various diverse coastal regions in the Gulf of Mexico, NSP in the United States has been reported only on the west coast of Florida (Sakamoto et al. 1987). Oysters and clams are the mollusks most often implicated in NSP (Baden, 1989).

In diarrheic shellfish poisoning (DSP), gastroenteritis may follow the consumption of mollusks (particularly mussels) contaminated with toxins produced by dinoflagellate species in the genus Dinophysis (Baden, 1989). In humans the most common symptoms of DSP are diarrhea, vomiting, abdominal pain, nausea, and headache. The incubation period may vary considerably, from less than half an hour to ten to fifteen hours (Underdal et al. 1985). Patients usually recover within 48 hours, though hospitalization may be necessary in some instances. No deaths have been reported (Underdal et al. 1985).

DSP was first observed in Japan (Yasumoto et al. 1984), and has been encountered in recent years in the following European countries as well: The Netherlands, Spain, and France, Norway, and Sweden (Krogh et al. 1985). A large outbreak of DSP occurred in Norway and Sweden in 1984-1985. An estimated 300-400 persons became ill after consuming freshly harvested and normally prepared mussels (Underdal et al. 1985). A finding of public health importance from this outbreak was the duration of mussel toxicity; although the dinoflagellate bloom terminated before the end of October 1984, mussels continued to be toxic through April 1985 (Krogh et al. 1985).

Amnesic shellfish poisoning (ASP) intoxication was first observed in eastern Canada in late 1987, when 130 cases and 2 deaths were reported following the ingestion of mussels (Gray, 1988). The toxin associated with this outbreak was identified as domoic acid. The source of the toxin has not been determined yet. Gastrointestinal and/or neurologic symptoms characterized the illness (Perl et al. 1988). Neurologic symptoms included confusion, memory loss, disorientation, and a facial grimace or chewing motion. Gastrointestinal symptoms included vomiting, abdominal cramps and diarrhea. The most serious illness occurred in elderly persons. No antidote is available, and supportive care, such as respirators, was required for some patients.

ALLERGIC REACTIONS

True allergic reactions to seafoods occur through an immunologic mechanism, with IgE-mediated allergies being the only thoroughly documented mechanism at this time (Taylor et al. 1989). Allergic reactions can occur to all classes of seafoods, including fishes, mollusks, and crustacea. Although not all species have been described as allergenic in the medical literature, there is no reason to doubt the potential allergenicity of all seafood species (Taylor et al 1989).

FREQUENCY OF MOLLUSCAN--BORNE OUTBREAKS

Only limited information exists from which to determine the relative importance of the role that seafood, in general, and mollusks, in particular, assume as vehicles in outbreaks of foodborne illness. However, the foodborne outbreak surveillance system in England and Wales, as well as that in the United States, suggests that seafood plays a relatively minor role as a vehicle in causing outbreaks of foodborne illness. In England and Wales between 1969-1978, a food vehicle was determined for 16.6% (1,078/6,500) of all foodborne outbreaks of bacterial etiology, seafood was implicated as the vehicle in only 19 (1.8%) of these 1,078 outbreaks in which a vehicle was identified (Vernon et al. 1974; Vernon, 1977; Hepner, 1980).

Although the foodborne outbreak surveillance data published annually by the Centers for Disease Control have limitations that prevent firm conclusions about the absolute incidence of outbreaks of foodborne illness by etiologic agent and implicated vehicle, it is noteworthy that for eight of the 10 years between 1972-1981, mollusks were the implicated vehicle in less than 2 percent of all reported outbreaks for which a vehicle was known (Anon, 1972-1981).

The magnitude of molluscan-borne outbreaks should not be minimized, however. Reports of over 5,000 cases of gastroenteritis believed to be of viral etiology were recorded for the period 1980-1989 by the Northeast Technical Services Unit, U.S. Food and Drug Administration, utilizing a variety of published and unpublished sources (Rippey, 1989). The ratio of total cases to those actually reported for all foodborne diseases has been estimated at 25:1 (Hauschild et al 1980). This being the case, one could estimate over 125,000 cases of viral-implicated gastroenteritis among shellfish consumers for the decade of the 1980s.

REFERENCES

Anon. (1925). Report of the Committee on Sanitary Control of Shellfish Industry in the United States. Pub Health Rep, Suppl no. 53, 1-17.

Anon. (1964). Shellfish-associated hepatitis. Part A: New Jersey and Pennsylvania . Part B: Connecticut. Part C: North Carolina. Hepatitis Surveillance Report No. 19, Atlanta, Centers for Disease Control, 39-40.

Anon. (1972-1981). Centers for Disease Control. Foodborne Disease Outbreaks Annual Summary

Anon. (1973). Surveillance of typhoid and paratyphoid fevers - 1971. World Health Organization. Weekly Epidemiological Record 24, 245-248.

Anon. (1981-1987). Landings for the United States, National Marine Fisheries Service, National Oceanic and Atmospheric Administration, United States Department of Commerce.

Anon. (1989). Centers for Disease Control. Toxigenic Vibrio cholerae O1 infection acquired in Colorado. MMWR 38, 19-20.

Appleton, H. and Pereira, M.S. (1977). A possible virus etiology in outbreaks of food-poisoning from cockles. Lancet i, 780-781.

Ayres, P.A. (1975). Mussel poisoning in Britain with special reference to paralytic shellfish poisoning. A review of cases reported 1814-1968. Environ Health, 261-265.

Baden, D.G. (1989). Public health problems of red tides. In: Tu AT, ed. Handbood of Natural Toxins. Vol. 3. Marine Toxins and Venoms. New York: Marcel Dekker, Inc., 259-277.

Baine, W.B., Mozzoti, M., Greco, D., Izzo, E., Zampieri, A., Angioni, G., Di Gioia, M.,Gangarosa, E. and Pocchiari, F. (1974). Epidemiology of cholera in Italy in 1973. Lancet ii, 1370-1374.

Black, R.E., Craun, G.F., and Blake, P.A. (1978). Epidemiology of common-source outbreaks of shigellosis in the United States, 1961-1975. Am J Epidemiol 108, 47-52.

Blake, P.A., Rosenberq, M.L., Costa, J.B., Ferreira, P.S., Guimares, C.L., and Gangarosa, E.J. (1977). Cholera in Portugal, 1974. I. Mode of transmission. Am J Epidemiol 105, 337-343

Blake, P.A., Merson, M.H., Weaver, R.E., Hollis, D.G., and Heublein, P C (1979) Disease caused by a marine Vibrio N Engl J Med 300:1-5.

Blake, P.A., Weaver, R.E., and Hollis, D.G. (1980a). Diseases of humans (other than cholera) caused by vibrios. Ann Rev Microbiol 34, 341-367.

Blake, P.A., Allegra, D.T., Snyder, J.D., Barrett, T.J., McFarland, L., Caraway, C.T , Feeley, J.C., Craig, J.P., Lee, J.V., Puhr, N.D., and Feldman, R.A. (1980b). Cholera-A possible endemic focus in the United States. N Engl J Med 302, 305-309.

Bostock, A.D., Mepham, P., Phillips, S. Skidmore, S., and Hambling, M.H. (1979) Hepatitis A infection associated with the consumption of mussels. J Infect 1, 171-177.

Dienstag, J.L., Lucas, C.R., Gust, I.D., Wong, D.C., and Purcell, R.H. (1976). Mussel-associated viral hepatitis, type A: serological confirmation. Lancet i, 561-564.

De Lorenzo, F., Manzillo, G., Soscia, M., and Balestrieri, G.G. (1974). Epidemic of cholera El Tor in Naples, I973. Lancet i:669.

Dismukes, W.E., Bisno, A.L., Katz, S., and Johnson, R.F. (1969). An outbreak of gastroenteritis and infectious hepatitis attrituted to raw clams. Am J Epidemiol 89, 555-561.

Dougherty, W., Altman, R. (1962). Viral hepatitis in New Jersey, 1960-1961. Am J Med 32, 704-716.

Feingold, A. (1973). Hepatitis from eating steamed clams [Letter]. JAMA 225, 526-527.

Gill, O.N., Cubitt, W.D, McSwiggan, D.A, Watney, B.M., and Bartlett, C.L.R. (1983). Epidemic of gastroenteritis caused by oysters contaminated with small round structured virus Br Med J 287, 1532-1534.

Gray, C. (1988). Mussel mystery: "The more you know, the more you don't know." Canadian Med Assn J 138:350-351.

Griffin, M.R., Dalley, E., Fitzpatrick, M., and Austin, S.H. (1983). Campylobacter gastroenteritis associated with raw clams. Med Soc NJ 80, 606-609.

Grohmann, G.S., Murphy, A.M., Christopher, P.J., Auty, E., and Greenberg, H.B. (1981). Norwalk virus gastroenteritis in volunteers consuming depurated oysters Aust J Exp Biol Med Sci 59, 219-228.

Gunn, A.D.G., Rowlands, D.F. (1963). A confined outbreak of food poisoning. Med Off 122, 75-79.

Hart, J.C. (1945). Typhoid fever from clams. Connecticut Monthly Health Bulletin 59, 289-292.

Hauschild, A.H.W., Bryan, F.L. (1980). Estimate of cases of food-and waterborne illness in Canada and the United States. J Food Prot 43: 435-440.

Heller, D., Gill, O.N., Raynham, E., Kirkland, T., Zadick, P.M., and Stanwell-Smith, R. (1986). An outbreak of gastrointestinal illness associated with consumption of raw depurated oysters. Br Med J 292, 1726-1727.

Hepner, E. (1980). Food poisoning and salmonella infections in England and Wales, 1976-1978. Publ Health, Lond 94:337-349.

Holmberg, S.D., and Farmer, J.J. (1984). Aeromonas hydrophila and Plesiomonas shigelloides as causes of intestinal infections. Rev Infect Dis 6, 633-639.

Hughes, J.M. (1979). Epidemiology of shellfish poisoning in the United States, 1971-1977. In: Taylor DL, Seliger HH, eds. Toxic Dinoflagellate Blooms. Amsterdam: Elsevier Publishing Co , 23-28.

Klontz, K.C., Tauxe, R.V., Cook, W.L., Riley, W.H., and Wachsmuth, I.K. (1987). Cholera after the consumption of raw oysters. Ann Intern Med 107, 846-848

Klontz, K.C., Lieb, S., Schreiber, M., Janowski, H.T., Baldy, L.M., and Gunn, R.A. (1988) Syndromes of Vibrio vulnificus infections. Clinical and epidemiologic features in Florida cases, 1981-1987. Ann Intern Med 109, 318-323.

Koff, R.S., Grady, G.F., Chalmers, T.C., Mosley, J.W., and Swartz, B.L. (1967). Viral hepatitis in a group of Boston hospitals. III. Importance of exposure to shellfish in a nonepidemic period N Engl J Med 276, 703-710.

Krogh, P., Edler, L., Graneli, E., and Nyman, U. (1985) Outbreak of diarrheic shellfish poisoning on the west coast of Sweden. In: Anderson D.M., A W. White, D.G. Baden, eds. Toxic Dinoflagellates. New York:Elsevier Publishing Co., 501-503.

Lin, F.Y.C., Morris, J.C., Kaper, J.B., Gross, T., Michalski, J., Morrison, C., Libonati, J.B., and Israel, E. (1986). Persistence of cholera in the United States: isolation of Vibrio cholerae O1 from a patient with diarrhea in Maryland. J Clin Microbiol 23, 624-626.

Linco, S.J., and Grohmann, G.S. (1980). The Darwin outbreak of oyster-associated viral gastroenteritis. Med J Aust 1, 211.

Lowry, A.W., Pavia, A.T., McFarland, L.M., Peltier, B.H., Barrett, T.J., Bradford, H.B., Quan, J.M., Lynch, J., Mathison, J.B., Gunn, R.A., and Blake, P.A. (1989). Cholera in Louisiana. Widening spectrum of seafood vehicles. Arch Intern Med 149, 2079-2084.

Lumsden, L.L., Hasseltine, H.E., Leake, J.P., and Veldee, M.V. (1925). A typhoid fever epidemic caused by oyster-borne infection (Ig24-25). Pub Health Rep, Supp. no. 50, 1-102.

Mason, J.O., and McLean, W.R. (1962). Infectious hepatitis traced to the consumption of raw oysters. Am J Hyg 1962;75:90-111.

McCollum, J.P.K., Pearson, R.C.M., Ingham, H.R., Wood, P.C., and Dewar, H.A. (1968). An epidemic of mussel poisoning in north-east England. Lancet ii, 767-770.

McIntyre, R.C., Tira, T, Flood, T., and Blake, P.A. (1979). Modes of transmission of cholera in a newly infected population on an atoll: implications for control measures. Lancet i, 311-314.

Medcof, J.C. (1985). Life and death with Gonyaulax: An historical perspective. In: Anderson DM, A.W. White, and D.G. Baden, eds. Toxic Dinoflagellates. New York:Elsevier Publishing Co., 1-8.

Mele, A., Rastelli, M.G., Gill, W.N., Di Bisceglie, D., Rosmini, F., Pardelli, G., Valtriani, C., and Patriarchi, P. (1989). Recurrent epidemic hepatitis A associated with consumption of raw shellfish, probably controlled through public health measures. Am J Epidemiol 130, 540-546

Meyer, K F., Sommer, H., and Schoenholz, P. (1928). Mussel poisoning. J Prev Med 2, 365-394.

Meyer, K.F. (1953). Food poisoning. N Engl J Med 249:843-852.

Morris, J.G., Wilson, R., Hollis, D.G., Weaver, R.E., Miller, H.G., Tacket, C.O., Hickman, F.W., and Blake, P.A. (1982). Illness caused by Vibrio damsela and Vibrio hollisae. Lancet i, 1294-1297.

Morris, J.G., Picardi, J.L., Lieb, S., Lee, J.V., Roberts, A., Hood, M., Gunn, R.A., and Blake, P.A. (1984). Isolation of nontoxigenic Vibrio cholerae O group 1 from a patient with severe gastrointestinal disease. J Clin Microbiol 19, 296-297.

Morris, J.G., and Black, R.E. (1985). Cholera and other vibrioses in the United States. N. Engl. J. Med. 312, 343-350.

Morse, D.L., Guzewich, J.J., Hanrahan, J.P., Stricof, R., Shayegani, M., Deibel, R., Grabau, J.C., Nowak, N.A., Herrmann, J.E., Cukorz, G., and Blacklow, N.R. (1986). Widespread outbreaks of clam-and oyster-associated gastroenteritis;. Role of Norwalk virus. N. Engl. J. Med. 314, 678-681.

Murphy, A.M., Grohmann, G.S., Christopher, P.J., Lopez, W.A., Davey, G.R., and Millsom, R H. (1979). An Australia-wide outbreak of gastroenteritis from oysters caused by Norwalk virus. Med J Aust 2, 329-333.

Ohara, H., Naruto, H., Watanabe, W., and Ebisawa, I. (1983). An outbreak of hepatitis A caused by consumption of raw oysters. J Hyg Camb 91, 163-165.

O'Mahoney, M.C., Gooch, C.D., Smyth, D.A., Thausell, A.J., Bartlett, C.L.R., and Noah, N.D. (1983). Epidemic hepatitis A from cockles. Lancet i:518-520.

Pavia, A.T., Campbell, J.F., Blake, P.A., Smith, J.D.L., McKinley, T.W., and Martin, D.L. (1987). Cholera from raw oysters shipped interstate [Letter]. JAMA 258, 2374.

Pennington, B.C., Stewart, W.B., Pollard, W.M., Marvel, P and De Silver, J F. (1902). Report on typhoid fever at Atlantic City to the Atlantic City Academy of Medicine. The Philadelphia Medical Journal, November 1, 634-35.

Perl, T.M., Bedard, L., Kosatsky, T., Hockin, J.C., and Remis, R.S. (1988). Gastrointestinal and neurologic illness related to mussels from Prince Edward Island: A new clinical syndrome associated with domoic acid. Presented at the 28th Interscience Conference on Antimicrobial Agents and Chemotherapy, Los Angeles, CA, October 21-24, abstract.

Pollitzer, R. 1959. Cholera (WHO Monographic Series No. 43). Geneva, World Health Organization, 859.

Ramsey, G.H., McGinnes, G.F., and Neal, P.R. (1928). An outbreak of typhoid fever and gastroenteritis attributed to the consumption of raw oysters. Pub Health Rep 43, 2395-2405.

Ratzan, K.R., Bryan, J.A., and Krackow, J. (1969). An outbreak of gastroenteritis associated with ingestion of raw clams. J Infect Dis 120, 265-268.

Reeve, G., Martin, D.L., Pappas, J., Thompson, R.E., and Greene, K.D. (1989) An outbreak of shigellosis associated with the consumption of raw oysters. N Engl J Med 321, 224-227.

Rindge, M.E. (1962). Infectious hepatitis in Connecticut. Connecticut Health Bulletin, May, 125-130.

Rippey, S.R. (1989) Shellfish-borne disease outbreaks. U.S. Public Health Service, Food and Drug Administration, Davisville, R.I.

Roos, B. (1956). Hepatitis epidemic conveyed by oysters. Svenska Lakartidningen 53, 989-1003.

Ruddy, S.J., Johnson, R.F., Mosley, J.W., Atwater, J.B., Rossetti, M.A., and Hart, J.C. (1969). An epidemic of clam-associated hepatitis. JAMA 208, 649-655.

Sakamoto, Y., Lockey, R.F., and Krzanowski, J.J. (1987). Shellfish and fish poisoning related to the toxic dinoflagellates. Southern Med J 80, 866-872.

Salmaso, S., Bonfiglio, B., Greco, D., Castellani-Pastoris, M., De Felip, G , Sitzia, G., Piu, G., Barra, L., Bracciotti, L.A., Congui, A., Angioni, G., Zampleri, A., and Baine, W.B. (1980). Recurrence of pelecypod-associated cholera in Sardinia. Lancet ii, 1124-1127.

Shimizu, Y. (1989). Toxicology and pharmacology of red tides: An overview. In: Okaichi T, Anderson DM, Nemoto T, eds. Red Tides. Biology, Environmental Science, and Toxicology. New York: Elsevier Science Publishing Co., 17-21.

Sommer, H., and Meyer, K.F. (1937). Paralytic shell-fish poisoning. Arch Path 24, 560-598.

Spellman, J.R., Levy, C.S., Curtin, J.A., and Ormes, C. (1986). Vibrio fluvialis and gastroenteritis. Ann Intern Med 105:294-295.

Tacket, C O., Brenner, F., and Blake, P.A. (1984). Clinical features and an epidemiolgical study of Vibrio vulnificus infections. J Infect Dis 149, 558-561.

Taylor, S.L , and Bush, R.K. (1989). Allergy by ingestion of seafoods. In: Tu AT, ed. Handbood of Natural Toxins. Vol. 3. Marine Toxins and Venoms. New York· Marcel Dekker, Inc , 149-184.

Truman, B.I., Madore, H.P., Menegus, M.A., Nitzkin, J.L., and Dolin, R. (1985). Snow Mountain agent gastroenteritis from clams. Am J Epidemiol 126, 516-525.

Turnbull, P.C.B., and Gilbert, R.J. (1982). Fish and shellfish poisoning in Britain. In Jellife EF, Jelliffe DB, eds. Adverse Effects of Foods. New York: Plenum Press, 297-306

Underdal, B., Yndestad, M., and Aune, T. (1985). DSP intoxication in Norway and Sweden, Autumn 1984-Spring 1985. In: Anderson DM, A.W. White, and D.G. Baden, eds. Toxic Dinoflagellates. New York:Elseviar Publishing Co., 489-494.

Vancouver, G. (1789). In: Robinson GG, and J. Robinson, eds. Voyage of Discovery to the North Pacific Ocean and Round the World. Vol. 2. London, 284-286.

Vernon, E., and Tillett, H. (1974). Food poisoning and salmonella infections in England and Wales, 1969-1972. Publ Hlth, Lond 88: 225-235.

Vernon, E. (1977). Food poisoning and salmonella infections in England and Wales, 1973-75. Publ Hlth, Lond 91/:225-235.

Yasumoto, T., Murata, M., Oshima, Y., Matsumoto, G.K., and Clardy, J (1984). Diarretic shellfish poisoning. In: Regalis EP, ed. Seafood Toxins. Washington, D.C.: American Chemical Society, 207-216.

BACTERIAL AGENTS IN SHELLFISH DEPURATION

Richard A. Souness and Graham H. Fleet

INTRODUCTION

In Australia, cultivation of the Sydney rock oyster (previously Crassostrea commercialis, now Saccostrea commercialis) represents 97% of total oyster production and has an annual Austriallan retail value in excess of $100 million. The state of New South Wales (NSW) is the principle producer of this oyster species where the main method of cultivation is by elevated sticks and trays in river estuaries. In several southern states of Australia small quantities of the Pacific oyster (Crassostrea gigas), are produced. At present, NSW growers are not allowed to farm this species as it is considered a noxious pest likely to replace S. commercialis.

The association of molluscan shellfish with outbreaks of human disease has been documented for over a century and is well reviewed (Wood, 1976; Brown and Dorn, 1977; Richards, 1987). The problem is intrinsically linked to the pollution of oyster farming waters with human sewage and associated pathogenic microorganisms. As pollution continues to increase, so does the risk to public health. Consequently, shellfish producers world-wide consider depuration as one of the main options for preserving the viability of their industry.

Compared with European and American experiences, the microbiological history of Australian oysters is relatively short (Table 1). This is probably a function of Australia's low population density and the concomitant lag in urban development adjacent to estuaries (Fleet, 1978a). Nevertheless, Australian consumers and the Australian oyster industry were abruptly awakened to the microbiological hazard of oysters in 1978, when a large outbreak of gastroenteritis involving some 2000 people was attributed to polluted oysters. In this case, the etiological agent was a Norwalk virus and the oysters had been harvested from waters that had suffered extreme sewage pollution due to persistent heavy rainfall (Murphy et al. 1979). The scale and seriousness of this incident, plus the economic losses that oyster producers experienced as a result of decreased consumer confidence in the product, provided impetus for the introduction of mandatory oyster purification in NSW.

Drawing from existing data on shellfish depuration in Europe and America, we developed depuration technology applicable to the requirements of the Sydney rock oyster and the local industry. Starting from laboratory experiments, the technology was developed through to a commercial-scale plant capable of handling 25,000 oysters per purification cycle. In considering design of the plant, a stacked, shallow tank configuration was selected as being best suited to industry needs, giving economy of space and ease of oyster handling. This technology has been used commercially in NSW for 11 years now, and has an outstanding record of public health safety. The purpose of this presentation is to give an overview of our research experiences, with special reference to the behaviour of bacteria during depuration.

TABLE 1. A microbiological history of Australian oysters

Microorganism	Counts	Reference
Coliforms	$80\text{-}10^5/100g$	Wells and Edwards, 1969
Escherichia coli	$0\text{-}1.6\times10^5/100g$	
Vibrio parahaemolyticus	$0\text{-}1.2\times10^4/100\ g$	Sutton, 1974
V. parahaemolyticus	$0\text{-}3.0\times10^5/100g$	Qadri, 1974
Fecal coliforms	$0\text{-}8.6\times10^3/100\ g$	
Salmonella	not detected	
Fecal coliforms	$0\text{-}5.0\times10^4/100g$	Qadri, Buckle and Edwards, 1975
Fecal coliforms	$<60\text{-}9.2\times10^3100/g$	Brown and McMeekin, 1977
E. coli	$0\text{-}17/g$	Fleet, 1978a
E. coli	$36\text{-}4.6\times10^3/100g$	Son and Fleet, 1980
Salmonella	not detected	
Bacillus cereus	$87\text{-}4.9\times10^2/g$	
V. parahaemolyticus	$17\text{-}4.25\times10^2/g$	
Clostridium perfringens	$7\text{-}30/g$	
V. cholerae	Detected in 1 sub-sample	Davey, Prendergast and Eyles, 1982
V. parahaemolyticus	$4.6/g$	
E. coli	$<1\text{-}2/g$	
Campylobacter jejuni	14% positive samples	Arumugaswamy and Proudford, 1987
E. coli		
V. cholera	$0\text{-}3/g$	Eyles and Davey, 1988
V. parahaemolyticus	$0\text{-}40/g$	
E. coli	up to $170/g$	
Salmonella	14% positive samples	

DEPURATION TECHNOLOGY

DESIGN PRINCIPLES

Water flow: There are two methods for moving water through a depuration plant, flow-through and recirculation. In the first system, water passes through the plant once and is discharged. With this method, the plant is exposed to fluctuations in the microbiological quality and physical attributes of the source water. Once drawn into the tanks of a recirculating system, water is subject to a controlled environment, where the costs associated with disinfection and maintaining water temperature and clarity are much reduced. Therefore, to have better control over the depuration environment, a recirculating system was chosen at the outset of our studies.

The faster rate of bacterial elimination from polluted shellfish that can be achieved in a recirculating system, as opposed to a flow-through system, was confirmed by (Sangrungruang, et al. 1989). They showed that the rate of reduction of coliform counts in oysters is greatly increased using recirculation, with reductions in counts being more than twice as fast as those in a flow-through system.

Water can be disinfected by a number of methods reviewed by Fleet (1978b) and Richards (1988). After examining the options available in relation to the local industry, ultraviolet (UV) light irradiation was selected as the means of water disinfection. UV irradiation has a potent, destructive effect on bacteria (Vasconcelos and Lee, 1972; Chang et al. 1985) and viruses (Hill et al. 1970). It has been shown to be an effective, cost efficient means for continuously disinfecting large volumes of seawater for oyster depuration systems (Kelly, 1961; Wood, 1961a). However, the efficiency of UV irradiation can be adversely affected by excessive water turbidity, fast flow-rates and poorly maintained UV lamps.

The sterilizing ability of the UV system is simply determined by measuring and comparing the viable count of total bacteria immediately before and after passage over the UV lamp. During pilot-plant depuration trials, we occasionally noted a gradual decrease in the sterilizing efficiency by the UV system over a 48 hr. period (Souness and Fleet, 1979). This decrease was indicated by a decline in the percent destruction of total bacteria by the UV sterilizer, from an initial value of 99.9% reducing to 95% after 48 hr. of water recycling. The significance of this reduced performance can be deceptive, as it does not necessarily imply a reduced ability of the UV system to kill pathogens such as Salmonella, Escherichia coli etc. Rather, other research in our laboratories (Pulsford unpublished data) showed that the apparent reduction in performance was due to a progressive increase in the total population of UV resistant species of bacteria. As the time of depuration increased, the proportion of Pseudomonas in the recycling tank water increased, so that by 48 hr Pseudomonas constituted 73 to 100% of the surviving bacteria. A small percentage of Vibrio sp. were also found at this stage. Vasconcelos and Lee (1972) also noted that Pseudomonas and Vibrio sp. were the main bacterial species found in UV treated seawater.

Pulsford (unpublished data) examined the bacterial composition of the slime that sometimes builds up in UV-based depuration systems. Pseudomonas and Vibrio sp. were prevalent in the slime, further suggesting the apparent UV resistance of these bacteria.

OPERATIONAL VARIABLES

Water temperature determines the efficiency of bacterial elimination from oysters during depuration by affecting:

1. oyster pumping rate
2. growth and survival of bacteria in the depuration water
3. integrity of the feces discharged by the oyster during depuration

Souness and Fleet (1979) assessed the effect of water temperature on oyster pumping rate, and hence the elimination of bacteria, by a dye uptake technique The pumping rate for S. commercialis peaked at 25°C, decreasing sharply above 30°C and below 15°C.

Using naturally contaminated Sydney rock oysters, the effect of two water temperature ranges (18-20°C and 27-29°C) on bacterial elimination was examined (Souness and Fleet 1979). Reductions in total bacteria, coliforms and E. coli in the oysters were faster at 18-20°C. At 27-29°C the total bacterial count of the tank water rapidly increased, even though water returning to the tank through the UV sterilizer showed a 99% destruction of incoming bacteria. It was concluded that at the higher temperature, multiplication of bacteria in the tank water was greater than their rate of destruction by the UV sterilizing unit.

Rowse and Fleet (1984) studied the effect of water temperature on the elimination of Salmonella charity and E. coli from the Sydney rock oyster during commercial depuration. Using both winter and summer harvested oysters they found purification to be inconsistent and incomplete at 13-17°C. At 18-22°C purification was rapid and consistent with levels of S. charity and E. coli being reduced to below 1 cell/g within 12 hr. They also noted that water temperatures above 22°C in summer induced oyster spawning. The recirculating water rapidly became turbid, rendering UV disinfection and depuration ineffective. In a study using artificially contaminated Sydney rock oysters, Arumugaswamy, et al. (1988) found that Campylobacter jejuni and Campylobacter coli were eliminated from levels of 6-8x10^2/g to undetectable levels within 48 hr at a water temperature of 18-20°C.

The levels of various bacterial species in the depuration water depend on their ability to survive and multiply at different temperatures. Rowse and Fleet (1982) showed that S. charity and E. coli did not survive well at 20 and 30°C in seawater, as is the case for most enteric bacteria. However, counts of total bacteria in the depuration water increased at higher temperatures which would imply ineffective depuration, if total bacterial counts of oysters are taken as the depuration index.

Discharged oyster feces contain viable bacteria entrapped in a mucous matrix (Haven et al. 1978) and there is some evidence to suggest that, at temperatures exceeding 25°C, the feces undergo chemical and microbial degradation leading to the release and multiplication of these bacteria (Goggins 1964) This phenomenon could also explain the buildup of bacterial numbers in depuration waters operating at elevated temperatures. Rowse and Fleet (1982) demonstrated the presence of viable bacteria, including E. coli and S charity, in the feces of the Sydney rock oyster; the level of bacteria in the feces being directly related to the level of oyster contamination These bacteria could be released from the feces into the depuration water and act as a source of recontamination of the oysters. The extent of release of these bacteria depends on the water temperature and the time of contact with the water. The release of bacteria from feces into the surrounding water is an important consideration when developing a depuration system utilizing shallow tanks. Precautions must be taken to avoid the physical disruption of feces and possible recontamination of the depuration water.

It is evident from this discussion that the water temperature required for optimal oyster pumping does not necessarily correspond with the temperature needed for efficient depuration. A water temperature less than optimal for pumping may be necessary to prevent excessive multiplication and build-up of bacteria in the depuration water. However, water temperature must not be allowed to decrease to a value below which efficient oyster pumping ceases. It is essential, therefore, to incorporate water temperature-control systems into depuration plant design This may include heating in winter and cooling in summer. For the Sydney rock oyster, the optimum temperature for depuration in a recirculating system is 18 to 20°C.

Water flow-rate: The recirculation rate of the water in the depuration plant has an important effect on the efficiency of oyster depuration. In particular, the recirculation rate should be such that the rate of kill of bacteria in the depuration water by UV irradiation, exceeds the rate of bacterial multiplication. If this does not occur, bacteria will accumulate in the water and reduce the efficiency of oyster cleansing. Ideally, the flow-rate should be expressed in terms of the number of times or cycles/hr that the total volume of water in the system is exposed to the UV sterilizer in order to prevent bacterial build-up in the water.

Souness and Fleet (1979) examined the impact of flow-rate on the bacteriological quality of depuration water and the rate of bacterial elimination by the Sydney rock oyster. A 20-fold build-up in bacterial numbers was observed in water (25°C) recirculated at 1 cycle/hr through the UV sterilizer while at 2 cycles/hr the total bacterial count increased by a factor of 10. When the water was recirculated at 3 cycles/hr, the total bacterial count of the water decreased. The latter set of conditions ensured rapid elimination of bacteria from oysters, including the reduction of coliforms and E. coli to undetectable levels within 48 hr.

A recirculation rate of 3-4 cycles/hr has been standard practice in our studies and, with other variables optimised, it gives consistent, effective depuration of a range of bacterial pathogens (Son and Fleet, 1980; Rowse and Fleet 1984). With slower flow-rates, as frequently cited in the literature (Richards, 1988), the potential exists for build-up in bacterial numbers within the system which may result in unacceptable depuration. For example, Barrow and Miller (1969) concluded that buildup of Vibrio sp. within one of two depuration plants they examined was due to the slow rate of water recirculation, allowing the bacteria to multiply at a rate much greater than their rate of destruction by the UV sterilizing unit. Oysters processed by this plant had Vibrio counts exceeding that at the beginning of depuration. Buisson,et al. (1981) also noted that fecal coliforms will build-up in depuration water if adequate flow-rates through the UV sterilizer were not maintained.

Water salinity and dissolved oxygen: Shellfish can adapt to moderate variations in both of these environmental parameters, but pumping activity may be temporarily affected during the process of acclimatisation. Many researchers (Wood, 1961b; Furfari, 1966 and 1976; Ayres, 1978, Haven et al. 1978; Souness, et al. 1979) have specified similar values for salinity and oxygen, irrespective of oyster species, for effective depuration. The accepted range of values are:

Salinity: $100 \pm 20\%$ of value at harvest area
Dissolved oxygen: 50-100% of saturation.

Rowse and Fleet (l984) examined the effect of water salinity on the elimination of E. coli and S. charity from the Sydney rock oyster. They found that sudden exposure to low water salinity (l 5-2.0%), relative to values in the harvest area, was stressful to the oysters, leading to a weakened physiological condition, ineffective depuration and even death, with mortality rates of 30% being common. Conversely, the elimination of these bacteria at high salinity (3.2-4.7%) was rapid and uniform, and no abnormal oyster mortality was noted during depuration. Rowse and Fleet (1984) also noted that high salinity was effective in preventing spawning of the Sydney rock oyster during depuration. As mentioned already, spawning can be a major problem in recirculating depuration systems (Souness, et al. 1979).

Oyster density: Excessive loading of oysters in depuration tanks leads to crowding and increased oxygen demand, the consequences of which are reduced oyster pumping and a diminished rate of bacterial elimination. Successful depuration trials using 2 and 4 oysters (S. commercialis)/L of water were conducted in a shallow tank system (Souness and Fleet, 1979). No significant differences in depuration rates were noted for the two oyster densities. At concentrations of 4 oysters /L or 500 oyster/m², the oysters formed a compact, single layer in the tanks. Successful depuration of oysters in a system operating at 4 oysters/L has been noted in other studies (Souness, et al 1979; Son and Fleet, 1980). Stacking or layering of oysters in tanks introduces the possibility that feces, discharged by oysters from the upper layers, will recontaminate oysters below.

DEPURATION PLANT DESIGN

A stacked tank format similar to several other designs (Ayres, 1978) was selected, but with a number of significant modifications. Previous stacked tank designs where water from the top tank discharges into the tank below and then to the next and so on, were considered unsatisfactory for Australia's high ambient temperatures because of the likelihood of excessive bacterial build-up in the lower tanks In our design (Figure 1), each tank was equipped with an individual supply of disinfected water passing through an inlet device that would ensure at least 3 exchanges of tank water every hour. A detailed description of the operation of the inlet device is given by (Souness, et al 1979).

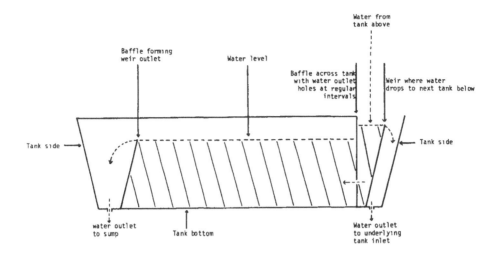

FIGURE 1. Cross-section showing details of depuration tank

Essentially, the flow-rate of water into each tank is controlled by the number and size of holes in the baffle and the pressure drop which is equal to the difference in height between the overflow weirs in the inlet and outlet sections of the tank. Since these parameters are fixed, the flow-rate through the tank is independent of external influence providing a predetermined, tamper-proof flow of water across the full width of the tank. The resulting plug flow ensures an absence of 'dead spots', where bacteria can accumulate, and turbulence, that could cause disruption of oyster feces. Water flowing over the outlet weir returns directly to the sump and then to the UV unit. The sump offers a convenient location for additional aeration and temperature control as well as sedimentation of suspended solids.

The tank configuration allowed for modular arrangement. In our case, each module consisted of 6 reinforced fiberglass tanks. When fully laden each tank contained 1,500 oysters and 350 L of water giving a depth of 3-4 cm of water over the oysters. The cascading effect of water at the inlet and outlet weir serves to aerate the water. The mean dissolved oxygen content of the water in each tank was at least 75% of saturation. Water cascading down through the system also introduced a cooling effect, so that the tank water was generally 2 to 3°C lower than the ambient temperature.

Depuration systems of this same design have been operating in NSW for 11 years, producing cleansed oysters that readily meet the microbiological standards required by Australian health authorities.

OPTIMAL CONDITIONS FOR DEPURATION OF THE SYDNEY ROCK OYSTER

Table 2 summarises the optimal conditions for effective depuration of the Sydney rock oyster. Although coliforms and E. coli were eliminated from oysters within 24 hr, we concluded that the depuration process should continue for 48 hr, thus offering a considerable margin of safety.

TABLE 2. Conditions necessary for the efficient depuration of the Sydney rock oyster.

Parameter	Value
Water temperature	18-22°C
Water flow-rate	3 cycles/h (minimum)
Salinity	100±20% of the value at harvest area
Dissolved oxygen	50-100% saturation
Oyster density	4 oyster/L (maximum) 500 oyster/m^2 (maximum)
Process time	not less than 48 hr

Using these conditions, Rowse and Fleet (1984) demonstrated the elimination kinetics of E. coli and S. charity from Sydney rock oysters in a commercial depuration plant. Table 3 shows that the greater proportion of these bacteria are eliminated from the oyster in the first 24 hr. After 48 hr, counts of these bacteria were below detectable levels.

TABLE 3. Elimination of Escherichia coli and Salmonella charity during commercial depuration.

Time	Percent reduction E. coli	S. charity
0	(378 cells/g)	(430 cells/g)
12	>99.9	99.90
24	>99.9	99.98
36	>99.9	>99 99
48	>99.9	>99.99

Source: Rowse, 1983.

BACTERIAL EFFICACY OF COMMERCIAL DEPURATION

An assessment of the commercial depuration plant resulting from our research demonstrated (Table 4) that polluted oysters could be cleansed to less than Australian health standards of 0.5 E. coli/g (NHMRC, 1987), within 48 hr. Occasionally, oysters were harvested with E coli counts as high as 110 cells/g and these were readily cleansed to undetectable levels within 48 hr. Rowse (1983) consistently attained reductions greater than 99.9% for E. coli and 99 99% for S charity, during commercial depuration.

Other pathogens that are eliminated from Sydney rock oysters are Bacillus cereus, Clostridium perfringens, V. parahaemolyticus (Son and Fleet 1980) and Campylobacter sp. (Arumugaswamy, et al 1987). However, Eyles and Davey (1984) questioned the purification of V. parahaemolyticus from the Sydney rock oyster.

TABLE 4.   Bacterial counts in Sydney rock oysters and tank water from a commercial
depuration plant

| Depuration time (h) | Bacterial count (cells/g or mL) | | | |
| | Total Count | | E. Coli | |
	Water (x10^3)	Oysters (x10^4)	Water	Oyster
0	20.6	25.5	0	31.7
12	5.4	-	0	-
24	1.6	8.1	0	0.9
48	1.0	1.64	0	0

Note: Counts are means of pooled samples from at least 6 tanks.
Source: Souness, 1979.

The mode of oyster contamination has been cited as a cause of inconsistent depuration results with suggestions that artificially contaminated oysters will eliminate bacteria at a faster rate than naturally contaminated oysters (Hoff and Becker, 1969; Eyles and Davey, 1984). However, Rowse (1983) showed that naturally and artificially contaminated Sydney rock oysters eliminated E. coli at a similar rate.

There is considerable evidence to suggest that oysters maintain an indigenous microbial population. During laboratory trials total bacterial counts in Sydney rock oysters did not fall below 10^4 cells/g (Son and Fleet, 1980).   Other investigators (Souness and Fleet, 1979; Pulsford, unpublished data) found that Sydney rock oysters rarely cleansed to bacterial counts below 10^4 cells/g during commercial depuration, and when oysters had initial bacterial counts below 10^4 cells/g, counts often increased during depuration to 10^4 cells/g. This phenomenon was not due to a failure in the depuration process but was possibly related to the maintenance of an indigenous microbial flora by the oyster. This view was also expressed by Vasconcelos and Lee (1972), who found that total bacterial counts of depurated oysters consistently tended towards 10^4 cells/g. Garland, et al (1982) addressed this question by examining oyster epithelial surfaces using electron microscopy, but their results did not support the concept of an indigenous flora.

The maintenance of an indigenous microbial flora by oysters suggests a selective mechanism by which they absorb and retain some microbial species in their digestive tract, but not other species. This close association of autochthonous marine bacteria with oysters may also be the cause of shellfish being unable to eliminate bacteria below a certain level. There are cases though, where marine bacteria, V. parahaemolyticus (Barrow and Millar, 1969; Son and Fleet, 1980) and V. vulnificus (Kelly and Dinuzzo, 1985), have been successfully eliminated from oysters.

The rapid initial decline in bacterial numbers followed by a more gradual decline, is a common feature of many depuration systems (Haven et al. 1978, Souness et al. 1979; Buisson, 1981; Son and Fleet, 1980; Greenberg, et al. 1982; Rowse and Fleet, 1984) and suggests a strong association between bacteria and oysters. Some researchers (Buisson, et al. 1981) explained this two-phase pattern of bacterial elimination as due to the varying degrees of adsorption of bacteria in the oyster digestive tract, with weakly attached bacteria being eliminated first. Comprehensive studies of the microbial ecology of shellfish in relation to their marine environment are needed to better understand this phenomenon.

CONCLUSION

Depuration is an effective method of eliminating human pathogenic microorganisms from shellfish when undertaken with equipment and conditions appropriate to individual species of shellfish Our studies have shown that there exists a number of critical points, requiring specific control to ensure reliable depuration of oysters. These points are:

1. Water flow: non-turbulent, 'plug' flow of water through tanks, with at least 3 exchanges/hr of total depuration water through the tanks and steriliser

2. Oyster loading: a single layer of oysters on raised trays in tanks to ensure uninterrupted feeding and prevent fouling of oysters with discharged feces

3. Water temperature: maintenance of water temperature to maximize oyster pumping rater while minimizing bacterial multiplication in the tank water

4   Oyster handling: gentle handling and culling of oysters after harvesting to ensure active feeding during depuration

5. Sterilizer capacity: sterilizer capable of dealing with anticipated extremes of water flow-rate, turbidity and microbial load

6. Water quality: dissolved oxygen and salinity values suited to individual oyster species and local conditions

7. Plant sanitation: regular cleaning and sanitiation of tanks, UV sterilizer and ancillary equipment.

Unfortunately, results of depuration studies are commonly reported without any indication of plant design and operating specifications, making comparisons between studies difficult, if not impossible. It is essential that a more uniform approach is taken for future depuration reaearch and that details of plant design and operating conditions be specified.

REFERENCES

Arumugaswamy, R.K. and Proudford, R.W. 1987. The occurrence of Campylobacter jejuni and Campylobacter coli in Sydney Rock Oyster (Crassostrea commercialis). Int J. Food Microbiol. 4: 101-104.

Arumugaswamy, R.K., Prouford, R.W, and Eyles, M.J. 1988. The response of Camplobactyer jejuni and Campylobacter coli in the Syndey Rock Oyster (Crassostrea commercialis), during depuration and storage. Inst. J. Food Microbiol. 7: 173-183

Ayres, P.A. 1978. Shellfish purification in installations using ultraviolet light Lowestoft, UK: MAFF. Direct. Fish. Res. Lab. Leafl. 43.

Barrow, G.I. and Miller, D.C. 1969. Marine bacteria in oysters purified for human consumption. Lancet 2:421-423.

Brown, L.D. and Dorn, C.R. 1977. Fish, shellfish and human health. J. Food Prot. 40:712-717.

Brown, R.K. and McMeekin, T.A. 1977. Microbiological aspects of oyster production in Southern Tasmania. Food Technol. Aust. 29:103-106.

Buisson D.H., Fletcher, G.C., and Begg, C.W. 1981. Bacterial depuration of the Pacific oyster (Crassostrea gigas) in New Zealand. N.Z.J. Sci. 24:253-262.

Chang, J.C , Ossoff, S F., Lobe, D.C., Dorfman, M.H., Dumais, C M , Qualla, R.G., and Johnson, J.D. 1985. UV inactivation of pathogenic and indicator microorganisms Appl. Environ. Microbiol. 49· 1361-1365.

Davey, G.R., Prendergast, J.K., and Eyles, M.J. 1982. Detection of Vibrio cholerae in oysters, water and sediment from the Georges River. Food Tech. Aust. 34:334-336.

Eyles, M.J. and Davey, G.R. 1984. Microbiology of commercial depuration of the Sydney rock oyster, Crassostrea commercialis. J. Food Prot. 47:703-706.

Fleet, G.H. 1978a. Protecting public from micro-biological pollution of oysters. Aust. Fish 37(12):18-20.

Fleet, G.H. 1978b. Oyster depuration - review. Food Tech. Aust. 30:444-454.

Furfari, S.A. 1966. Depuration plant design. U.S. Dept. Health, Education and Welfare. Washington, Public Health Service. Publ. No. 999-FP-119p.

Furfari, S.A. 1976. Shellfish purification: a review of current technology. FAO Tech. Conf. on Aquaculture, Publ. FIR: AQ/Conf/76/R.11. Kyoto. 16 p.

Garland, C.D., Nash, G.U., and McMeekin, T.A. 1982. Absence of surface-associated microorganisms in adult oysters (Crassostrea gigas). Appl. Env. Microbiol 44:1205-1211.

Goggins, P.L. 1964. Depuration in Maine. In Houser, L.S. Fifth Natl. Shellfish Sanitaiton Workshop. U.S. Dept of Health, Education and Welfare. Washington, p. 78-92.

Greenberg, E.P., Dubois, M., and Palhof, B. 1982. The survival of marine vibrios in Mercenaria mercenaria, the hardshell clam. J. Food Safety 4:113-123.

Haven, D.S., Perkins, F.O., Morales-Alamo, R., and Rhodes, M.W. 1978. Bacterial depuration by the American oyster (Crassostrea virginica) under controlled conditions. Virginia Institute of Marine Science, Gloucester Point, VA, USA. Special Scientific Report No. 88.

Hill, W.F., Hamblet, F.E., Benton, W.H., and Akin, E.W. 1970. Ultraviolet devitalization of eight selected enteric viruses in estuarine water. Appl. Microbiol 19:805-812.

Hoff, J.C. and Becker, R.C. 1969. The accumulation and elimination of crude and clarified poliovirus suspensions by shellfish. Amerc. J. Epid. 90:53-61.

Kelly, C.B. 1961. Disinfection of sea water by ultraviolet radiation. Am. J. Pub. Health, 51:1670-1680.

Murphy, A.M., Grohman, G.S., Christopher, P.J., Lopez, W.A., Davey, G.R., and Millsom, R.H. 1979. An Australia-wide outbreak of gastroenteritis from oysters caused by Norwalk virus. Med. J. Aust., 2:329-333.

NHMRC (National Health & Medical Research Council). 1979. Code of hygienic practice for oysters and mussels for sale for human consumption. Australian Government Publishing Service, Canberra. 20.

Qadri, R.B. 1974. Microbiology of shellfish with particular reference to aspects of public health. PhD thesis, University of NSW.

Qadri, R.B., Buckle, K.A., and Edwards, R.A. 1975. Sewage pollution in oysters grown in the Georges River - Botany Bay area. Food Technol. Aust. 27:236-242.

Richards, G.P. 1987. Shellfish associated enteric virus illness in the United States, 1934-1984. Estuaries 10:84-85.

Richards, G.P. 1988. Microbial purification of shellfish: a review of depuration and relaying. J Food Prot. 1:218-251.

Rowse, A.J. 1983. The elimination of Salmonella charity and Escherichia coli from the rock oyster Crassostera commercialis during commercial purification MSc thesis, University of NSW.

Rowse, A.J. and Fleet, G.H. 1982. Viability and release of Salmonella charity and Escherichia coli from oyster feces. Appl. Envir. Microbiol. 44:544-548.

Rowse, A.J. and Fleet, G.H. 1984. Effects of water temperature and salinity on elimination of Salmonella charity and Escherichia coli from Sydney rock oysters (Crassostrea commercialis). Appl. Environ. Microbiol. 48:1061-1063.

Sangrungruang, K., Sahavacharin, S., and Ramanudom, J. 1989. Depuration of some economically important bivalues in Thailand. ASEAN Fd. J. 4:101-106.

Son, N.T. and Fleet, G.H. 1980. Behaviour of pathogenic bacteria in the oyster, Crassostrea commercialis, during depuration, relaying and storage. Appl. Environ. Microbiol. 40:994-1002.

Souness, R.A. 1979. Depuration of the Sydney rock oyster, Crassostrea commercialis. MAppSc thesis, University of NSW.

Souness, R.A., and Fleet, G.H. 1979. Depuration of the Sydney rock oyster, Crassostrea commercialis. Food Tech. Aust. 31:397-404.

Souness, R.A., Bowre, R.G., and Fleet, G.H. 1979. Commercial depuration of the Sydney rock oyster, Crassostrea commercialis. Food Tech. Aust. 31:531-537.

Sutton, R.G. 1974. Some quantitative aspects of Vibrio parahaemolyticus in oysters in the Sydney area. In: Fujino, T., G. Sakaguchi, R. Sakazaki, and Y. Takeda. ed. International Symposium on Vibrio parahaemolyticus. Saikon Publ. Tokyo, p. 71-76.

Vasconcelos, G.J. and Lee, J.S. 1972. Microbial flora of Pacific oysters (Crassostrea gigas) subjected to ultraviolet-irradiated seawater. Appl. Microbiol. 23:11-16.

Wells, G.C. and Edwards, R.A. 1969. A survey of sewage pollution in Georges River oysters. Food Tech. Aust. 212:616-619.

Wood, P.C. 1961a. The principles of water sterilization by ultra-violet light and their application in the purification of oysters. Ministry of Agriculture, Food and Fisheries. Fisheries Investigation Series II. 23:1-47.

Wood, P.C. 1961b. The production of clean shellfish. J. Roy. Soc. Health, 81:173.

Wood, P.C 1976. Guide to shellfish hygiene. WHO Offset Publication No. 31. WHO Geneva, 80 p.

HUMAN ENTERIC VIRUSES AND DEPURATION OF BIVALVE MOLLUSKS

Mark D. Sobsey and Lee-Ann Jaykus

INTRODUCTION

When bivalve molluscan shellfish are exposed to fecally contaminated waters, they can accumulate enteric bacteria and viruses during normal filter-feeding and related activities. As a result, human consumption of raw or partially-cooked bivalves contaminated with enteric viruses has caused infectious hepatitis and viral gastroenteritis. Due to continued development and usage of the coastal zone, the quantity of productive shellfish acreage subjected to fecal contamination continues to increase in many areas, and this has become a growing public health and economic concern. Sanitary standards for the evaluation, classification and certification of shellfish and their harvest waters have been established in the U.S. and other countries (Food and Drug Administration, 1989). These shellfish sanitation practices and standards, based on identification and control of sources of fecal contamination and on the densities of total and fecal coliform bacteria in shellfish and their harvest waters, have been effective in protecting the public from shellfish-associated illness due to enteric bacteria, but they are thought to be inadequate indicators of enteric viral contamination. Many documented outbreaks of hepatitis A and viral gastroenteritis have been linked to the ingestion of contaminated shellfish, including outbreaks attributed to shellfish subjected to depuration (controlled purification) or relaying. In some outbreaks,the contaminated shellfish were harvested from waters meeting present bacterial standards.

The processes of depuration (controlled purification) and relaying (shellfish transfer to "clean" estuarine environments for a period of time prior to harvest) are important measures to further improve the virological and microbiological quality of shellfish harvested from approved and restricted waters and perhaps to recover shellfish in polluted (prohibited) waters for ultimate use food commodities The effectiveness of depuration and relaying processes for eliminating enteric bacteria has been well documented and these processes are used routinely now in some countries (U S Food and Drug Administration, 1989) However, their effectiveness in eliminating viral contamination is less well documented and has been questioned Indicator bacteria and some enteroviruses appear to be readily eliminated by shellfish subjected to purification processes, but less is known about the reductions of hepatitis A virus (HAV) and most agents of viral gastroenteritis. The rate and extent of viral reduction in shellfish subjected to depuration and relaying and the factors influencing viral depuration efficiency must be better understood and documented. In this paper the following topics concerning viral depuration and relaying are reviewed: the important human enteric viruses potentially present in fecally contaminated shellfish, the illnesses they cause and their sources; the epidemiological evidence, especially in the U.S., that shellfish have caused viral illness; the factors influencing viral persistence in shellfish habitats; the factors influencing viral uptake and persistence in shellfish; the factors influencing the subsequent elimination of viruses under depuration and relaying conditions; and the needs for additional research and field information on virus occurrence and behavior in depurated and relayed shellfish.

ENTERIC VIRUSES

The enteric viruses, like other viruses, are obligate, intracellular parasites at the genetic level that can infect and cause diseases in their cellular hosts. The simplest viruses consist of a nucleic acid core (either single- or double-stranded DNA or RNA) surrounded by a protein coat (the capsid) comprised of multiple copies of one or several different polypeptides. Some viruses, but not the important shellfish-borne enteric viruses, are also surrounded by a lipoprotein envelope. The outer protein coat or capsid of the enteric viruses protects the nucleic acid from adverse environmental conditions and also functions in virion attachment to cells and host specificity. The nucleic acid is responsible for viral infectivity and the viral replication processes in the host cell that result in the production of progeny viruses.

Enteric viruses can be defined as those that enter the body via the oral route and infect the alimentary canal (gastrointestinal tract). They multiply initially in the gut and are excreted in the feces at concentrations as high as 10^9 infectious units per gram. All enteric viruses, with the exception of coronaviruses, lack a lipoprotein envelope, and because of their protein outer surface (the capsid) they are resistant to the acids, enzymes and other degradative agents present in the stomach and elsewhere in the intestinal tract. As shown in Table 1, there are more than 100 types of enteric viruses that are taxonomically grouped into several viral families and genera. With respect to transmission by fecally contaminated shellfish, the most important human enteric viruses on the basis of documented shellfish-borne illness are hepatitis A virus (HAV; presently classified as enterovirus 72), Norwalk virus, and the other small, round, Norwalk-like viruses causing acute gastroenteritis (caliciviruses, astroviruses, and possibly parvoviruses and picornaviruses). Infection with some of these viruses is invariably localized in the gut (e.g., Norwalk virus), but some others, including HAV, produce generalized or systemic disease. Localized gastrointestinal infections by some of these enteric viruses are frequently asymptomatic.

The main mode of transmission of enteric viruses is the so-called fecal-oral route, and person-to-person contact is the most common means of spread in the population. However, these viruses are environmentally important because they may also be transmitted via drinking water, bathing water and foods (particularily bivalve molluscan shellfish) that have become contaminated with sewage or other sources of human (and perhaps animal) feces and other excreta.

Table 1. Human Enteric Viruses

Virus or Group	No. Types	Major Illnesses
Enteroviruses		
Polioviruses	3	Poliomyelitis; aseptic meningitis
Coxsackieviruses A	24	Aseptic meningitis; herpangina; rash
Coxsackieviruses B	6	Aseptic meningitis; myocarditis; pericarditis
Echoviruses	34	Aseptic meningitis; rash; GI
Hepatitis A Virus (HAV)	1	Infectious hepatitis
Other enteroviruses	>4	Encephalitis; hemorragic conjunctivitis
Adenoviruses	41	Acute GI (types 40/41); upper resp.
Reoviruses	3	Mild GI and UR illness
Rotaviruses	>4	Acute GI
Norwalk and other small, round GI viruses (calici-, astro-, parvo-?)	several	Acute GI
Enteric non-A, non-B (HEV)	1?	Infectious hepatitis (calicivirus?)
Coronaviruses, enteric	1?	UR illness; GI?; necrotizing colitis?

Hepatitis A Virus.

Hepatitis A virus (HAV), presently classified as enterovirus 72 within the family Picornaviridae, is a non-enveloped icosahedral virion, 27-28nm in diameter, containing single-stranded, plus-sense RNA, three major capsid polypeptides (VP1, VP-2 and VP-3) and probably a fouth capsid polypeptide (VP-4). Like other enteroviruses, HAV has a buoyant density in CsCl of 1.33-1.34 g/ml and a sedimentation coefficient of 156 to 160 S (Gust and Feinstone, 1988; Hadler and Margolis, 1989;Hollinger and Ticehurst, 1990). HAV is stable at pH 3 and resistant to inactivation by organic solvents such as ether, chloroform, and fluorocarbons, as are the other enteroviruses. However, HAV is more stable at elevated temperatures than are most other enteroviruses. It is stable at 60°C for one hour, and temperatures above 60°C are necessary to destroy infectivity within a short period of time (Parry and Mortimer, 1984; Siegl et al., 1984). At room temperature or lower,the virus remains infectious for several weeks, even under dessicating conditions (Sobsey et al., 1988).

The predominant mode of HAV transmission is person-to-person via the fecal-oral route, although consumption of contaminated water and food (including shellfish), is also important. In developing countries, where sanitation and hygienic conditions are relatively poor, HAV infection is highly endemic and is usually acquired in early childhood; these infections are typically inapparent. In developed countries HAV outbreaks continue to occur despite high levels of hygine Hepatitis A virus is the primary agent of what used to be called infectious hepatitis. Type A hepatitis is an acute disease involving the liver. Case severity ranges from asympomatic infection to GI illness, to acute, jaundiced infection involving necrosis of liver hepatocytes. Infectious hepatitis has an incubation period of 15-45 days and the illness may last several weeks to months; mortality is <1%. The virus is shed in the feces of infected individuals up to two weeks prior to the onset of clinical symptoms, has been detected in the feces of approximately 50% of patients within the 1st week after the onset of symptoms and may be present for up to several weeks (Gust and Feinstone, 1988, Hollinger and Ticehurst, 1990). To date, only one serotype of HAV has been identified, and its distribution is worldwide. However, based upon nucleotide sequence and antigenic information, different strains of HAV can be distinguished (Jansen et al , 1990). HAV infection confers lifelong immunity, but in developed countries like the U.S , the majority of children and young adults have never been infected and, therefore, are susceptible if exposed to the virus.

HAV could not be cul tivated in cell culture until 1979, when Provost and Hilleman (1979) finally propagated the virus in primary marmoset liver cells and fetal rhesus kidney cell cultures Lack of visible cytopathic effects and late appearance of viral antigen post-infection necessitated the development of assays such as the radioimmunofocus assay (RIFA) for the enumeration of virus infectivity in cell culture (Lemon et al., 1983). More recently, cytopathic variants of HAV have been isolated in cell cultures (Anderson, 1987; Cromeans et al., 1987).

Other Enteroviruses.

The other enteroviruses, including polioviruses, coxsackie viruses A and B, echoviruses, and newer enteroviruses not classified in these aforementioned categories, are non-enveloped, 27 nm diameter, icosahedral particles containing single-stranded, plus-sense RNA.Their protein coats (capsids) consist of four major polypeptides (VP-1 to VP-4), and mature virions have a buoyant density in CsCl of 1.34g/ml. These viruses are relatively heat-resistant, acid-resistant (pH 3 for 1 to 3 hours), and resistant to many common laboratory disinfectants, including 70% alcohol, 5% lysol, ether, deoxycholate,and other lipid solvents. Enteroviruses are inactivated by dessication, heat (50°C for 1 hour), and certain chemicals (e.g.,formaldeyhde and free residual chlorine).

Of the enteroviruses, the polioviruses have received the most attention because of their ability to cause poliomyelitis, a severe paralytic disease with a relatively high mortality rate in older children and adults. With the development and widespread use in developed countries of killed vaccines or live vaccines, poliomyelitis has all but disappeared in these countries, except for rare live vaccine-associated cases. Childhood poliomeylitis is more common in developing countries where control by widespread vaccination has not yet occurred. In developed countries polioviruses have not disappeared because recipients of the live vaccine can fecally excrete high virus concentrations (10^6 infectious units per gram) for periods of days to weeks after vaccination. Consequently, vaccine strains of poliovirus are plentiful in sewage and fecally contaminated waters

The other enteroviruses, including coxsackie viruses and echoviruses, continue to be important human pathogens. They cause such severe illnesses as: aseptic meningitis, a non-paralytic central nervous system illness resembling poliomyelitis; herpangina (type A coxsackie viruses), an acute illness characterized by fever, lesions in the mouth, loss of appetite, vomiting and abdominal pain; pleurodynia(type B coxsackieviruses), an illness with symptoms of severe pleuritic pain, headache and fever, and carditis (type B coxsackievirus) which can lead to permanent cardic abnormalities Less severe illnesses caused by the enteroviruses include rashes, conjunctivitis, respiratory illnesses of the "common cold" type and mild febrile illnesses with or without diarrhea.

Hepatitis E Virus; Enterically Transmitted Non-A, Non-B Hepatitis

Recently, it has been established that infectious heptitis disease can be caused by a distinct enteric virus unrelated to hepatitis A virus and now designated hepatitis E virus (HEV) or enteric non-A, non-B hepatitis virus. Clinically and epidemiologically, the disease transmitted by this virus is very similar to hepatitis A However, the disease is marked by a very high mortality rate in pregnant women and has a geographic distribution somewhat different than hepatitis A, with high prevalence in Asia, Africa, Mexico and Central America HEV has recently been characterized, and it appears to be a calicivirus (Hollinger, 1990; Reyes et al , 1990). The virus is a non-enveloped, icosahedral particle about 33 nm in diameter. It contains a single-stranded, plus-sense, RNA genome of about 7.6 kilobases and the genome has been cloned and at least partially sequenced (Reyes et al., 1990).HEV is an important agent of water-borne epidemics of hepatitis in some parts of the world, and it has been implicated in shellfish-borne illness (Caredda et al., 1981; 1985; 1986; Torne et al., 1988).

Reoviruses

The reoviruses, which morphologically resemble the rotaviruses (see below) and belong to the same family (Reoviridae) have been known for some time, and they are readily detected in quantity in sewage and fecally contaminated water by cell culture infectivity. However, they are of only minor importance as agents of water- and shellfish-borne gastroenteritis, and their role in enteric illness in general is uncertain. They occasionally cause mild respiratory and enteric disease, but because they have been isolated from healthy people as well, they are not considered major enteric viral pathogens.

AGENTS OF EPIDEMIC VIRAL GASTROENTERITIS

Gastroenteritis is second only to viral respiratory disease as the most prevalent viral illness, with about one episode per person per year in the U.S. population. The illness usually has a rapid onset and is self-limited, but it can be severe and even fatal in infants, geriatrics, the malnourished or the debilitated Symptoms of gastroenteritis vary with the virus type but include one or more of the following: diarrhea, nausea, vomiting, low grade fever, abdominal cramps, headache, anorexia, myalgia and malaise. At least three main groups of gastroenteritis viruses are known to be important in human disease: (i) the rotaviruses, (ii) the small, round viruses typified by the Norwalk virus and (iii) the enteric adenoviruses (types 40 and 41). Another group of viruses suspected of causing gastroenteritis is the coronaviruses, but their role is uncertain. They are more important in respiratory illness.

Rotaviruses

The rotaviruses are members of the Reoviridae family. The virions are about 70 nm diameter and consist of a double layer capsid composed of several distinct polypeptides, 11 discrete segments of double-stranded RNA, each encoding for distinct viral proteins, and an internal virion polymerase. The group A rotaviruses, of which there are at least 5 serotypes, are major causes of gastroenteritis and possess a common antigen (Kapikian and Chanock, 1989; 1990a). There are least two other groups of rotaviruses infecting humans, designated B and C, but they are uncommon in the U.S.

Rotavirus gastroenteritis has an incubation period of 1-3 days and an acute onset marked by severe watery diarrhea, vomiting (a hallmark symptom) and low-grade fever. Illness typically lasts 5-8 days and is self-limiting Viruses are fecally shed in high concentrations (up to 10^{11} particles per gram) for about 1.5 weeks, beginning at onset of illness. As with all viral gastroenteritis, the primary threat to health is severe dehydration caused by excessive diarrhea and vomiting, and therefore, fluid and electrolyte replacement is most important in managing the disease. Rotavirus infection produces both circulatory (humoral) and intestinal (secretory) immunity. Individuals with serum antibodies are generally immune to reinfection with the same serotype, but may be infected with different serotypes.

The rotaviruses are found worldwide, cause peak infection in children ages 4 months through 3 years, and >90 percent of children are infected by age 5. Adult reinfection is possible at high doses and asymptomatic infection is frequent in both healthy children and previously infected adults. Symptomatic infection is common in debilitated children and older adults (age >60 years) Rotaviruses account for about one-third to one-half of hospitalized cases of diarrhea in children aged 6-24 months, and infections are most frequent in winter and early spring in temperate climates. Transmission is usually person-to-person by the fecal oral route, and feces may contain 10^{11} virus particles/gram. The high risk groups for rotavirus infections are: diaper-age children, contacts of diaper-age children, adult travelers, military personnel, and institutionalized geriatrics. Epidemic transmission of rotaviruses occurs in nurseries and day care centers, via contaminated drinking water and possibly by airborne transmission indoors. Although rotaviruses have been detected in contaminated coastal waters, they have not been strongly implicated in shellfish-borne gastroenteritis.

Norwalk Virus and the Norwalk-type Viruses

Norwalk virus is the prototype of several small, round, structured, gastroenteritis viruses (Kapikian and Chanock, 1989; 1990b). The virion is about 27 nm diameter and appears to contain a single capsid protein of about 60,000 molecular weight. Recent success in cloning and characterizing the Norwalk virus nucleic acid shows that it is single-stranded, plus-sense RNA of about 7 5 kilobases. Based upon this information and serological studies on the antigenic relatedness of Norwalk virus to other small, round viruses, it is probably a calicivirus.

Norwalk gastroenteritis has an incubation period of 1-3 days, an acute onset with symptoms of mild to moderate diarrhea lasting 1-2 days, vomiting (often frequent), fever, headaches, myalgias and malaise. The virus is fecally shed from onset of illness to about 3 days thereafter. As with other viral gastroenteritis, dehydration is the most serious threat to health and is managed by fluid and electrolyte therapy. Immunity to Norwalk virus is paradoxical because adults who have been infected are usually susceptible to reinfection several months or years later, despite serum antibodies Adults who lack serum antibodies are often resistant to infection, even after repeated exposures to the virus

Norwalk virus is distributed worldwide, and about one-half to two- thirds of U.S. adults have antibodies. Norwalk infection usually occurs in older children and adults, and infection in infants and young children is rare. The level of Norwalk virus that is fecally shed by infected individuals is uncertain because there are no laboratory host systems to detect and quantify virus infectivity. Epidemic transmission of Norwalk virus is common, especially common source outbreaks due to contaminated drinking water and uncooked or improperly handled foods (especially shellfish and salads), and there have been outbreaks in schools, social gatherings and restaurants where contaminated food was consumed, nursing homes, summer camps and cruise ships. The role of Norwalk virus in endemic acute gastroenteritis of adults is uncertain, but may be appreciable.

Other Small, Round Gastroenteritis Viruses

Several other small, round viruses causing gastroenteritis have been identified. These viruses can be morphologically classified into two groups: "structured" viruses and "featureless" viruses (Appleton, 1987; Appleton and Pereira, 1977; Appleton et al., 1981). The structured viruses have characteristic surface morphologies by electron microscopy and appear to be caliciviruses and astroviruses. The caliciviruses are about 31-35 nm in diameter, probably contain single-stranded, plus-sense RNA and contain one virion (capsid) protein (Cubitt, 1987). The astroviruses are about 29-30 nm in diameter and are thought to contain plus-sense RNA. Their capsid proteins are of uncertain number and size (Kurtz and Lee, 1987). The "featureless" gastroenteritis viruses have been tentatively considered picornaviruses or parvoviruses, but their taxonomy is uncertain They are 25-30 nm diameter and have an indistinct surface when viewed by electron microscopy. They have been implicated in shellfish-borne and other food-borne gastroenteritis.

The caliciviruses and astroviruses are probably worldwide in distribution and infect all age groups. Transmission is primarily person-to-person via the fecal-oral route, and infection occurs early in life (by age 4); more than two-thirds of older children and adults have immunity. Epidemic transmission of these viruses is documented by community and nosocomial outbreaks in nurseries, schools and geriatric homes, as well as outbreaks associated with water and shellfish (Cubitt, 1987; Kurtz and Lee, 1987).

Enteric Adenoviruses

The enteric adenoviruses, serotypes 40 and 41 of the human adenoviruses, resemble other adenoviruses that cause primarily upper respiratory illness. They are icosahedral, about 85 nm diameter, possess several major capsid proteins (hexons, pentons and fibers), several additional minor capsid proteins, internal core proteins, and have linear, double-stranded DNA as their nucleic acid. The enteric adenoviruses have a worldwide distribution and produce infection early in life (1-16 months) with little or no seasonal fluctuation. These viruses are responsible for 4 to 10% of infantile diarrhea and epidemics have occurred in hospital nurseries. These is no documented shellfish-borne disease due to these viruses, but adenoviruses have occasionally been detected in shellfish habitats.

Enteric Coronaviruses

The enteric coronaviruses are round, oval or pleomorphic in shape, enveloped with club-shaped outer spikes, 80-150 nm in diameter and contain single-stranded, plus-sense RNA. These viruses are probably found worldwide and they have been implicated in gastroenteritis and necrotizing enterocolitis of infants. They have not been implicated in shellfish-borne gastroenteritis.

Detection of Gastroenteritis Viruses

It is ironic that the most important human gastroenteritis viruses, the rotaviruses and Norwalk-type viruses, can not be cultivated readily or at all in cell cultures. Diagnosis of infection and illness is based on detection of viral antigens in stools, or, less frequently, on diagnostic rises in serum antibodies The first technique for detection of rotaviruses and Norwalk-type viruses was immune electron microscopy, but this technique is too cumbersome and insensitive for detecting these viruses in environmental samples, including shellfish. Later, solid-phase radioimmunoassays and enzyme immunoassays were developed to detect the antigens or antibodies of these viruses. However, these techniques are also too insensitive to detect the viruses at the low concentrations at which they would be present in shellfish or other environmental samples. Human rotaviruses can be cultivated in certain cell cultures using specialized techniques such as proteolytic enzyme treatment of the virus specimen and host cell cultures, centrifuging the specimen onto the cell layers, immunofluorescence to detect individual foci of infected cells, or roller tubes cell cultures to promote the development of cytopathic effects. These techniques have been applied successfully to the detection of rotaviruses in environmental waters, but they are more tedious than the cell culture methods for detecting and quantifying the enteroviruses, such as the polioviruses. Laboratory techniques are not available to detect Norwalk and other Norwalk-type viruses by their infectivity.

ENTERIC VIRUS CONTAMINATION OF SHELLFISH HABITATS

Sources of Enteric Virus Contamination

The continued and widespread disposal of sewage and other fecal wastes to coastal estuarine and marine waters is the main source of enteric viruses in shellfish environments. Fecal waste sources include raw sewage discharges, sewage treatment plant effluents, faulty septic (on-site) waste disposal systems, boat waste discharges, and land runoff. In the U.S. more than 8 billion gallons of sewage, of which only about one-half receives secondary treatment, are discharged per day into coastal waters (Bitton, 1980). Although virus densities in raw sewage vary greatly with season and geographic location, it is estimated that typical concentrations in raw sewage range from about 10^2-10^5 infectious units per liter of sewage (Bitton, 1980; Rao and Melnick, 1986). Conventional sewage treatment processes reduce the amounts of viruses in wastewater to varying degrees (Feachem et al., 1983; Rao and Melnick, 1986). Primary sedimentation removes about 50% of the viruses in wastewater, primarily by sedimentation of viruses in settleable solids. Secondary (biological) treatment is capable of reducing viruses in primary sewage to a greater extent. Trickling filters typically achieve a 50-90% removal of viruses, mainly by adsorption to the biological slime layer and subsequent inactivation therein. Virus removal by activated sludge treatment has been reported as high as 90- 99%, and it occurs by virus adsorption to the activated sludge floc and inactivation caused by microbes in the activated sludge Disinfection of treated sewage effluents can inactivate viruses, but the virucidal effectiveness of wastewater disinfection depends upon the type and concentration of disinfectant used, the contact time, and the quality of the effluent. High levels of organic matter and suspended solids in effluent limit the efficiency of disinfection. Viruses tend to be more resistant to inactivation by chlorine and other disinfectants than are coliform bacteria Overall, a typical sewage treatment system consisting of primary sedimentation, biological treatment (e.g. activated sludge or trickling filtration), and disinfection is not likely to reduce raw sewage virus concentrations by more than 99%.

Persistence of Enteric Viruses in Shellfish Habitats

Virus survival in shellfish environments depends upon a variety of physical, chemical, and biological factors (Rao and Melnick, 1986). Physical factors affecting virus survival include temperature, UV radiation, and adsorption to particulate matter. Temperature is one of the most important environmental influences on virus survival, and it is well established that viruses survive longer at lower temperatures. Exposure to high temperatures results in damage to both the viral nucleic acid and protein coat, thereby producing virus inactivation. UV radiation in sunlight near the surface of a body of water can also result in virus inactivation by damaging viral nucleic acid. Virus adsorption to organic and inorganic particulate matter present in sewage, effluents and natural waters may promote virus survival by shielding the virus from biological and physiochemical factors causing inactivation. Solids-associated viruses tend to accumulate in sediments where they may survive longer than in the water column (LaBelle and Gerba, 1980, Rao et al., 1984).

However, temperature, pH, ionic strength, and the presence of organic matter influence the nature and extent of virus adsorption to solids. Storms, wave action, currents, tides, and dredging can resuspend sediment-associated viruses. Shellfish harvested after periods of heavy rain or flooding have been implicated in a number of viral disease outbreaks, which may be related to less effective sewage treatment, increased runoff of fecal contamination into coastal waters and resuspension of sediment-associated viruses.

Chemical factors influencing virus persistence in shellfish waters include pH, salinity, and the presence of certain cations and organic chemicals. Enteric viruses are generally stable in the pH range of natural waters, pH 5 to 9. However, certain salts, including those found in sea water enhance inactivation of enteric viruses (Block,1983). Organic matter tends to protect viruses from inactivation by reacting with disinfectants and other strong oxidants having virucidal activity, by absorbing virucidal UV radiation and by precipitating in seawater and thereby providing particulate material with which the viruses may become protectively associated. However, readily degradable organic matter may stimulate natural microbial activity in aquatic environments that is biologically antagonistic to viruses (Ward et al., 1986).

A number of biological factors influence virus inactivation in water. One of these is the type of virus itself, because specific viruses differ in their survival in the environment. For example, HAV survives much longer than poliovirus in water and wastewater (Sobsey et al.,1986). Naturally occurring microbes in water, especially bacteria, possess virucidal capabilities (Herrmann and Cliver, 1973, Herrmann et al., 1974; Shuval et al., 1971; Ward et al., 1986). Although biological inactivation mechanisms are not fully understood, proteolytic enzymes may be involved in enterovirus inactivation (Block,1983; Ward et al., 1986).

Occurrence of Enteric Viruses in Shellfish and Their Habitat

Viral contamination of shellfish and their estuarine and coastal marine environment is well documented. Since 1965, when enteric virus isolation from shellfish was first reported, viruses have been detected in many types of edible bivalves (Gerba and Goyal, 1978; DeLeon and Gerba, 1990), even when no viruses were detectable in their overlyingwaters (Goyal et al., 1979). Table 2 summarizes data on virus isolation from shellfish for the period 1965-1989. Enteric virus isolation has been reported for Crassostrea virginica, Mercenaria mercenaria, Mercenaria californicus, Mytilus galloprovincialis and Mytilus edulis. Additionally, Goyal et al. (1984) have reported the isolation of enteroviruses from blue crabs at a sludge dump site in the northern Atlantic

The types of viruses found in shellfish are those with direct or indirect association with the alimentary tract of man or other homeothermic animals and whose characteristics permit survival and transmission by feces. Those isolated include polioviruses, echoviruses, coxsackieviruses, reoviruses, adenovirus, rotaviruses and various bacteriophages. Other viruses have been isolated but could not be identified. The virus type most frequently recovered from shellfish is poliovirus, whose prevalence is explained by active immunization programs using live, attenuated vaccines in developed nations.

A number of studies report virus isolations from shellfish harvested from open areas (Ellender et al., 1980; Vaughn et al 1979b; 1980; Wait et al. 1983), although at lower frequency than from shellfish taken from prohibited areas. Virus isolation rates for shellfish taken from both approved and prohibited beds range from 0% to 40%, but rates for approved shellfish generally fall below 20% those for prohibited shellfish frequently exceed 25%. In all cases where enteric viruses have been detected in shellfish, levels of contamination are relatively low, ranging from 1 to approximately 200 PFU (plaque forming units) per 100 grams of shellfish.

The isolation of enteric viruses in shellfish harvested from approved waters has prompted investigation into the relationship between viral contamination of shellfish and overlay water. Metcalf and Stiles (1965; 1968) examined both oysters and overlay water and concluded that the

Table 2. Virus Isolation from Shellfish

Shellfish Species	Location	Virus Type	Isolation Frequency	Comments	Reference
C. virginica	N.H. estuaries	E-9 C-B4 Unk.	6/10 (60%) pools	Waters polluted with raw sewage	Metcalf and Stiles, 1965
Mussels	Genoa Harbor, It.	P-3	-	-	Petrilli and Crovari, 1965
Mussels	Bari & Parma mkts., It.	E-3,-9,-13	50 total samples	-	Bellelli and Leogrande, 1968
C. virginica	N.H. estuaries	P-1,-2,-3 CB-2,3,4 E-9	114/459 25% of pools	4 year study	Metcalf and Stiles, 1968
M. mercenaria C. virginica	R.I. estuary	-	33% of clams 55% of oysters	Prohibited area	Liu et al., 1968
Mytilus galloprovincialis	Leghorn Coast, It.	E-5,-6,-8,-12 CA-18	5/68 (7.3%) of pools	Purified mussels put 3 days in massively-polluted seawater	Bendinelli & Ruschi, 1969
C. virginica	N. Atlantic estuary	1 entero. 30 coliphages	59/130 (45%)	Treated effluent discharged to estuary	Gerba and Goyal, 1978
Oysters; Mussels	Poitiers, Fr.	CA-16 most common	7/70 (10%) oyster 2/10 (20%) mussel pools	Tested by flaccid paralysis production in mice	Denis, 1973
C. virginica	Galveston Bay, TX.	P-1,-2	28.6%	-	Metcalf et al. (1972)

Table 2. Virus Isolation from Shellfish, Continued

Shellfish Species	Location	Virus Type	Isolation Frequency	Comments	Reference
C. virginica	Galveston Bay LA. coast Japanese imports	P-1,-3 E-4	2/17 (12%)-TX. 1/24 (4%)-LA. 1/1 (100%)-Jap.	Positive samples in TX. fr. approved waters	Fugate et al., 1975
M. cali-fornicus	CA. coast	unspecified	18/39 (46%)	Beds nr. outfall discharges of 1° & 2° sewage effl.	Gerba and Goyal, 1978
C. virginica	Great Bay, N.H. estuary	Coliphages; enteric viruses unspecified	80/158 (51%) coli-phage; 12/158 (7.6%) enteric virus pools	estuary drains 7 rivers having 1° treated sewage	Vaughn and Metcalf, 1975
M. mercenaria	Chesapeake & Chincoteague Bays, MD.	chlamydiae phages	-	Viruses detected by election microscopy	Harshbarger et al., 1977
C. virginica	Galveston Bay, TX.	P-1,-2 E-1	14/40 (35%) pools	2/10 fr. open areas 12/30 fr. closed areas	Goyal et al., 1979
C. virginica	Pass Christian Reef, Graveline Bayou, MS. estuaries	P-1,-2 E-24, & not identified	32/109 (29.3%)	2/22 fr. open areas 30/87 fr. closed areas	Ellender et al., 1980
C. virginica	N.C. coast	-	3/47 (6.4%) pools	0/23 fr. open areas 3/24 fr. closed areas	Sobsey et al., 1980
M. edulis	Intertidal zone, Br. coast	Adeno-virus, C-B3 Reovirus	4/6 (67%) pools	Waters grossly sewage-polluted	Watson et al., 1980

Table 2. Virus Isolation from Shellfish, Continued

Shellfish Species	Location	Virus Type	Isolation Frequency	Comments	Reference
C. virginica M. mercenaria	Great So. Bay, Oyster Bay & Raritan Bay, off N.Y. & N.J. coasts	P-1,-2 E-2,-15,-20,-23 CB-3 & not indentified	oysters: 5/16 (31.3%) clams: 13/22 (59.1%) pools	Oysters - 2/8 fr. open areas 3/8 fr. closed areas Clams - 2/5 fr open areas 11/17 fr. closed areas	Vaughn et al., 1979, 1980
M. mercenaria	N.C. coast	P-1 E-7,-27 Unidentified	9/28 (32.1%) pools	3/13 fr. open areas 6/15 fr. closed areas	Wait et al., 1983
C. virginica	Terrebonne Paris, LA.	-	3/41 (7.3%) pools	Viruses in oysters fr. closed areas only	Cole et al., 1986
M. arenaria M. mercenaria	Mass. coast	None detected	0/50 (0%) soft shell clams, 0/21 (0%) hard shell clams, 0/14 (0%) oyster pools	10 samples fr. open areas; 75 samples fr. restricted areas	Khalifa et al., 1986
C. virginica	5 states, unspecified	Not specified	5/81 (6.2%)	0/31 fr. open areas 5/50 fr. closed areas	Richards and Hetrick, 1988
Mussels; oysters- species not specified	S. Afr.	P-2 CB-1 E-3,-7 Reo 1,3 Adeno 31	8/28 (29%) enteroviruses, 22/28 (79%) reoviruses	Waters with varying pollution levels	Grabow et al., 1989

coexistence of viruses in overlay water and oysters occurred in only 43% of the cases Goyal et al (1979) reported parallel isolations in only 50% of the cases. When viruses were isolated simultaneously from water and oyster samples, the types of viruses identified in the two samples frequently differed (Goyal et al , 1979).Other investigators have reported generally poor correlations between the presence of phages or enteric viruses in shellfish samples and parallel samples of sediment and/or overlay water (Gerba et al., 1980; Vaughn and Metcalf, 1975; Vaughn et al., 1980). Cole et al. (1986) reported no in parallel isolation of virus in 41 samples of oysters and overlay water. Investigators have concluded that viral presence in overlay water is not an adequate predictor of viral contamination of shellfish. Discrepancies between viral contamination of water and shellfish are attributable to discontinuous pollution patterns and viral persistence in shellfish after disappearance from overlay water (Metcalf and Stiles, 1968, Goyal et al., 1979). It has been suggested that the only reliable way in which to determine the virological quality of shellfish is to test the shellfish meat itself (Metcalf and Stiles, 1968).

It has been assumed that shellfish are passive carriers of enteric viruses, and viral replication in shellfish has never been demonstrated (Metcalf and Stiles, 1968). Liu et al. (1966a; 1966b) observed that virus in the digestive tract of quahogs was neither intracellular nor adsorbed to cells. Chang et al. (1971) inoculated proflavine-labeled light sensitive virus into northern quahogs and later examined them for light-sensitive progeny. The failure to detect such progeny provided convincing evidence of the absence of viral replication in contaminated shellfish.

Relationships Between Viruses in Shellfish and Bacteriological Standards

Methods for detecting enteric viruses and pathogenic bacteria in water and shellfish are complex, expensive, and time consuming. Fecal coliform bacteria are used as a sanitary quality indicator for shellfish and their harvesting waters   This index is considered appropriate because fecal coliforms are normal inhabitants of the gastrointestinal tract of warm-blooded animals and are excreted in the feces in large numbers; their presence is taken as evidence of recent fecal pollution (Gerba and Goyal, 1978). Some states in the U.S continue to use total coliforms for classification of shellfish harvesting waters, despite their lesser specificity for detecting coliforms of fecal origin. There has been increasing concern about the adequacy of bacteriological standards as indicators of the virological quality of shellfish. Reports of viral isolation from shellfish are well documented, and coliform bacteria are known to be more sensitive to disinfection, natural inactivation, and depuration than are viruses (Gerba and Goyal, 1978).

Investigators have examined the relationships between viral contamination of shellfish and several bacteriological indicators. No statistically significant relationships have been found consistently between virus presence in clams and oysters and a variety of bacteriological and physicochemical quality parameters of water and shellfish (Cole et al ,1986; Ellender et al., 1980; Fugate et al., 1975; Goyal et al., 1979; Gerba et al., 1980; Grabow et al., 1989; Kilgen et al., 1984; Vaughn et al., 1980, Wait et al., 1983). Bacteriological indicators examined include total and fecal coliforms in overlay water, as well as total coliforms, fecal coliforms, $E.$ coli, fecal streptococci, Vibrio parahemolyticus, and aerobic plate count in shellfish. Goyal et al. (1979) were able to isolate viruses from oyster samples when the total coliform MPN was as low as 3/100 ml water. Ellender et al. (1980) reported viral isolation from clams at fecal coliform levels less than 20 per 100 grams of shellfish However, Richards and Hetrick (1988) have reported a correlation between virus and fecal coliform levels in shellfish They consistently isolated viruses from clams and oysters each time fecal coliform levels exceeded 130 per 100 grams of shellfish.

Because the current bacteriological standards for determining the sanitary quality of shellfish and shellfish-growing waters do not adequately reflect the occurrence of enteroviruses, the use of coliphages has been suggested as an alternate indicator of enteric viral contamination (Kott et al , 1974). Coliphages are fecally excreted by a proportion of humans and are found at relatively high levels in raw sewage (10^2-10^5 infectious units/ml) and treated effluents. Their sources, size, survival in the environment and reductions by sewage treatment, including chlorine disinfection, are similar to certain enteric viruses. The use of coliphages as indicators is valid only if their presence accurately and consistently indicates enteric viral contamination (Kott et al., 1974). Vaughn and Metcalf (1975) reported the results of parallel examination of sewage effluent, sediment, shellfish, and water for coliphages and enteric viruses. The majority of enteric virus isolations (63%) were observed in samples yielding no coliphages. In controlled experiments, oysters were found to accumulate 5-30 times more coliphages than model enteric viruses. Shifts in dominant coliphage type occurred during the course of the study and evidence for coliphage replication in estuarine waters was presented. Vaughn and Metcalf (1975) concluded that the application of coliphages as an index of shellfish hygiene was not recommended. Kennedy et al. (1986) reported wide ranges of coliphage recovery from fresh oysters and poor correlation between coliphage counts and E. coli levels. Burkhardt et al. (1989) recently reported that male-specific coliphages were unable to replicate within whole clams, although replication in clam homogenates was observed. As yet,there have not been systematic studies on the relationships between human enteric viruses and male-specific coliphages in shellfish to determine the validity of this group of phages as viral indicators.

EPIDEMIOLOGY OF SHELLFISH-ASSOCIATED ENTERIC VIRAL DISEASE

Since 1956, when the transmission of viral illness due to the consumption of contaminated shellfish was first recognized with the report of a large outbreak of shellfish-associated hepatitis A (Roos,1956), considerable epidemiological evidence linking viral illness to shellfish consumption has accumulated Richards (1985) reported over 100 U.S. outbreaks, which alone have accounted for more than 1000 cases of hepatitis A. Appleton (1987) reported 96 outbreaks of shellfish-associated illness of suspected viral etiology in England and Wales between 1965 and 1986. Hepatitis A caused 11 of these outbreaks, viral gastroenteritis occurred in 57, and 26 were unclassified. While improved sanitation requirements have controlled the transmission of most shellfish-associated bacterial diseases in recent decades, it is apparent that outbreaks of hepatitis A and viral gastroenteritis continue to occur. Also important are endemic or sporadic cases of shellfish-acquired viral illness that are not detected as outbreaks, yet caused by shellfish consumption (Gerba, 1988) Epidemiological evidence clearly indicates that shellfish can serve as an effective vehicle for the transmission of viral hepatitis and gastroenteritis.

A summary of U S. shellfish-associated outbreaks and cases of hepatitis and gastroenteritis adapted from Rippy (1989) is shown in Table 3. For most of these outbreaks, viral etiology could not be ascertained but was the most probable cause. European surveillance data on shellfish-borne outbreaks is provided by Gunn and Rowlands (1969) and Appleton (1987) Since the early 1970's when identification methods for hepatitis A virus and several viral agents responsible for gastroenteritis were first developed, clear epidemiological evidence has been established linking the consumption of contaminated shellfish to the transmission of hepatitis A, Norwalk virus, Snow Mountain agent and several other small, round, gastroenteritis viruses (astroviruses, caliciviruses, and parvoviruses) (Appleton, 1987, DeLeon and Gerba,1990). Tables 4 and 5 summarize the data implicating viruses directly in shellfish-borne hepatitis and gastroenteritis outbreaks, respectively, and include reports from the U.S., Europe, and Australia.However, it should be noted that detection methods for enteric viruses are sophisticated and largely unavailable. When this is considred along with the typical difficulties in investigating foodborne disease outbreaks, it can be concluded that the number of unreported outbreaks and cases of shellfish-borne viral disease in the general population far exceeds the number reported (Richards, 1985).

Besides viral contamination and viral illness associated with raw shellfish, there is evidence that cooking shellfish may not sufficiently inactivate the viruses present   Morse et al. (1986) reported that while most of the 1017 cases of shellfish-transmitted gastroenteritis occurring during an 8-month period in New York in 1982 were caused by the consumption of raw clams and oysters, there was an appreciable attack rate (26%) among those who ate only steamed clams. It has also been reported that 7-13% of poliovirus added to oysters survived cooking by steaming, frying, and baking (DiGirolamo et al., 1970). Although steaming for only about 60 seconds causes soft-shelled

84

Table 3.   Reported Outbreaks and Cases of Shellfish-associated Gastroenteritis
            and Hepatitis in the United States, 1979-1988[*]

YEAR	GASTROENTERITIS		HEPATITIS	
	Outbreaks	Cases	Outbreaks	Cases
1988	37	42	1	51
1987	20	24	0	0
1986	83	>247	0	0
1985	32	218	2	2
1984	52	474	0	0
1983	68	>2376	2	5
1982	17	1858	1	11
1981	13	225	1	9
1980	15	>155	1	1
1079	14	81	2	18
Totals	351	>5700	10	97

[*]Adapted from Rippey (1989).

clams to open (and hence typically be eaten), 4-6 minutes of steaming is required to reach 100°C (Koff and Sear, 1967).   Temperatures normally achieved during steaming are inadequate to completely inactivate HAV (Peterson et al , 1978; Koff and Sear, 1967).

ACCUMULATION OF VIRUSUS BY SHELLFISH

Shellfish Feeding and Related Activities.

Bivalves are laterally compressed and possess a shell with two valves, hinged dorsally, that completely enclose the body. They are plankton feeders and extract food and large volumes of seawater from their environment. Potential food particles present in the ingested water are trapped on a mucus sheet secreted by the gills and transported along the gills by ciliary and muscular activity towards the mouth.   In the mouth region, particles are sorted by ciliated labial palps. Accepted particles are passed through the mouth and enter the stomach and digestive tract, while rejected particles are moved past the gill sand are ejected as pseudofeces. The particles ingested by this filter feeding process are phytoplankton and bacteria, but they may also include viruses and a variety of chemical toxins (dinoflagellate toxins, heavy metals, radiochemicals and hydrocarbons) (Earampamoorthy and Koff, 1975).

The volume of water pumped by bivalves varies with species and with environmental conditions such as temperature, salinity, and turbidity. Oysters can filter seawater up to a rate of 40 L/hour, and thus may be exposed to considerable quantities of microorganisms in polluted water.

Mechanisms of Virus Accumulation by Shellfish.

At least two possible mechanisms of virus accumulation by shellfish have been proposed One mechanism suggests that viruses adsorbed to acceptable food particles are transported along the gills to the mouth and digestive system. The other mechanism suggests that viruses present

Table 4.   Reported Viral Gastroenteritis Outbreaks Attributable to Shellfish

Year	Shellfish	Location	No. Cases	Agent	Reference
1976-1977	Cockles	Chelmsford & Southampton, U.K.	797	SRVs[**]	Appleton & Pereira (1977)
1978	Oysters	Georges R., Aus.[*]	2000+	Norwalk	Murphy et al. (1979)
	Oysters	Georges R., Aus.[*]	5	Norwalk	Eyles et al. (1981)
	Oysters	Georges R., Aus.[*]	150	Norwalk	Lince & Grohmann (1980) Murphy et al. (1979)
1980	Oysters	Florida	6	Norwalk	Gunn et al. (1982)
1981	Clams	N.Y.[*], England[*]	210	Not specified	Guzewich & Morse (1986), Richards (1985)
1982	Clams	N.Y.[*], Mass.[*], N.C.[*], R.I.[*]	441	Norwalk	Guzewich & Morse (1986), Morse et al. (1986) Richards (1985)
	Clams	N.Y.[*]	659	Norwalk	Guzewich & Morse (1986), Mores et al. (1986) Richards (1985)
	Oysters	N.Y.[*], Mass.[*]	230	Norwalk	Guzewich & Morse (1986), More et al. (1986) Richards (1985)
	Oysters	Louisiana[*]	472	Norwalk like	Richards (1985)
1983	Clams	N.J., England[*]	135+	Not specified	Richards (1985)
	Clams	N.J., England[*]	400+	Not specified	Richards (1985)
	Clams	N.J., England[*]	1100+	Not specified	Richards (1985)
	Clams	N.Y.	84	Snow Mtn.	Truman et al. (1987)
	Oysters	London, England	181+	SRVs	Gill et al. (1983)
1984	Oysters	Florida[*]	93	Not specified	Richards (1985)
	Shellfish, unspecified	New York	256	Not specified	Guzewich & Morse (1986)
1985	Shellfish, unspecified	New York	98	Not specified	Guzewich & Morse (1986)
1985-1986	Oysters	England	NR	SRVs & parvoviruses	Appleton (1987)
	Cockles & mussels	England	NR	Calici- & parvoviruses	Appleton (1987)
	Cockles	England	NR	Astro- parvoviruses	Appleton (1987)
	Oysters & mussels	England	NR	SRVs & parvoviruses	Appleton (1987)
	Cockles	England	NR	SRVs & parvoviruses	Appleton (1987)
	Oysters	England	NR	Parvoviruses	Appleton (1987)

[*] Indicates source of contaminated shellfish.
[**] SRVs = small, round viruses.

Table 5. Shellfish-Borne Hepatitis A Outbreaks

Year	Shellfish	Location	# Cases	Reference
1955	Oysters	Sweden	629	Roos (1956)
1961	Oysters	AL., MS.[*]	84	Mason & McLean (1962)
	Clams	N J., N.Y.[*]	459	Dougherty & Altman (1962)
	Clams	CT.	15	Goldfield (1976), Richards (1985)
	Oysters	AL.	31	Goldfield (1976), Richards (1985)
1962	Clams	N.Y.	3	Goldfield (1976), Richards (1985)
1963-64	Clams	PA., N.J.[*]	252	CDC (1964)
	Clams	CT., R.I.[*]	123	CDC (1964), Ruddy et al. (1969)
	Oysters	N.C.	3	CDC (1964)
	Clams	N.Y.	43	Richards (1985)
	Clams	Wash. DC, N.J.[*]	3	Richards (1985)
	Oysters	Br. Col., Can.	2	Goldfield (1976)
1966	Clams	N.J., MA.[*], MD.[*]	4	Dismukes et al. (1969)
	Clams	MA.	3	Richards (1985)
	Clams	N.Y., MA.	4	Richards (1985)
1967	Clams & oysters	TX.	3	Goldfield (1976)
1968	Clams	N.Y.	3	Richards (1985)
1969	Clams	N.Y.	6	Richards (1985)
	Oysters	FL.	13	Richards (1985)
1971	Clams	MA.	5	CDC (1972)
	Clams	R.I.	3	Feingold (1973)
1972	Clams	FL.[*], MA.	2	Richards (1985)
1973	Oysters	TX., GA., LA.[*]	293	Portnoy et al. (1975)
	Clams	MN., FL.	1	Guidon & Pierach (1973)
1976	Mussels	Australia	7	Deinstag et al. (1976)
1977	Unspecified	Washington	17	Richards (1985)
1978	Cockles	England	41	Bostock et al. (1979)
1979	Unspecified	USA unspecified	8	Richards (1985)
	Oysters	AL., GA., FL.[*]	10	Smith et al. (1979)
1982	Clams	N.Y.[*], MA.[*] N.C.[*], R.I.[*]	11	Richards (1985)
1984-85	Mussels & clams	Livorna, It.	75+	Mele et al. (1989)

[*]Source of contaminated shellfish.

hemolymph has also been documented and further supports the association of viruses with phagocytes (DiGirolamo et al., 1975; Liu etal., 1966a; Metcalf et al., 1979; 1980).

Studies have demonstrated that many common species of shellfish rapidly accumulate viruses when exposed to contaminated water. Virus uptake has been documented for oysters (Crassostera virginica, C. gigas, C. glomerata, Ostera edulis, O. lurida), clams (M. mercenaria, Mya arenaria, Tapes japonica), mussels (Mytilus edulis, M. galloprovincialis) cockles (Cerastoderma edule) and Hawiian bivalves of the genus Pinnidae and Isognomonidae. Various crab species (Pachygrapsus and Hemigrapsus species, Cancer magister, Cancer antennarius, Callinectes sapidus) and the Florida conch and sea hare (Gerba and Goyal, 1978) have also been found to accumulate virus. Reviews of virus accumulation by shellfish have been presented by Gerba and Goyal (1978), Richards (1988), and DeLeon and Gerba (1990).

Kinetics and Extent of Virus Accumulation by Shellfish

Table 6 provides summary data of virus accumulation by shellfish from 1964 to present. Shellfish were exposed to known concentrations of various viruses in aquaria with either static or flowing water systems. In most studies, the viruses were not only taken up, they were significantly accumulated by the shellfish. Accumulation was frequently to a higher level than surrounding water. Reported accumulation factors varied from less than one-fold to over 1000-fold The highest uptakes were 900-fold and 1500-fold as reported by Hoff and Becker (1969) and Canzonier (1971), respectively Most accumulation ranged from 10- to 100-fold (Hamblet et al , 1969, Liu et al., 1966a; Metcalf et al., 1979; 1980; Mitchell et al , 1966; Seraichekas et al., 1968). In some cases, virus accumulation did not exceed the exposure level (Duff, 1967; Hedstrom and Lyke, 1964, Metcalf and Stiles, 1965; et al., 1987; Scotti et al., 1983). This low level of accumulation may be due to suboptimal conditions for metabolic activity (Hamblet et al., 1969).

Maxiumum uptake of viruses by shellfish takes place within a few hours of exposure to contaminated water Investigators have reported appreciable uptake by 1-3 hours (Hedstrom and Lyke, 1964; Liu et al.,1966a; Metcalf and Stiles, 1965). Mitchell et al. (1966) found that oysters concentrated polioviruses to 27 times their concentration in seawater after only 1 hour exposure Others have reported slightly longer exposures of 4-16 hours to obtain maximum uptake (Metcalf etal., 1979). In flow-through systems shellfish can concentrate viruses to higher levels in their tissues than in the surrounding water(Hamblet et al., 1969; Hoff and Becker, 1969; Liu et al., 1966b,Metcalf et al , 1979; Mitchell et al., 1966). However, a threshold or plateau of virus titers in shellfish has been observed (Gaillot et al.,1988; Liu et al , 1966a; Mitchell et al., 1966). This has been explained as an equilibrium between uptake and elimination, indicating that viral concentration by shellfish is a dynamic process (Mitchell etal , 1966) and that virus levels in shellfish are maintained as long as sufficient viruses are present in the surrounding water. As the concentration of viruses in water decreases, the viral content of the shellfish also decreases (Gaillot et al , 1986; Hedstrom and Lyke, 1964; Liu et al., 1966a; Metcalf and Stiles, 1965). Other researchers have found, however, that shellfish may not attain high virus levels when uptake is performed in static systems (Hedstrom and Lycke, 1964; Metcalf and Stiles, 1965), probably because static systems do not simulate natural conditions as well as flow-through systems do.

Factors Influencing Virus Uptake by Shellfish

The degree of viral uptake and accumulation by shellfish may depend upon several factors, including the hydraulic characteristics of the exposure system and exposure time, virus type, viral concentration in water, temperature, presence of particulate matter or turbidity, shellfish species, differences in individual animals, food availability, pH, and salinity. Because uptake of viruses by shellfish is dependent upon active feeding, any factor affecting the physiological activity of the animal can influence accumulation.

Table 6. Virus Accumulation by Shellfish

Shellfish Species	Virus Type	System	Exposure Time	Concentration Factor	Comments	References
O. edulis	P-3	Circ.	2 hr.	1-fold	Increasing viral titers on prolonged exposure in water	Hedstrom and Lyke, 1964
C. virginica	P-1 CB-3	Circ.	24 hr.- 7 days	1-fold	Min. 4.5 log $TCID_{50}$ in water for uptake. Min. virus titer increase in tissues on prolonged exposure	Metcalf and Stiles, 1965
M. mercenaria	P-1	Static	1 hr.- 6 days	30-fold in 6 days	1-4 hrs. max. uptake; titer in meat maintained if enough virus in water	Liu et al., 1966a
M. mercenaria	P-1	Flowing	1-48 hrs.	Up to 1000-fold	Uptake more efficient if low virus levels in water; more virus uptake with better environmental conditions	Liu et al., 1966b
C. virginica	Phage	-	-	10-fold	-	Hoff et al., 1965
C. virginica	P-1	Flowing	1-22 hr.	10-27-fold	Low water contamination	Mitchell et al., 1966
M. edulis	P-3 CA-8	Circ.	18-36 hr.	<1-fold	-	Duff, 1967

Table 6. Virus Accumulation by Shellfish, Continued

Shellfish Species	Virus Type	System	Exposure Time	Concentration Factor	Comments	References
M. mercenaria	P-1	-	4 hr.- 3 days	-	4 hr. max.; contamination maintained for 3 days	Liu et al., 1967
C. virginica	P-1	Flow-through	6-24 hrs.	5-18-fold	4.2-4.6-fold @ high turb. 9.2-18.1-fold @ low turb.	Hamblet et al., 1969
M. mercenaria	P-1	Flowing	20-72 hrs.	10-100-fold	Assayed individual shellfish	Seraichekas et al., 1968
O. lurida C. gigas T. japonica	P-1	Flowing	24 hrs.	<1-900-fold	0.4-3.6-fold for clarified virus; 10-900-fold for crude virus	Hoff and Becker, 1969
M. mercenaria	S-13 coliphage	Flowing	16-231. hrs.	2-1500-fold	Assayed digestive glands 2-1100-fold in clam pools 100-1500-fold in individuals	Canzonier, 1971
Pachygrapsus sp. Hemigrapsus sp. (crabs)	P-1	-	12-48 hr.	<1-fold	25% uptake in 12 hr. 63% uptake in 48 hr.	DiGirolamo et al., 1972[a]
Cancer magister Cancer antennarius (crabs)	Coliphage T4	-	12-48 hr.	<1-fold	58% uptake in 24 hr. 73% uptake in 48 hr.	DiGirolamo et al., 1972[b]
C. gigas O. lurida	P-1	Static	48 hr.	<1-fold	C. gigas-46% uptake in 12 hr. 88% uptake in 48 hr. O. lurida-86% uptake in 12 hr. 94% uptake in 24 hr.	DiGirolamo et al., 1975

Table 6. Virus Accumulation by Shellfish, Continued

Shellfish Species	Virus Type	System	Exposure Time	Concentration Factor	Comments	References
C. glomerata	Reo-3 SFV*	Static	-	-	radio-labelled virus	Bedford et al., 1978
M. arenaria	P-2	Agitated water	6 hr.	<1-34.5-fold	<1-28.7-fold virus uptake with or without cornstarch and kaolinite; 19.5-34.5-fold uptake of feces-associated virus	Metcalf et al., 1979; 1980
C. virginica	P-1	Static	6-24 hrs.	<1	low uptake, <5% in 24 hr.	Jensen, 1980
Callinectes sapidus (blue crab)	P-1 Coliphage MS-2	-	-	<1	Max. uptake in 2 hr.	Hejkal & Gerba, 1980
C. virginica C. gigas M. mercenaria	P-1	Flowing	48-432 hr.	-	No uptake in up to 120 hr. at water levels <0.01 pfu/ml	Landry et al., 1982
C. virginica M. mercenaria	P-1	Flowing	48-168 hr.	-	Little uptake of sediment-bound virus	Landry et al., 1983
C. gigas	Cricket paralysis virus	Agitated water	6-16 hr.	1-fold	Radio-labelled virus	Scotti et al., 1983

Table 6.   Virus Accumulation by Shellfish, Continued

Shellfish Species	Virus Type	System	Exposure Time	Concentration Factor	Comments	References
Pinnidae Isognomonidae	P-1 Rota-SA-11	-	24-72 hr.	-	Max. uptake in 24-72 hr.	Dow et al.,
M. edulis	-	-	20 hr.	1-fold	Max. uptake in 24 hr.	Croci et al., 1984
Cerastoderma edule	P-1 HAV	Aerated water	6-48 hr.	≤1-fold	P-1 uptake 1-fold in 6-24 hr. HAV uptake <1-fold in 24-48 hr. Uptake in presence of normal fecal extract	Millard et al., 1987
M. edulis	P-1	Flowing	1 hr.- 10 days	<1-5-fold	<1-fold uptake in 1-6 hr. 1.8-5 fold uptake in 35-75 hr.	Gaillot et al., 1988

Static Versus Continuous-flow Exposure. As noted above, the type of exposure system used in uptake studies affects the level of accumulation. Static or stationary water systems generally result in lower uptakes than do flowing systems (Hamblet et al., 1969; Liu etal., 1966a; 1966b). It has been suggested that flowing seawater more closely simulates the natural environment of shellfish and is thus more conducive to feeding. Static systems do not optimize other physiological conditions such as dissolved oxygen, food availability, and metabolic waste dilution which may also be important in facilitating shellfish feeding (Hamblet et al., 1969).

Shellfish Species and Variability among Individual Animals: Interspecies differences in virus uptake rates have been noted by DiGirolamo et al. (1975) who found that viral accumulation occurred four times faster in Olympia oysters as compared to Pacific oysters. Also important is the difference in virus accumulation by individual specimens of the same species. Seraichekas et al. (1968) noted that individual variability for clams reached 10-100 fold, while Canzonier (1971) found bioaccumulation that ranged from 100-1500 fold for individual clam digestive glands. Variable viral uptake rates by individual animals has also been reported for soft shell clams and Pacific oysters (Metcalf et al., 1979, Scotti et al., 1983). It is suggested that many factors influence the uniformity of viral uptake by individual animals, including the physiological condition of the shellfish, size and weight of the specimen, localized hydrographic conditions and overcrowding (Metcalf et al., 1979; Scotti et al., 1983; Serichekas et al., 1968). However, one study reported that the weight of oysters did not appreciably affect the amount of virus accumulated (Scotti et al., 1983).

In addition to mollusks, crustaceans also accumulate viruses. DiGirolamo et al (1972) found that poliovirus uptake by West Coast crabs varied from 28-63% within 12 and 48 hours of exposure to contaminated water, respectively. Coliphage uptakes ranged from 58- 73%. When allowed to feed on virally-contaminated mussels, the crabs accumulated 74-94% of the viruses present in the shellfish. Hejkal and Gerba (1980) reported that accumulation of viruses by blue crabs began within two hours of exposure to contaminated seawater, with final concentrations rarely exceeding those present in the water. Highest virus concentrations were found in the hemolymph and digestive tracts.

Type of Virus: Differences in accumulation rates have been noted for different viruses in comparative studies (Bedford et al., 1978; Canzonier, 1971; Duff, 1967). It is hypothesized that this is due to virus surface properties which differ among virus types. DiGirolamo et al (1977) reported that attachment of viruses to shellfish mucus was due to ionic bonding of the virus to the negatively-charged sulfate radicals of mucus. The strongest positively-charged viruses should bind most efficiently. Surface characteristics also influence the ability of viruses to associate with particulates and sediments, which also may affect the efficiency of viral uptake by shellfish (Canzonier, 1971)

Most viral uptake studies in shellfish have been done using strains of poliovirus, other enteroviruses or coliphages, but hepatitis A virus, reovirus, Semliki Forest virus, Cricket Paralysis Virus and rotavirus have also been studied (Bedford et al., 1978; Dow et al., 1984; Scotti et al., 1983; Werner, 1983). Millard et al. (1987) investigated the accumulation of HAV by cockles. Filter feeding cockles were exposed for 24 to 48 hours to a concentration of 1.6×10^4 RFU/ml in seawater supplemented with normal fecal extract. The virus was recovered from shellfish extracts at levels as high as 9.4×10^4 RFU/ml and was present primarily in digestive tract tissue.

Virus Concentration in Water. Investigation of the relationship between the level of viral contamination in water and viral accumulation by shellfish has produced somewhat conflicting results. Early studies indicated that large amounts of viruses in seawater were required to demonstrate viruses in oysters (Hedstrom and Lyke, 1964; Metcalf and Stiles, 1965). More recent studies confirm that the greatest amount of bioaccumulation of viruses in other shellfish species occurs in the presence of the greatest numbers of contaminating viruses in water (Bedford et al., 1978, Metcalf et al., 1979). However, these studies used very high viral titers in waters. Using lower titers, Gaillot et al. (1988) observed that maximum numbers of viruses were recovered from mussels when the viral concentration in water was higher (390 PFU/ml) as opposed to lower (64-71 PFU/ml), but appreciable uptake was also noted at very low virus levels in water (3-8 PFU/ml).

Liu et al. (1966b) found that comparatively higher uptake of viruses in quahogs was achieved at low (10-45 PFU/ml) versus high (1500-4000 PFU/ml) virus levels in seawater At low virus levels of 10 PFU per ml seawater, they observed as many as 9000 PFU of virus/gram in digestive tracts of quahogs. They concluded that viral uptake by shellfish was more efficient when surrounding water was contaminated with low levels of virus. Canzonier (1971) also found virus accumulation factors exceeding 1000-fold for hard shell clams contaminated with S-13 coliphage at relatively low levels of 1-8 PFU/ml of water.

The aforementioned studies report virus accumulation by shellfish at levels in water far exceeding those typically found in environmental waters. Investigation of uptake efficiency at more realistic contamination levels was undertaken when sporadic isolation of enteric viruses from shellfish was reported at water column levels as low as 0.009. 0.001, and 8.34×10^{-5} PFU/ml (Goyal et al., 1979; Metcalf et al., 1979; Vaughn et al., 1979b, 1980). Landry et al. (1982) investigated poliovirus accumulation by clams and oysters at water levels of 0 002 to 0.18 PFU/ml of both feces-associated and monodispersed virus. These levels corresponded to those reported for light to moderately polluted waters. They concluded that viral accumulation by shellfish may not be efficient at water column concentrations less than 0.01 PFU/ml. However, upon prolonged exposure to water levels of 0.1 PFU/ml, an equilibrium between accumulation and depuration of viruses by shellfish was reached such that a steady state of viral carriage existed. The authors suggested that sediments may serve as a potential source of virus for accumulation, but subsequent study revealed low viral accumulation in the presence of artificially contaminated sediments (Landry et al., 1983). The importance of contamination of sediments to the accumulation of viruses by shellfish remains uncertain.

Temperature: The importance of temperature was documented by Metcalf and Stiles (1968) who demonstrated that oysters in polluted water at temperatures below 7·C failed to accumulate viruses. This was attributed to the cessation of pumping and feeding activities at low temperatures At temperatures exceeding 7·C, accumulation was noted and the accumulation rate may be temperature dependent. Meinhold (1982) found that the maximum uptake of poliovirus by the Eastern oyster occurred in 5 hours at 6·C, 2-3 hours at 17·C, and 1-3 hours at 28·C. However, Bedford et al. (1978) reported that the rate of accumulation of radioactively-labelled reovirus by rock oysters was not appreciably affected by temperatures of 15 to 25·C.

Suspended Solids or Turbidity: There is considerable evidence documenting that virus association with solids in water increases the extent of viral uptake by shellfish Hoff and Becker (1969) found that crude virus preparations containing cellular debris were concentrated 40 to 60 times greater by clams and oysters than were clarified virus suspensions. Metcalf et al (1979, 1980) investigated the bioaccumulation of poliovirus 2 by the softshell clam using stock virus, solids-associated virus, and feces-associated natural virus More viruses were bioaccumulated when suspended solids were present, with maximum efficiency for feces-associated virus It was concluded that the efficiency of bioaccumulation depends in part on the type of suspended solids. Millard et al. (1987) also found that incorporation of normal fecal extract into virally contaminated water resulted in greater viral uptakes by cockles. Investigators agree that the state in which viruses exist in natural waters is an important consideration in assessing viral uptake by shellfish (Hoff and Becker, 1969, Millard et al., 1987).

At least three explanations have been put forth to explain the improved uptake of solids-associated virus by shellfish (Metcalf et al., 1979). Particulate matter may stimulate shellfish feeding or alternatively, solids-associated virus may have a greater liklihood of ingestion and adsorption to mucus sheaths. It is also possible that viruses are afforded greater protection from inactivation in water by the presence of solids, thereby increasing their persistence water and sediments

However, excess turbidity or suspended particulate matter can inhibit viral uptake by shellfish. Hamblet et al. (1969) reported 18-fold uptake in oysters subjected to low turbidity (16-24 ppm) seawater but only 5-fold accumulation in high turbidity (54-77 ppm) seawater. They hypothesized that high turbidity clogged the gills and palps, thereby interfering with pumping, feeding, and filtration.

Salinity and pH: Differences in ionic concentrations (salinity) or pH may affect virus uptake by influencing the bond between viruses and shellfish mucus. DiGirolamo et al. (1977) observed that decreasing salinity from 28 to 7 ppt resulted in a 20% increase in poliovirus binding by mucus. The efficiency of poliovirus binding by mucus was also pH-dependent They hypothesized that this phenomenon occurs due to competition between cations and viral capsids for mucus anions.

Other Factors: Investigators have proposed other factors influencing viral uptake by shellfish but their importance remains uncertain. Bedford et al (1978) showed that reovirus uptake by rock oysters was influenced by the presence of food material such as algae in the surrounding water. Limited comparative uptake data indicate that bacteria are accumulated faster than viruses (Bedford et al., 1978).

ELIMINATION OF VIRUSES FROM SHELLFISH· DEPURATION AND RELAYING

Both the uptake and elimination of viruses by shellfish under natural and controlled conditions are dependent on a number of physical, chemical and biological factors Because shellfish accumulate and eliminate viruses as a result of their natural physiological activities, factors influencing these activities will influence viral accumulation and elimination. In addition, virus-related factors and the effects of the design and operating characteristis of depuration systems that influence shellfish activity will affect virus elimination. Some of the specific factors to be considered are. type of shellfish, variability of individual animals, degree of viral contamination, type of virus, temperature, salinity, turbidity, and virus association with particles.

While virus levels in shellfish tissues increase at increasing virus concentrations in surrounding water, they decrease with decreasing viral content of water (Hedstom and Lycke, 1964, Liu et al., 1966b, Metcalf and Stiles, 1965), and when shellfish contaminated with viruses they will eliminate viruses as a result of the natural processes of pumping and elimination of waste

Mechanisms of Virus Elimination from Shellfish

Elimination of viruses and other microbial contaminants from shellfish appears to be a result of active elimination as a primary mechanism (Metcalf, 1987) and physical inactivation of viruses as well (Canzonier, 1971). Much of the reduction of microbial contaminants relies on the ability of the animals to eliminate contaminating microorganisms from their digestive tracts through normal feeding, digestion and excretion. As proposed by Canzonier (1971), there may be two separate aspects of viral elimination from shellfish. In one of these, active elimination occurs by the normal excretion of waste products. Viruses transported through the alimentary canal are expelled from the anus in the form of a fecal ribbon, and virus excretion may also occur by ejection as part of the pseudofeces, particularily when the shellfish is not pumping efficiently. Elimination of viruses by these process should be rapid and efficient. A second aspect of virus elimination by shellfish concerns those viruses that become sequestered in other shellfish tissues and spaces and are apparently more resistant to elimination. The elimination of these viruses may be independent of feeding and related activities and may instead may be a function of virus inactivation processes in the shellfish. Removal of these viruses is less efficient and may account for low level viral persistence in shellfish. Evidence for such phenomena is provided by the studies of Hay and Scotti (1986) who found that most labelled viruses were removed effectively by depuration, but that the small proportion of initial viruses persisting after 64 hours were intracellularly adsorbed and associated with connective tissue and epithelial cells surrounding the digestive tract.

Physical inactivation of viruses may occur independently of shellfish activity and involve inactivating factors such as temperature, pH, and salinity. Davis (1986) suggested that physical inactivation was responsible for the majority of HAV reduction in the Eastern oyster in laboratory-scale depuration experiments, while both active elimination and physical inactivation were responsible for poliovirus reduction. It has also been suggested that enzymatic activities of mussel tissues may contribute to virus reductions (Duff, 1967).

Depuration and Relaying

There is considerable interest in and practice of methods to reduce viral and other microbial contaminants in shellfish by exploiting their processes of elimination Two types of active elimination practices are widely used: depuration and relaying. Depuration, or controlled purification, is the process of reducing the levels of bacteria and viruses in contaminated, live shellfish by placing them in a controlled water environment. Depuration is typically done in tanks provided with a supply of uncontaminated, often disinfected, seawater under specified operating conditions. Shellfish remain in the tanks for a period of time sufficient to allow them to purge themselves of contaminants, though the duration of this time period varies. The seawater used is sometimes treated by such processes as sedimentation, filtration, ozonation or UV irradiation to maintain acceptable quality.

Relaying, or natural purification, refers to the transfer of shellfish from contaminated growing areas to approved areas for natural biological cleansing using the ambient environment as a treatment system.

These reclamation processes are currently being used in a number of countries. While shellfish reclaimed by depuration or relaying are considered acceptable for consumption based on bacterial quality standards, the viral quality of shellfish may not be acceptable and additional precautions are used in some countries in an effort to assure virological quality. For example, in the U.S. only shellfish taken from no worse than moderately contaminated waters classified as "restricted" can be depurated or relayed. This is because of concerns about the effectiveness of depuration or relaying to extensively reduce viruses to negligable levels in heavily contaminated shellfish that may initially contain high concentrations of viruses if they were taken from grossly contaminated waters Some outbreaks of viral hepatitis and gastroenteritis have been attributed to inadequately depurated shellfish taken from heavily contaminated waters (Food and Drug Administration, 1983; Gill et al , 1983; Grohmann et al., 1981).

Numerous studies have been done to determine the effectiveness of virus removal by depuration and relaying. Table 7 summarizes the data from these investigations Reviews are provided by Gerba and Goyal (1978), Richards (1988), and DeLeon and Gerba (1990). Virus reduction by controlled purification and relaying has been reported for oysters (C. virginica, C gigas, O edulis, O. lurida) (Cook and Ellender, 1986; DiGirolamo et al., 1975; Hamblet et el, 1969; Hedstrom and Lyke, 1964; Hoff and Becker, 1969; Gerba and Goyal, 1978; Metcalf and Stiles, 1968; Metcalf et al., 1973; Mitchell et al , 1966; Scotti et al., 1983; Sobsey et al., 1980b, Sobsey et al., 1988), clams (M. mercenaria, M campechiensis, M arenaria, T. japonica) (Canzonier, 1971; DiGirolamo et al., 1975; Docs and Noss, 1989; Hoff and Becker, 1969; Gerba and Goyal, 1978, Liu et al., 1967a, 1976b, Metcalf et al., 1979; Seraichekas et al., 1968), and mussels (M edulis) (Crovari, 1958; Power and Collins, 1989; 1990). Most studies have used poliovirus strains, although other picornaviruses (coxsackie viruses, echoviruses, Cricket Paralysis Virus and HAV), rotaviruses and coliphages have been used (Canzonier, 1971; Docs and Noss, 1989; Gerba and Goyal, 1978; Liu et al , 1967a; Metcalf and Stiles, 1968; Metcalf et al., 1973; Sobsey et al., 1980b; 1988; Scotti et al., 1983, Werner, 1983).

Rates and Extents of Viral Elimination by Shellfish

As shown by the results of studies summarized in Table 7, the rate and extent of viral reduction from shellfish is highly variable and may depend on a variety of influencing factors, many of which will be discussed below. However, some general patterns of viral elimination are often observed. Maximum rates of viral elimination generally occur during the first 24 to 48 hours of depuration, regardless of initial level of contamination (Docs and Noss, 1989, Hamblet et al , 1969; Hedstrom and Lyke, 1964; Hoff and Becker, 1969; Mitchell et al., 1966; Metcalf et al., 1979, Seraichekas et al., 1969). Investigations have shown that 90%-95% of polioviruses were eliminated from eastern oysters within 6-8 hours of depuration, with 99.9% eliminated within 24 hours (Hamblet et al., 1969, Mitchell et al, 1966). Somewhat slower rates were reported by Seraichekas et al. (1969) who found >90% reductions of viruses after 48 hours of depuration. Similarly, Sobsey

Table 7. Virus Elimination by Shellfish

Shellfish Species	Virus Type	Water System	Time	Comments	Reference
M. edulis	P-2	Intermittent flow	48 hr.	-	Crovari, 1958
O. edulis	P-3	Flow-through	24-100+ hr.	<1 log reduction in 24 hr. 2 log reduction in 100 hr. Incomplete after 100 hr.	Hedstrom & Lyke, 1965
C. virginica	P-1 CB-3 E-6	Natural waters	4 months	Lab-contaminated oysters in estuary water at 5°C	Metcalf and Stiles, 1968
C. virginica	P-1	Flowing	8-96 hr.	95% reduction in 8 hr. 99% reduction in 24 hr. Incomplete after 96 hr.	Mitchell et al., 1965
M. mercenaria	P-1,-3 CB-4	Flowing	72-96 hr.	Reduction complete in 72-96 hr.	Liu et al., 1967
C. virginica M. mercenaria	Staph. phage	-	Up to 100 hr.	Reduction incomplete in 100 hr.	Gerba & Goyal, 1978
C. virginica	P-1	Flow-through	6-48 hr.	>90% reduction in 6 hr. >99.9% reduction in 24 hr. Complete in 48 hr.	Hamblet et al., 1969
M. mercenaria	P-1	Flowing	48-72 hr.	>90% reduction in 48 hr. Incomplete in 72 hr.	Seraichekas et al., 1968

Table 7. Virus Elimination by Shellfish, Continued

Shellfish Species	Virus Type	Water System	Time	Comments	Reference
O. lurida C. gigas T. japonica	P-1	Flowing	48-96 hr.	Reduction complete in. 48 hr. for clarified virus Reduction in 72-96 hr. for crude virus	Hoff & Becker, 1969
M. mercenaria	Coli-phage S-13	Flowing	48-168 hr.	Reduction incomplete in 6+ days if initial contamination level was low	Canzonier, 1971
C. virginica	CB-3	-	-	Reduction complete in 7 days at 9-13°C; reduction complete in 3 days at 21-12°C	Metcalf et al., 1973
C. virginica O. lurida	P-1	Static	120 hr.	16-21% virus remaining remaining at 120 hr.	DiGirolamo et al., 1975[a]
O. lurida	P-1	Flowing	72 hr.	24% virus remaining at 24 hr. <10% virus remaining at 72 hr.	DiGirolamo et al., 1975[b]
M. arenaria	P-2	Flowing	2-6 days	Reductions of feces-associated viruses: 80-87% in 2 days, 92-95% in 4 days, and 98-99% in 6 days	Metcalf et al., 1979; 1980
C. virginica	Bovine entero-virus type 1	Relay	30 days	>99.99% reduction in 30 days in winter and spring 80% reduction in 30 days in summer and fall	Sobsey et al., 1980
C. gigas	Cricket paralysis virus	Flowing	10 days	Incomplete reduction after 10 days	Scotti et al., 1983

Table 7. Virus Elimination by Shellfish, Continued

Shellfish Species	Virus Type	Water System	Time	Comments	Reference
C. virginica	P-1	Relay	45 days	Incomplete reduction after 45 days	Cook & Ellender, 1986
C. virginica	P-1 HAV	Flowing	2-5 days	HAV - >10% persistence in 5 days P-1 - <1% persistence in 5 days	Sobsey et al., 1988
M. campehciensis	F-specific coliphage	Relaying	14 days	Rapid decreases in 48 hr.	Docs & Noss, 1989
M. edulis	P-1; wild-type coliphage	Recircu-lating water	52 hr.	P-1 reduction uniform for 52 hr.; final 1.86 log reduction; phage reduction variable; incomplete reduction of both viruses in 52 hr.	Power & Collins, 1989; 1990

et al. (1980b) found poliovirus reduction in oysters to be rapid and extensive, with less than 3% of the viruses remaining after 3 days.

After the first 24-48 hours, virus depletion rates are much slower (Hoff and Becker, 1969, Metcalf et al , 1979). Some investigators report complete removal of detectable virus after depuration for 48-96 hours (Hamblet et al., 1969; Hoff and Becker, 1969; Liu et al, 1967a). However, numerous investigators report incomplete virus removal (Canzonier, 1971, DiGirolamo et al , 1975; Hedstrom and Lyke, 1964; Mitchell et al., 1966; Metcalf et al , 1979; 1980; Power and Collins, 1989; Scotti et al.; 1983; Seraichekas et al., 1968; Sobsey et al, 1988) Early studies reported viral persistance after depuration for 100 hours (Hedstrom and Lyke, 1964; Hoff and Becker, 1969; Liu et al , 1967a). Two studies demonstrated incomplete virus removal after 6 days of depuration (Canzonier, 1971, Metcalf et al., 1979, 1980). Scotti et al. (1983) reported the greatest degree of persistence, with depuration of experimentally contaminated oysters always incomplete after 10 days. Lewis et al. (1986) found that for mussels naturally contaminated with enteroviruses, at least 22% of the initial viruses persisted for up to 8 days of depuration in a static system with daily water changes.

Shellfish Activity

Shellfish activity may influence viral reduction by shellfish, and it can be assessed in several ways, including pumping, metabolism, filtration, and feeding. Pumping or water transport refers to the rate of water movement through the gills. Metabolic rate and oxygen consumption are equivalent terms and refer to energy metabolism per unit time. Filtration refers to the removal of suspended particles from the water column. Feeding involves the sorting, acceptance, and assimilation of food particles occurring subsequent to filtration. While these activities are important to the elimination of viruses and other microbial contaminants from shellfish, it is not clear that they can be relied upon as monitoring tools to assess the effectiveness of viral elimination Recently, efforts were made to evaluate the predictive ability of hard clam siphon extension activity, which is associated with feeding activity and perhaps elimination of microbes, to indicate viral and bacterial depuration efficiency (Cantelmo and Carter, 1987; Carter and Cantelmo, 1987). Poliovirus depuration from experimentaly contaminated clams was somewhat more extensive in animals with high rather than low siphon extension activity However, siphon extension activity did not predict depuration of C perfringens spores and varied among individual animals, seasonally, and with exposure to different temperatures prior to depuration. Therefore, the ability of this and other parameters of shellfish activity to predict viral depuration efficiency remains uncertain

Type of Shellfish and Individual Variation

Depuration rates are affected by type of shellfish. For example, under identical conditions, the Eastern oyster can eliminate poliovirus to undetectable levels in 24 hours, while 48 hours is required for the soft shell clam (Hoff and Becker, 1969). However, variability in depuration efficiency of individual shellfish may be of greater importance. It has been observed that during the latter phases of depuration (48-72 hours), while the majority of shellfish were free of viruses, a few still harbored minimal amounts of viral contaminants (Metcalf et al., 1979, 1980; Seraichekas et al , 1969) Scotti et al. (1983) found a 100-fold range in depuration efficiency for virus-infected Pacific oysters when assaying individual shellfish. Variable individual depuration rates may reflect variations in viral uptake or physiological inactivity of the shellfish (Scotti et al., 1983; Seraichekas et al., 1969). Metcalf et al. (1980) concluded that depuration and relaying of shellfish should reduce microbial contamination but cannot guarantee all shellfish to be virus-free.

Type of Virus

Rates of virus elimination by shellfish differ with virus type. Differences in surface charge may cause viruses to be eliminated at different rates. Viruses with a strong positive surface charge may attach more effectively to the negatively-charged sulfate radical of shellfish mucus and therefore depurate more slowly (Duff, 1967, DiGirolamo et al , 1975). Power and Collins (1989) reported differences in both the relative rates and patterns of elimination of coliphage ϕA1-5a and poliovirus by mussels. They suggested that the similarities in size and shape of these viruses indicate that elimination may not be a function of virus size and shape alone. Sobsey et al. (1987, 1988) compared the depuration of HAV and poliovirus in experimentlally-contaminated oysters. HAV reduction was slow, with up to 18% viral persistance after five days Poliovirus persistance was less than 1% after five days. Other investigations have shown that poliovirus depuration proceeds quickly, with 80-99% removal in 48 hours (Davis, 1986; Hoff and Becker, 1969; Liu et al., 1967a; Metcalf et al , 1979). After the initial drop, low levels of poliovirus may persist for up to six days (Hoff and Becker, 1969). Virus-specific differences in depuration patterns have led investigators to caution the use of model viruses for the prediction of overall viral depuration efficiency by shellfish (Power and Collins, 1989; Sobsey et al., 1987; 1988).

Initial Levels of Viral Contamination

The efficiency of depuration is highly dependent upon the initial contamination level of the shellfish. Less time is required for virus elimination from lightly contaminated shellfish (Canzonier, 1971; Cook and Ellender, 1986; Liu et al , 1967b; Metcalf and Stiles, 1968, Metcalf et al , 1979). Liu et al (1967b) found that hard shell clams contaminated with 10_2 PFU of poliovirus per gram of meat required 24 hours to eliminate them to nondetectable levels, whereas 72 hours were required at contamination levels of 10^3 PFU/gram. No apparent difference in rate of elimination was seen Mesquita (1988) exposed mussels to a coliphage at low water titers for long times and high water titers for short times. Depuration patterns for both contamination schemes were similar, however exposure to high titers for short times showed more rapid phage elimination. Both exposure treatments resulted in prolonged retention of a portion of the phage.

Physical Factors Influencing Depuration of Viruses

Temperature: Temperature exerts a major effect on shellfish activity. Within the physiological range of tolerance of the shellfish, feeding (and hence virus accumulation and elimination) normally increases with temperature. Loosanoff (1958) reported a steady increase in pumping rate of Eastern oysters up to 30°C. Lower temperatures result in a slower rate of shellfish feeding and hence slower virus removal (Cook and Ellender, 1986). There appears to be a threshold temperature below which no depuration occurs, presumably due to shellfish inactivity. This temperature is estimated to be 7-10°C (Cook and Ellender, 1986; Metcalf and Stiles, 1965). Metcalf showed that oysters kept in water of <5°C depurated virtually no viruses after four months. Liu et al. (1967a) found that an initial poliovirus 1 or 3 concentration of about 10^2 PFU/g in Mercenaria mercenaria was reduced in 24 hours by >99% at 18-20°C, <90% at 13-14oC and only about 50% at 5-8°C. Meinhold (1982) demonstrated statistically significant differences in the rates of poliovirus 1 elimination from Crassostrea virginica at 6, 12, and 28°C, with the most rapid elimination occurring at 28°C. For simian rotavirus SA-11, elimination from Crassostrea virginica was shown to be more rapid at 28°C than at 6°C, while no significant difference was demonstrated between 17 and 28°C (Werner, 1983). In studies using coliphage S-13 in Mercenaria mercenaria the rate of virus elimination was more rapid at 24°C than at 16°C (Canzonier, 1971). In contrast, Sobsey et al (1987) found that temperature had only a minor affect on elimination of poliovirus 1 and hepatitis A virus by Eastern oysters in a laboratory scale depuration system. Rates of poliovirus 1 elimination were progressively slower at the lower temperatures, with reductions of 97% or more achieved by 1, 2 and 3 days at temperatures of 23, 17 and 12°C, respectively. Hepatitis A virus reductions were minimal (<90%) after 5 days at all three temperatures.

Turbidity and Particulate Matter: Because turbidity of shellfish waters influences feeding and water transport, it also may efffect depuration. Jorgensen (1955) exposed Ostrea virginica to both 0.1 and 1.0 g/l of silt, calcium carbonate, and kaolinite and observed that pumping rates decreased by approximately 50% and 80%, respectively. Loosanoff and Engle (1947) reported that oysters fed effectively only if the surrounding water contains small quantities (<2 million/ml) of suspended food material (Chlorella sp. cells) and that a high concentration (>5 million/ml) resulted in low rates of feeding. They proposed that at high turbidity, oyster activity is directed primarily to cleansing the gills to prevent possible suffocation, resulting in minimal feeding and in the formation of large amounts of pseudofeces. The production of pseudofeces, however, also contributes to turbidity.

Variations in turbidity or association of viruses with particles do not appear to appreciably affect the rate of virus elimination by shellfish, at least at turbidities likely to be encountered in depuration systems. Hamblet et al. (1969) found that poliovirus elimination by Crassostrea virginica occurred at similar rates in waters of high (54-80 ppm) and low (8-21 ppm) turbidity; poliovirus was eliminated by about 3 logs (99.9%) despite turbidity levels as high as 80 mg/l. Crude and clarified poliovirus was eliminated at similar rates by Olympia oysters, Pacific oysters, and Manila clams (Hoff and Becker, 1969), although crude virus persisted somewhat longer than did clarified virus in Olympia oysters and Manila clams. Meinhold (1982) reported that the association of poliovirus with estuarine sediment during uptake did not affect their elimination rates.

Chemical Factors Influencing Depuration of Viruses

Salinity: Within limits, shellfish can adapt to changes in water salinity. Loosanoff (1952) (as cited by Furfari, 1966) found that Crassostrea virginica adapted to a salinity of 27 PPT were affected as follows: at salinity decreases to 20, 15 and 5 PPT, pumping for 6 hours was 76, 11 and 0 4% of normal, respectively. Galtsoff (1964) found no significant difference in rate of oxygen consumption by Crassostrea virigincia when the salinity was varied from 31.6 to 24.1 ppt. As cited by Furfari (1966), Goggins and colleagues found in pilot depuration plant studies that soft-shell clams taken from a salinity of 18 PPT to a salinity of 30 PPT required more than 12 hours to overcome the stresses of acclimation. Hard-shell clams acclimated to a salinity of 31 PPT were found to depurate poliovirus 1 rapidly and extensively in water of salinities 23-31 PPT, while little depuration occurred at salinities of 17-21 PPT (Liu et al., 1967a). All shellfish depuration activity ceased when water salinity was reduced by 50-60% of its original value. Sobsey et al. (1987) found that salinities of 8-28 PPT, had no appreciable effect on poliovirus 1 elimination by Eastern oysters, although after only 1 day of depuration, virus reduction was least at the lowest salinity. However, the same oysters depurated HAV more efficiently at salinities of 28 PPT than at 8 or 18 PPT (Sobsey et al., 1987). Haven et al. (1977) reported that mean salinities of 14- 21.4 PPT had no effect on coliform bacteria depuration of Chesapeake Bay oysters, which successfully live over a salinity range of 3-32 PPT

pH and Oxygen Availability: Although pH and oxygen availability of water can influence shellfish activity, the effects of these parameters on viral depuration by shellfish has not been studied extensively. The normal pH range of seawater is 7.5-8.4, and several studies have reported that lowering the pH to below this range decreases pumping activity and oxygen consumption of shellfish (Galtsoff, 1964). In recent studies in the authors' laboratory there were no appreciable differences in the reductions of hepatitis A virus, poliovirus 1 and male-specific coliphage MS-2 by Eastern oysters or hard-shell clams when depuration water was at pH 7.0 or 8.0 (unpublished results).

Design Factors Influencing Depuration of Viruses

Static Versus Flowing Depuration Systems: Depuration occurs more rapidly and efficiently in flowing versus stationary or static depuration systems (DiGirolamo et al., 1975; Liu et al., 1967b; Mitchell et al., 1966). DiGirolamo et al. (1975) depurated West Coast oysters in both static and flowing systems; the flowing system yielded a 99% reduction of polioviruses after 72 hours, whereas up to 21% of the viruses remained after 120 hours in the static system. Mitchell et al. (1966) reported that a static system limits some essential conditions for shellfish activity including dissolved oxygen, food, and dilution of metabolic waste. Because virus elimination is dependent upon active feeding, they suggested that the flowing system better simulates the natural environment, which promotes increased feeding and results in faster depuration rates for contaminated shellfish (Mitchell et al., 1966).

Other Design Considerations. In addition to flowing as opposed to static systems to maximize viral and other microbial reductions in shellfish by depuration, other design factors are important to the effectiveness of the process. Loading (the quantities of shellfish in a unit of space), water distribution, flow rate, flow patterns (e.g., turbulent versus laminar flow), tank depth and depth of shellfish in containers are all likely to influence viral reductions by depuration. However, most of these factors have not been rigorously studied with respect to their influence on viral depuration efficiency.

Effects of Other Factors on Depuration of Viruses

Investigators have proposed other factors affecting viral elimination rates from shellfish, including availability of food and the effects of shellfish spawning. The presence or absence of food (algae) had no effect on HAV or poliovirus elimination rates in oysters (Sobsey et al., 1987). Power and Collins (1990) found that food availability affected depuration rates only in filtered seawater, in which case elimination was faster. In another study Power and Collins (1989) reported an adverse effect of spawning on poliovirus and coliphage elimination by mussels, with slower and somewhat less extensive elimination from spawning than from non-spawning mussles. Only limited studies have been reported on the effects of disinfection of depuration water on virus reductions. Hedstrom and Lyke (1964) demonstrated a failure to disinfect viruses in oyster tissues by water chlorination.

VIRUS REDUCTIONS BY RELAYING

The effectiveness of relaying depends on many of the same factors as depuration. Cook and Ellender (1986) found that the rate at which microorganisms are reduced is slower in relayed than in depurated oysters. This is attributable to natural fluctuations in temperature and salinity, differential concentration of enteric viruses, and mechanical disturbance of the animals. Elimination proceeded faster at higher water temperatures, up to 31°C. At temperatures below 10°C, oyster activity and enteric microbe elimination were slowed (Cook and Ellender, 1986). Sobsey et al (1980b) studied the extent of enterovirus elimination in relayed oysters and found that during winter and spring months, virus elimination exceeded 99.9% in 30 days. Most rapid rates occurred at water temperatures of 0-25°C During the summer and fall months, persistence of up to 20% of initial viruses was observed after 30 days Cook and Ellender (1986) reported that oysters with a high level of initial contamination frequently failed to eliminate all viruses after 45 days of relaying Salinity of the water had little effect provided it was sufficient to maintain metabolic function. Oysters under severe physiological stress failed to eliminate viruses as quickly as healthy oysters (Cook and Ellender, 1986).

COMPARATIVE ELIMINATION OF VIRUSES AND BACTERIA BY SHELLFISH

The relationship of viral elimination to bacterial elimination in shellfish subjected to purification has been investigated. While one study has reported similar patterns and rates of removal for viruses and bacteria (Hoff and Becker, 1969), the vast majority conclude that such parallels do not exist (Burkhardt et al., 1988; Canzonier, 1971, Cook and Ellender, 1986, Hedstrom and Lyke, 1964, Power and Collins, 1989, 1990; Scotti et al, 1983; Sobsey et al, 1980b) Generally, viral elimination rates are slower than those observed for bacteria Canzonier (1971) found that elimination of E. coli from clams proceeded much more rapidly than did elimination of phage S-13. Sobsey et al. (1980b) found that elimination of E. coli and enterococci in relayed oysters was much more extentive than was elimination of an enterovirus. Scotti et al (1983) concluded that viral and bacterial depuration rates were unrelated for C. gigas, with considerable persistence of Cricket Paralysis Virus and yet extensive reduction of E. coli.

Power and Collins (1989; 1990) reported that not only the relative rates but also the patterns of elimination of E. coli, coliphage, and poliovirus during depuration of mussels were significantly different. This observation suggests that the mechanisms effecting elimination may vary from one microorganism to another and may depend on phenomena such as differential tissue adsorption, interstitial sequestering, and spontaneous/hematocyte-induced inactivation (Canzonier, 1971; Power and Collins, 1989; 1990; Scotti et al., 1983). It is generally agreed that the lack of correlation between viral and bacterial removal in shellfish means that bacterial eliminination is not a reliable index of viral elimination during depuration and relaying (Canzonier, 1971; Cook and Ellender, 1986; Scotti et al., 1983; Sobsey et al., 1980b).

VIROLOGICAL ASPECTS OF COMMERCIAL DEPURATION AND RELAYING OF SHELLFISH

Depuration and relaying of shellfish have been recognized reclamation options throughout the world for many years (Furfari, 1966). In the U.S. depuration plants treating soft- and hard-shell clams and oysters have operated in a number of states, including Maine, Massachusetts, New Jersey, New York, South Carolina, Florida, and California. Depuration plants are reported in New South Wales, Australia, France, Italy, Spain and the United Kingdom. Depuration practices may vary appreciably in different countries, and some practices have led to viral illness from depurated shellfish. In some countries shellfish to be depurated are taken from contaminated waters regardless of water quality. This is of public health concern because it has been shown that heavily contaminated shellfish eliminate viruses more slowly than do lightly contaminated ones (Mitchell et al., 1966; Seraichekas et al., 1968; Metcalf et al., 1979; and Liu et al., 1967b). The importance of the initial levels of fecal contamination in shellfish to be depurated is highlighted by several documented outbreaks of shellfish-borne viral illness. In 1983 over 2500 people in New York and New Jersey became ill from consuming clams imported from England. These clams had been harvested from a grossly contaminated (i.e. prohibited) area, were relayed to a contaminated area, and were then inadequately purified for a 2-day period in depuration tanks (Food and Drug Administration, 1983). Indeed, depurated English shellfish have been linked to a number of other outbreaks of viral disease. In 1983, 181 persons in Great Britain were stricken with Norwalk gastroenteritis from eating Pacific oysters that had been depurated for 72 hours (Gill et al., 1983). A 1967 outbreak of non-bacterial (presumably viral) gastroenteritis in the U.K. involved a group of people who consumed raw oysters depurated for 36-48 hours (Gunn and Rowlands, 1969).

Depurated oysters in Australia have also been the cause of human viral illness Following an Australia-wide outbreak of Norwalk virus gastroenteritis linked to the consumption of raw oysters and involving over 2000 persons in 1978 (Murphy et al., 1979), a study was conducted in which human volunteers test-consumed depurated oysters. All oysters were commercially depurated for 48 hours or relayed to pollution-free natural waters for 7 days, after which all shellfish meats contained less than 230 total coliforms/100g. During the 17-week study, 52 out of 1382 volunteers became ill with Norwalk gastroenteritis following ingestion of the shellfish (Grohmann et al., 1981).

From these reports it is apparent that depuration and relaying are not always capable of completely eliminating viral contamination in shellfish. Despite these outbreaks of viral illness associated with depurated and relayed shellfish, experience has shown that depuration and relaying can produce shellfish products that are apparently of negligible risk of causing viral illness in consumers. However, greater efforts are needed to document by microbiological-virological and epidemiological studies the effectiveness of depuration and relaying to reduce viral contamination of shellfish such that the purified products are of low risk in causing viral illness.

SUMMARY

In this review we have attempted to summarize the state of knowledge and understanding of viral contamination of bivalve molluscan shellfish as food commodities, the role of these seafoods in the transmission of enteric viral disease, the uptake and elimination of viruses by shellfish, including their mechanisms and influencing factors, and the control of viral contamination of shellfish by depuration and relaying. It is apparent that there is considerable knowledge and understanding of these important public health issues of shellfish sanitation and safety However, it is also clear that there are considerable deficiencies in our information and understanding of viral contamination of shellfish and its control by depuration and relaying. Additional research must be done and actual practice must be further improved to decrease viral contamination of depurated and relayed shellfish and thereby reduce the risks of shellfish-borne viral illness.

Specifically, there are research needs for better understanding of the uptake, persistence and elimination of particular viruses in shellfish, with special emphasis on the behavior of the most important shellfishborne viral pathogens that have been studied either inadequately or not at all Hence, studies are needed on shellfish uptake and elimination of hepatitis A virus, Norwalk virus and the other small, round, gastroenteritis viruses responsible for epidemic shellfish-borne illness Research is needed to establish the reasons why some viruses are eliminated efficiently by shellfish while other viruses are not. It is necessary to determine if such differences in viral persistence in shellfish are related to the specific properties of the viruses, the shellfish species or to other factors. There also must be greater understanding of the reasons for low level persistence of viruses in shellfish, even after prolonged purification efforts. Research is needed to determine if and how such prolonged viral persistence relates to the physical state of the viruses, the locations of the viruses in specific shellfish tissues, and the mechanisms responsible for virus transport to and retention in these tissues. Only by a better understanding of these phenomena will it be possible to devise improved strategies to achieve more effective elimination or inactivation of viruses in depurated shellfish.

Because depuration practices vary widely throughout the world and even within the same country, there is a need to systematically identify through applied research those depuration and relaying practices and conditions that maximize virus reductions. Because of the complexities of the mechanisms and processes of virus reduction in shellfish, greater emphasis must be placed on defining those design, operating and other process conditions which achieve maximum and optimum virus reductions in shellfish by depuration and relaying.

Because of the difficulties in identifying and quantifying enteric viruses in shellfish for routine monitoring and surveillance and because of the unreliability of conventional bacterial indicators to predict viral contamination of shellfish, research is needed to develop and evaluate improved indicators of viruses in shellfish and reliable monitoring strategies to determine the effectiveness of depuration and relaying to reduce viral contamination to acceptable levels.

REFERENCES CITED

Alter, M.J., R.J Gerety, L A. Smallwood, R.E. Sampliner, E. Tabor, F. Deinhardt, G. Frosner, and G.M Matanoski. 1982 Sporadic non-A non-B hepatitis: Frequency and epidemiology in the urban U.S. population J. Infect Dis., 145 886-893.

Anderson, D.A. 1987. Cytopathology, plaque assay, and heat inactivation of hepatitis A virus strain HM175. J Med. Virol., 22:35-44.

Appleton, H. 1987. Small round viruses: Classification and role in food-borne infections. In: Novel Diarrhoea Viruses. Ciba Foundation Symposium. Wiley, Chinchester, U.K., pp. 108-125.

Appleton, H., S.R. Palmer, and R J. Gilbert. 1981. Foodborne gastroenteritis of unknown aetiology A virus infection? Br. Med. J., 282:1801-1802

Appleton, H. and M.S. Pereira. 1977. A possible virus aetiology in outbreaks of food poisoning from cockles. Lancet i:780-781.

Bedford, A.J., G. Williams, and A.R. Bellamy. 1978. Virus accumulation by the rock oyster Crassostrea glomerata. Appl. Environ. Microbiol., 35:1012-1018.

Bellelli, E. and G. Leogrande. 1967. Ricerde bacteriologiche e virologiche sui mitili. Ann. Sclavo., 9:820-828.

Bendinelli, M. and A. Ruschi. 1969. Isolation of human enteroviruses from mussels. Appl. Microbiol., 18:531-532.

Bitton, G 1980 Introduction to Environmental Virology. John Wiley and Sons, New York, 326 pp.

Block, J.-C. 1983. Viruses in environmental waters. In: G. Berg (ed.) Viral Pollution of the Environment, CRC Press, Boca Raton, FL., pp. 117-145.

Bostock, A.D., P. Mepham, and S. Phillips. 1979. Hepatitis A infection associated with the consumption of mussels. J. Infection, 1:171-177.

Burkhardt III, W., S.R. Rippey, and W.D. Watkins. 1989. The behavior of coliphage and F + Escherichia coli within whole clams and their homogenates Abstr Ann Meet Am Soc. Mibrobiol., 89:348.

Burkhardt III, W., S R. Rippey, and W.D. Watkins. 1988. A comparison of classical and alternative indicators used for assessing the purification rates of depurating shellfish. Abstr Ann Meet. Am. Soc Microbiol 88:291.

Cantelmo, F.R. and T.H. Carter. 1987. Assessment of hard clam siphon extension activity as an indicator of commercial depuration efficiency. In· Oceans '87 Proceedings, Vol. 5; Coastal and Estuarine Pollution. The Marine Technology Society, Washington, D.C , pp. 1735-1739

Canzonier, W J 1971. Accumulation and elimination of coliphage S- 13 by the hard clam, Mercenaria mercenaria. Appl. Microbiol., 21:1024-1031.

Caredda, F. S. Antinori, T. Re, C. Pastecchia, and M. Maroni 1986. Acute non-A non-B hepatitis after Typhoid Fever Br. Med. J., 292:1429.

Caredda, F., S. Antinori, T. Re, C. Pastecchia, C. Zavaglia, and M. Maroni. 1985 Clinical features of sporadic non-A, non-B hepatitis possibly associated with faecal-oral spread. Lancet, ii:444-445.

Caredda, F., A.D. Monforte, E. Rossi, S. Lopez, and M. Maroni. 1981. Non-A non-B hepatitis in Milan. Lancet,. ii:48.

Carter, T.H. and F R. Cantelmo. 1987. Viral and Clostridium perfringens content of hard clams during commercial depuration. In: Oceans '87 Proceedings, Vol. 5, Coastal and Estuarine Pollution. The Marine Technology Society, Washington, D.C., pp 1723-1727.

Caul, E.O 1987. Astroviruses· human and animal. In: Novel Diarrhoea Viruses. Ciba Foundation Seminar. Wiley, Chichester, U.K. pp. 120-125.

(CDC) Centers for Disease Control. 1982. Epidemiologic notes and reports-Enteric illness associated with raw clam comsumption-New York. Morbidity and Mortality Weekly Report, 31:229-451.

(CDC) Centers for Disease Control. 1979. Epidemiologic notes and reports-Viral hepatitis outbreaks-Georgia, Alabama. Morbidity and Mortality Weekly Report, 28 581.

(CDC) Centers for Disease Control. 1972. Shellfish-associated hepatitis-Massachusetts. Morbidity and Mortality Weekly Report, 21:20.

Chang, P.W., O.C. Liu, L T. Miller, and S.M. Li. 1971. Multiplication of human enteric viruses in Northern quahogs. Proc. Soc. Exp. Biol. Med., 136:1380-1384.

Cole, M.T., M.B. Kilgen, L.A. Reily, and C.R. Hackney. 1986. Detection of enteroviruses, bacterial indicators and pathogens in Louisiana oysters and their overlying waters J. Food Protection, 49:596-601.

(CDC) Communicable Disease Center. 1964. Epidemics of infectious hepatitis. Part A. New Jersey and Pennsylvania. Part B. Connecticut Hepatitis Surveillance Report 18 , pp.14-29

(CDC) Communicable Disease Center. 1964. Shellfish-associated hepatitis:Part A. New Jersey and Pennsylvania. Part B. Connecticut. Part C. North Carolina. Hepatitis Surveillance Report 19 , pp. 30-40.

Cook, D.W. and R.D. Ellender. 1986. Relaying to decrease the concentration of oyster-associated pathogens. J. Food Protection, 49.196-202.

Croci, L , G.D. Filip, S. Gizzarelli, L. Toti, S. Arigento, S. Amato, F. Lombardi, F. Novello, R. Santoro. 1984. Methodology for the determination of of polioviruses in edible Lamellibranch molluses. Microbiol. Alimento Nutrition, 2:85-90.

Cromeans, T., M.D. Sobsey, and H.A. Fields. 1987. Development of a plaque assay for a cytopathic, rapidly replicating isolate of hepatitis A virus. J. Med Virol., 22 45-56

Crovari, P. 1958 Some observations on the depuration of mussels infected with poliomyelitis virus. Ig , Mod. 5·22-32

Cubitt, W D 1987 The candidate caliciviruses. In: Novel Diarrhoea Viruses. Ciba Foundation Symposium. Wiley, Chichester, U.K. pp. 126-143.

Davis, A L 1986. Elimination of hepatitis A virus and poliovirus 1 from the eastern oyster, Crassostrea virginica. Master's Thesis, Department of Environmental Sciences and Engineering, University of North Carolina, Chapel Hill, N.C.

DeLeon, R. and C.P. Gerba. 1990. Viral disease transmission by seafood. In. J.O. Nraigu and M.S. Simmons (eds), Food Contamination from Environmental Sources. John Wiley and Sons, Inc., New York pp 639-662.

Denis, F.A. 1973 Coxsackie group A in oysters and mussels. Lancet, i:1262.

Dienstag, J.L., I.D. Gust, C.R. Lucas, D.C. Wong, R.H. Purcell. 1976. Mussel-associated viral hepatitis type A: Serological confirmation. Lancet, i:561-564.

DiGirolamo, R., J. Liston, and J. Matches. 1977. Ionic binding, the mechanism of viral uptake by shellfish mucus. Appl. Environ. Microbiol., 33:19-25.

DiGirolamo, R , J. Liston, and J. Matches. 1975. Uptake and elimination of poliovirus by West Coast oysters. Appl. Environ. Microbiol., 29:260-264.

DiGirolamo, R., L Wiczynski, M. Daley, and F. Miranda. 1972. Preliminary observations on the uptake of polioviruses by West Coast shore crabs. Appl. Microbiol., 23:170-171.

Dismukes, W E , A L. Bison, S. Katz, and R.F. Johnson. 1969. An outbreak of gastroenteritis and infectious hepatitis attributed to raw clams. Am. J. Epidemiol., 89:555-561.

Docs, J. and C.I. Noss. 1989. Effects of relaying on indicator bacteria and bacteriophage levels in Mercenaria campechiensis from the Indian River. Abstr. Ann. Meet. Am. Soc. Microbiol. 89 370.

Dougherty, W.J. and R. Altman. 1962. Viral hepatitis in New Jersey 1960-1961. Am J Med, 32:704-716.

Dow, M A., R. Fujioka, and P.C. Loh. 1984. Recovery of enteric viruses from Hawaiian marine waters using indigenous bivalve molluscs. Abstr. Ann. Meet. Am. Soc. Microbiol. 84:215.

Duff, M.F. 1967. The uptake of enteroviruses by the New Zealand marine blue mussel, Mytilus edulis aoteanus. Am. J. Epidemiol., 85:486-492.

Ellender, R.D., J.B. Mapp, B.L. Middlebrooks, D.W. Cook, and E.W. Cake. 1980. Natural enterovirus and fecal coliform contamination of Gulf Coast oysters. J. Food Protection. 43:105- 110.

Earampamoorthy, S. and R.S. Koff. 1975. Health hazards of bivalve-mollusk ingestion. Ann. Internal Med., 83:107-110.

Eyles, M J. and G R. Davey. 1984. Microbiology of commercial depuration of the Sydney rock oyster, Crassostrea commercialis. J. Food Protection, 47:703-706.

Eyles, M.J. 1983. Assessment of cooked prawns as a vehicle for transmission of viral disease J Food Protection, 46:426-428.

Eyles, M.J., G.R. Davey, and E J. Huntley. 1981. Demonstration of viral contamination of oysters responsible for an outbreak of viral gastroenteritis. J Food Protection, 44:294-296

Feachem, R G., D J Bradley, H. Garelick and D.D. Mara. 1983. Sanitation and Disease. Health Aspects of Excreta Disposal and Wastewater Management. John Wiley and Sons, New York.

Feingold, A.O 1973. Hepatitis from eating steamed clams. J. Am. Med. Assoc., 225:526-527.

Food and Drug Administration. 1983. England Shellfish Program Review - 1983. Food and Drug Administration, North Kingstown, R.I., 68 pp.

Food and Drug Administration. 1989. National Shellfish Sanitation Program Manual of Operations. Part I and Part II. Center for Food Safety and Applied Nutrition, Washington, D.C

Fries, C R. and M R. Tripp 1970. Uptake of viral particles by oyster leukocytes in vitro J. Invertebrate Pathology, 15:136-137.

Fugate, K.J., D.O Cliver, and M.T. Hatch. 1975. Enteroviruses and potential bacteria indicators in Gulf Coast oysters. J. Milk Food Technol., 38:100-104.

Furfuri, S.A. 1966. Depuration Plant Design. Public Health Service Publication 999-FP-7, Washington, D.C. 119 pp.

Gaillot, D., D. Terver, C. Finance, and L. Schwartzbrod. 1988. Improved method for contamination of mussels by poliovirus. Int. J. Food Microbiol., 6:333-339

Galtsoff, P.S. 1964 The American Oyster, Crassostrea virginica, U S. Fish and Wildlife Service, Fishery Bulletin, 64:1-480.

Gerba, C P 1988 Viral disease transmission by seafoods. Food Technol., 42.99-103

Gerba, C.P., Goyal, S.M., I. Cech, and G.F. Bodgan. 1980. Bacterial indicators and environmental factors as related to contamination of oysters by enteroviruses. J Food Protection, 43:99-101.

Gerba, C.P. and S.M. Goyal. 1978. Detection and occurence of enteric viruses in shellfish: A review. J. Food Protection, 41:743-754.

Gill, O.N., W.D. Cubitt, N.A. McSwaggan,B.M. Watney, and C.L.R. Bartlett 1983. Epidemic of gastroenteritis caused by oysters contaminated with small, round structured viruses. Br. Med. J., 287:1532-1534.

Goldfield, M. 1976. Epidemiological indicators for transmission of viruses by water. In: G Berg, H.L. Bodily, E.H. Lennette, J.L. Melnick, and T.G. Metcalf (eds.) Viruses in Water. American Public Health Association, Washington, D.C., pp.70-85.

Goyal, S.M. 1984. Viral pollution of the marine environment. CRC Critical Rev. Environ. Contr., 14:1-32.

Goyal, S.M., W.N. Adams, M.L. O'Malley and D.W. Lear. 1984. Human pathogenic viruses at sewage sludge disposal sites in the middle Atlantic region Appl. Environ Microbiol., 48:758-763.

Goyal, S.M., C.P. Gerba, and J.L. Melnick. 1979. Human enteroviruses in oysters and their overlying waters. Appl. Environ Microbiol., 37:572-581

Grabow, W.O K., G K. Idema, P. Coubrough, and B.W. Bateman. 1989. Selection of indicator systems for human viruses in polluted seawater and shellfish. Water Sci. Technol., 21:111-117.

Grohmann, G.S., A.M. Murphy, P.J. Christopher, E. Auty, and H.B. Greenberg 1981. Norwalk virus gastroenteritis in volunteers consuming depurated oysters. Australian J. Exp. Biol Med. Sci., 59 219-228.

Guidon, L. and C.A. Pierach 1973. Infectious hepatitis after ingestion of raw clams. Minn. Med , 56:15-19.

Gunn, R A , H.T Janowski, S. Lieb, E C. Prather, and H.B. Greenberg. 1982. Norwalk virus gastroenteritis following raw oyster consumption. Am. J. Epidemiol., 115:348-351.

Gunn, A.D.G. and D.F. Rowlands. 1969. A confined outbreak of food poisoning which an epidemiological exercise attibuted to oysters. Medical Officer, 122:75-79.

Gust, I.D. and S.M. Feinstone. 1988. Hepatitis A. CRC Press, Boca Raton, Fl., 239 pp

Guzewich, J.J. and D.L. Morse. 1986. Sources of shellfish in outbreaks of probable viral gastroenteritis Implications for control. J. Food Protection, 49:389-394.

Hadler, S.C. and H.S. Margolis. 1989. Viral hepatitis. In. Viral Infections of Humans. Epidemiology and Control A.S Evans (ed.) 3rd ed. Plenum Publishing Corp., New York. pp. 351-391

Hamblet, F.E., W.F. Hill, E.W. Akin, and W.H. Benton. 1969. Oysters and human viruses: Effect of seawater turbidity on poliovirus uptake and elimination. Am. J. Epidemiol , 89:562-571.

Harshbarger, J.C., S.C. Shang, and S.V. Otto. 1977. Chlamydiae (with phages), mycoplasmas and rickettsiae in Chesapeake Bay bivalves. Science, 196:666-668.

Haven, D.S., F.O. Perkins, R. Morales-Almo and M.W. Rhodes. 1978. Bacterial Depuration by the American Oyster (Crassostrea virginica) under Controlled Conditions Vol. I. Biological and Technical Studies. Special Scientific Report No 88, Virginia Institute of Marine Science, Gloucester Point, Va.

Hay, B. and P. Scotti. 1986. Evidence for intracellular absorption of virus by the Pacific oyster, Crassostrea gigas. New Zealand Journal of Marine and Freshwater Research, 20.655-659.

Hedstrom, C.E. and E. Lyke. 1964. An experimental study on oysters as virus carriers. Am. J. Hyg., 79:134-142.

Hejkal, T.W. and C.P. Gerba. 1980. Uptake and survival of viruses in the blue crab, Callinectes sapidus. Abstr. Ann. Meet. Am. Soc. Microbiol., 80:210.

Herrmann, J.E. and D.O. Cliver. 1973. Degradation of coxsackievirus type A9 by proteolytic enzymes. Infect. Immun., 7:513-517.

Herrmann, J.E., K D. Kostenbader, Jr. and D.O. Cliver. 1974. Persistence of enteroviruses in lake water. Appl. Microbiol., 28·895-896.

Hoff, J C and R.C. Becker. 1969. The accumulation and elimination of crude and clarified poliovirus suspensions by shellfish. Am. J. Epidemiol., 90:53-61.

Hoff, J.C., W. Jakubowski, and W.J. Beck. 1965. Studies on bacteriophage accumulation and elimination by the Pacific oyster (Crassostera gigas). In: Proc. 1965 Northwest Shellfish Sanit. Res. Planning Conf., Publ. Health Serv. Publ. No. 999-FP-6, pp. 74-90.

Hollinger, F.B 1990 Non-A, Non-B Hepatitis Viruses. In: Fields Viology, Vol. 2, B N. Fields and D.M Knipe (eds.). Raven Press, New York. pp. 2239-2273.

Hollinger, F.B. and J. Ticehurst. 1990. Hepatitis A virus. In: Fields Viology, Vol 1, B.N. Fields and D M. Knipe (eds.) Raven Press, New York. pp. 631-667.

Jansen, R.W, G. Siegl and S.M. Lemon. 1990. Molecular epidemiology of human hepatitis A virus defined by an antigencapture polymerase chain reaction method. Proc. Natl. Acad. Sci. USA, 87:2867-2871.

Jensen, H.R., Jr. 1980. The development of an acid precipitation procedure for concentrating viruses from artifically and naturally contaminated oysters. Master's Thesis. Department of Environmental Sciences and Engineering, University of North Carolina, Chapel Hill, North Carolina.

Jorgensen, C.B 1955. Quantitative aspects of filter feeding in invertebrates Biological Reviews, 30:391-454.

Kapikian, A.Z and R.M. Chanock. 1989. Viral gastroenteritis. In· A.S. Evans (ed.) Viral Infections of Humans. Epidemiology and Control, 3rd ed. Plenum Publishing Corp., New York. pp. 293-340.

Kapikian, A.Z. and R.M. Chanock. 1990a. Rotaviruses. In: Fields Virology, 2nd ed. Raven Press, New York. pp. 1353-1387.

Kapikian, A.Z. and R.M Chanock. 1990b. Norwalk Group of Viruses. In: Fields Virology, 2nd ed , Raven Press, New York. pp. 671-693.

Kennedy Jr., J.E., C.I. Wei, and J.L. Oblinger. 1986. Distribution of coliphages in various foods. J. Food Protection, 49:944-951.

Khalifa, K.I., B. Werner, and R. Timperi Jr. 1986. Non-detection of enteroviruses in shellfish collected from legal shellfish beds in Massachusetts. J. Food Protection, 49:971-973.

Kilgen, M.B , M. Cole, C. Hackney, D. Sbaih, and L. Reily. 1984. Relationship between enteric viruses and bacteriological indices in Louisiana oysters and waters Abstr Ann Meet. Am Soc Microbiol., 84:214

Koff, R.S. and H S. Sear. 1967. Internal temperature of steamed clams. New Eng. J Med., 276:737-739.

Koff, R.S., G.F. Grady, T.C. Chalmers, J.W. Mosley, B.L. Swartz, and the Boston Inter-Hospital Liver Group. 1967. Viral hepatitis in a group of Boston hospitals. III. Importance of exposure to shellfish in a nonepidemic period. New Eng. J. Med., 276:703-710.

Kott, Y., N. Roze, S. Sperber, and N. Betzer. 1974. Bacteriophages as viral pollution indicators Water Res., 8:165-171.

Kurtz, J.B. and T.W. Lee. 1987. Astroviruses: human and animal. In: Novel Diarrhoea Viruses, John Wlley and Sons, New York. pp. 92-101.

LaBelle, R.L and C.P. Gerba. 1980. Influence of estuarine sediment on virus survival under field conditions Appl. Environ. Microbiol., 39:749-755.

LaBelle, R.L., C.P. Gerba, S.M. Goyal, J.L. Melnick, I. Cech, and G. Bogdan. 1980. Relationship between environmental factors, bacterial indicators, and the occurrence of enteric viruses in estuarine sediments. Appl. Environ. Microbiol., 39:588-596.

Landry, E.F., J.M Vaughn, T.J. Vicale, and R. Mann. 1983. Accumulation of sediment-associated viruses in shellfish. Appl. Environ. Microbiol., 45:238-247.

Landry, E F., J M Vaughn, T.J. Vicale, and R. Mann. 1982. Inefficient accumulation of low levels of monodispersed and feces-associated polioviruses in oysters. Appl. Environ. Microbiol., 44:1362-1369.

Larkin, E.P. and D.A. Hunt. 1982. Bivalve mollusks: Control of microbiological contaminants. Bioscience, 32:193-197.

Lemon, S.M., L.N. Binn, R.H. Marchwicki. 1983. Radioimmunofocus assay for quantitation of hepatitis A virus in cell cultures. J. Clin. Microbiol., 17:834-839.

Lewis, G., M.W. Loutit and F.J. Austin. 1986. Enteroviruses in mussels and marine sediments and depuration of naturally accumulated viruses by green lipped mussels (Perna canaliculus), N.Z.J. Mar. Freshwater Res., 20:431-438.

Linco, S.J. and G.S. Grohmann. 1980. The Darwin outbreak of oyster-associated viral gastroenteritis. Med. J. Australia, 1:211-213.

Liu, O.C., H.R. Seraichekas, D.A. Brashear, W.P. Hefferman, and V.J. Cabelli. 1968 The occurrence of human enteric viruses in estuaries and shellfish. Bact. Proc., 68 151

Liu, O.C., H.R. Seraichekas, and B.L. Murphy. 1967a. Viral depuration of the Northern quahog. Appl Microbiol , 15:307-315

Liu, O C., H.R. Seraichekas, and B L Murphy. 1967b. In. G.Berg (ed.) Transmission of Viruses by the Water Route. Wiley Interscience, New York, New York. pp. 419-437.

Liu, O.C., H.R. Seraichekas, and B.L. Murphy. 1966a. Viral pollution of shellfish I. Some basic facts of uptake. Proc. Soc. Expt. Biol. Med., 123:481-487.

Liu, O.C., Seraichekas, H.R., and B.L. Murphy. 1966b. Fate of poliovirus in Northern quahogs Proc. Soc. Expt Biol. Med., 121:601-607.

Loosanoff, V.L 1952. Behavior of oysters in water of low salinities Convention address, National Shellfish Sanitation Association, Atlantic City, N.J.

Loosanoff, V.L 1958. Some aspects of behavior of oysters at different temperatures. Biological Bull., 114:57-70.

Loosanoff, V.L. and J.B. Engle. 1947. The effect of different concentrations of microorganisms on the feeding of oysters (Ostrea virginica). Fishery Bulletin 42, US Fish and Wildl. Serv , 51:31-57.

Mackowiak, P A , C.T. Caraway, and B.L. Portnoy. 1976. Oyster-associated hepatitis: Lessons from the Louisiana experience. Am. J. Epidemiol , 103:181-191.

Mason, J.O. and W.R. McLean. 1962. Infectious hepatitis traced to the consumption of raw oysters- an epidemiologic study. Am. J. Hygiene. 75:90-111.

Meinhold, A.F. 1982. Uptake, elimination, and tissue distribution of poliovirus in the American oyster. Masters Thesis. Department of Environmental Sciences and Engineering University of North Carolina, Chapel Hill, North Carolina.

Mele, A., M.G. Rastelli, O.N. Gill, D. DiBisceglie, F. Rosmini, G. Pardelli, C Valtriani, and P Patriarchi. 1989. Recurrent epidemic hepatitis A associated with consumption of raw shellfish, probably controlled through public health measures Am J Epid., 130:540-546

Melnick, J.L and Gerba, C.P. 1980. The ecology of enteroviruses in natural waters CRC Crit Rev Environ Control, 10:65

Mesquita, M.M.F. 1988. Effects of seawater contamination level and exposure period on bacterial and viral accumulation and elimination processes by Mytilis edulis. Water Sci Technol , 20::265-270

Metcalf, T.G., D. Eckerson, E. Moulton, and E.P. Larkin. 1980. Uptake and depletion of particulate-associated polioviruses by the soft shell clam. J. Food Protection, 43 87-88

Metcalf, T.G., B Mullin, D. Eckerson, E. Moulton, and E.P. Larkin 1979 Bioaccumulation and depuration of enteroviruses by the soft shelled clam, Mya arenaria Appl. Environ. Microbiol , 38:275-282.

Metcalf, T.G , L W Slanetz, and C.M Bartley. 1973. Enteric pathogens in estuary waters and shellfish. In. C.O. Chichester and H.D. Graham (eds.) Microbial Safety of Fishery Products Academic Press, New York pp. 215-234.

Metcalf, T.G , J.M. Vaughn, and W.C. Stiles. 1972. Virus enumeration and public health assessments in polluted surface water contributing to transmission of virus in nature. In· J.F. Malina and B.P. Sagik (eds.) Virus Survival in Water and Wastewater Systems Center for Research in Water Resources, Austin, Texas. pp. 57-70.

Metcalf, T.G. and W.C. Stiles. 1968 Enteroviruses within an estuarine environment Am J. Epidemiol , 88:379-391.

Metcalf, T.G. and W.C. Stiles 1967. Survival of enteric viruses in estuary waters and shellfish. In G. Berg (ed.) Transmission of Viruses by the Water Route. Interscience, New York pp 439-447.

Metcalf, T.G. and W.C. Stiles 1965. The accumulation of enteric viruses by the oyster Crassostrea virginica. J Infect. Dis., 115:68-76.

Millard, J., H Appleton, and J V Parry. 1987. Studies on heat inactivation of hepatitis A virus with special reference to shellfish. Epidemiol. Infect., 98:397-414.

Mitchell, J.R , Presnell, M.W., E.W. Akin, J.M. Cummins, and O.C. Liu. 1966. Accumulation and elimination of poliovirus by the Eastern oyster. Am. J. Epidemiol., 84:40-50.

Morse, D.L., J.J. Guzewich, H.P. Hanrahan, R. Stricof, M. Shayegani, R. Deibel, J. Grabau, N.A. Nowak, J.E. Herrmann, G. Cukor, and N.R. Blacklow. 1986. Widespread outbreaks of clam and oyster-associated gastroenteritis- role of Norwalk virus. New Eng. J. Med., 314:678-681.

Murphy, A.M , G.S Grohmann, R.J. Christopher, W.A. Lopez, G.R. Davey, and R.H Millsom 1979 An Australian-wide outbreak of gastroenteritis from oysters caused by Norwalk virus. Med J Australia, 2:329-333.

Murray, J C 1990 The effect of temperature and salinity on the depuration of hepatitis A virus and other microbial indicators from clams. Master's Thesis. Department of Environmental Sciences and Engineering, University of North Carolina, Chapel Hill, North Carolina

O'Mahoney, M.C., C.D. Gooch, D.A. Smyth, A.J. Thrussell, C.L.R. Bartlett, and N.D. Noah. 1983. Epidemic hepatitis A from cockles. Lancet, i:518-520.

Parry, J V. and P P. Mortimer. 1984. The heat sensitivity of hepatitis A virus determined by a simple tissue culture method. J. Med. Virol., 14:227-238.

Peterson, D A , L.G Wolfe, E P. Larkin, and F.W. Deinhardt 1978. Thermal treatment and infectivity of hepatitis A virus in human feces. J. Med. Virology, 2:201-206.

Petrilli, F. and P. Crovari 1965. Aspetti dell'inquinamento delle acque marine con partiolare riguardo alla situazione in liguera. G. Ig. Med. Prevent., 8:269-311.

Pietri, Ch., B. Hughes, J.M. Crance, D. Puel, C. Cini, and R. Deloince. 1988. Hepatitis A virus levels in shellfish exposed in a natural marine environment to the effluent from a treated sewage outfall. Water Sci. Technol., 20:229-234.

Portnoy, B.L , P A. Mackowiak, C.T. Caraway, J.A. Walker, T.W. McKinley, C A Klein. 1975 Oyster-associated hepatitis- Failure of shellfish certification programs to prevent outbreaks. J. Am. Med. Assoc., 233:1065-1068.

Power, U.F.and J.K Collins. 1990. Elimination of coliphage and E. coli from mussels during depuration under varying conditions of temperature and salinity and food availability. J. Food Protection, 53:208-212.

Power, U.F. and J.K. Collins. 1989. Differential depuration of poliovirus, Escherichia coli, and a coliphage by the common mussel, Mytilus edulis. Appl. Environ. Microbiol , 55:1386-1390.

Provost, P.J and M.R. Hilleman. 1979. Propagation of human hepatitis A virus in cell culture in vitro Proc Soc Exp. Biol. Med., 160:213-221.

Rao, V C and J L Melnick. 1986 Environmental Virology. American Society for Microbiology, Washington, D.C. 88 pp.

Rao, V.C., K.M. Seidel, S.M. Goyal, T.G. Metcalf, and J.L. Melnick. 1984. Isolation of enteroviruses from water, suspended solids, and sediments from Galveston Bay: Survival of poliovirus and rotavirus adsorbed to sediments. Appl. Environ. Microbiol., 48:404-409.

Ratzan, K.R., J.A. Bryan, and J. Krackow. 1969. An outbreak of gastroenteritis associated with ingestion of raw clams. J. Infect. Dis., 120:265-268.

Reyes, G.R., M.A. Purdy, J.P. Kim, K.-C. Luk, L.M. Young, K.E. Fry and D.W. Bradley. 1990. Isolation of a cDNA from the virus responsible for enterically transmitted Non-A, Non-B hepatitis. Science, 247:1335-1339.

Richards, G.P. 1988. Microbial purification of shellfish: A review of depuration and relaying. J.Food Protection, 51:218-251.

Richards, G.P. and F.M. Hetrick. 1988. Detection of enteroviruses in oysters from the environment. Abstr. Ann. Meet. Am. Soc. Microbiol., 88:290.

Richards, G.P. 1985. Outbreaks of shellfish-associated enteric viruses illness in the United States· Requisite for development of viral guidelines. J. Food Protection, 48:815-823.

Rippey, S.R. 1989 Shellfish Borne Disease Outbreaks Northeast Technical Services Units, Food and Drug Administration, Davisville, R.I.

Roos, R. 1956. Hepatitis epidemic conveyed by oysters. Svenska Lakartidningen, 53:989-1003.

Ruddy, S.J., R.F. Johnson, J.W. Mosley, J.B. Atwater, M.A. Rossetti, and J.C. Hart. 1969. An epidemic of clam-associated hepatitis. J. Am. Med. Assoc., 208:649-655.

Scotti, P.D, G.C. Fletcher, D.H. Buisson, S. Fredericksen. 1983. Virus depuration in the Pacific oyster (<u>Crassostrea virginica</u>) in New Zealand. N.Z. J. Science, 26.9-13.

Seraichekas, H.R., D.A. Brashear, J.A. Barnick, P.F. Carey, and O.C. Liu 1968. Viral depuration by assaying individual shellfish. Appl. Microbiol., 16:1865-1871,

Shuval, H.I, A. thompson, B. Fattal, S. Cymbalista and Y. Weiner. 1971. Natural virus inactivation processes in seawater. J. Sanit Eng. Div. Proc. Am Soc. Civ. Eng 97:587-600.

Siegl, G., M. Weitz and G. Kronauer. 1984. Stability of hepatitis A virus. Intervirology, 22:218-226

Sobsey, M.D., P.A. Shields, F.H. Hauchman, R.L. Hazard and L.W. Caton, III. 1986. Survival and transport of hepatitis A virus in soils, groundwater and wastewater. Wat. Sci. Technol., 18:97-106.

Sobsey, M.D., P.A Shields, F.S. Hauchman, A.L. Davis, V.A. Rullman, and A. Bosch. 1988 Survival and persistence of hepatitis A virus in environmental samples. In: A.J Zuckerman (ed.) Viral Hepatitis and Liver Disease. Alan R. Liss, Inc., New York, pp. 121-124.

Sobsey, M.D., C.R. Hackney, R.J Carrick, B. Ray, and M L. Speck. 1980a Occurrence of enteric bacteria and viruses in oysters. J. Food Protection, 43:111-113.

Sobsey, M.D., R. Carrick, C. Hackney, and B. Ray. 1980b. Enterovirus elimination from oysters in estuary water. Abstr. Ann. Meet. Am. Soc. Microbiol., 80·178.

Sobsey, M.D., A.L. Davis and V.A. Rullman. 1987. Persistence of hepatitis A virus and other viruses in depurated Eastern oysters. Oceans '87 Proceedings, Vol. 5, Coastal and Estuarine Pollution, Marine Technology Society, Washington, D C pp 1740-1745

Torne, J., R. Miralles, S. Tomas, and P. Saballs. 1988. Typhoid fever and acute non-A non-B hepatitis after shellfish consumption. Eur. J. Clin. Microbiol. Infect. Dis., 7:581-582

Truman, B.I., H.P. Madore, M.A. Memegus, J.L. Nitzkin, and R. Dolin. 1987. Snow Mountain agent gastroenteritis from clams. Am. J. Epidemiol., 126:516-525.

Vaughn, J.M., E F. Landry, T.J. Vicale, and M.C. Dahl. 1980. Isolation of naturally occurring enteroviruses from a variety of shellfish species residing in Long Island and New Jersey embayments. J. Food Protection, 43:95-98.

Vaughn, J.M. and T.G. Metcalf. 1975. Coliphages as indicators of enteric viruses in shellfish and shellfish-raising waters. Water Res., 9:613-616.

Vaughn, J.M , E.F Landry, T J. Vicale, and M.C. Dahl. 1979a. Modified procedure for the recovery of naturally accumulated poliovirus from oysters. Appl. Environ. Microbiol., 38:594-598.

Vaughn, J.M., E F. Landry, M.Z. Thomas, T.J. Vicale, and W.F. Penello. 1979b. Survey of human enterovirus occurrence in fresh and marine surface waters on Long Island Appl Environ Microbiol., 38:290-296.

Wait, D.A , C R Hackney, R.J. Carrick, G. Lovelace, and M.D. Sobsey. 1983. Enteric bacterial and viral pathogens and indicator bacteria in hard shell clams. J. Food Protection, 46:493-496.

Ward, R.L., D.R. Knowlton and P.E. Winston. 1986. Mechanism of inactivation of enteric viruses in fresh water. Appl. Environ. Microbiol., 52:450-459.

Watson, P.G., J.M. Inglis, and K.J. Anderson. 1980. Viral content of a sewage-polluted intertidal zone. J. Infection, 2:237-245.

Werner, K. 1983 Uptake and Elimination of Simian Rotavirus SA-11 by the Eastern Oyster, Crassostrea virginica. Master's Thesis, Department of Environmental Sciences and Engineering, University of North Carolina, Chapel Hill, N.C.

VIBRIOS IN DEPURATION

Gary E. Rodrick and Keith R. Schneider

INTRODUCTION

Shellfish such as the eastern oyster (Crassostrea virginica) and the southern quahog (Mercenaria campechiensis) are filter feeders. They pump large volumes of seawater and may accumulate and concentrate bacteria and/or other contaminants if present in their seawater environment   Depuration is a natural process of purification by which the shellfish cleanse themselves   In depuration, shellfish are placed in "clean" recirculating seawater that has been treated with a disinfectant or disinfecting process. The accumulated pollutants are eliminated from the organism by mechanisms that are not well known.

The lack of correlation between fecal coliform standards and the presence of indigenous pathogenic Vibrios (Rodrick et al., 1983) has caused many problems for the regulatory agencies and shellfish industry.   These bacteria have been linked to the consumption of raw or improperly cooked shellfish (Blake et al. 1979)   The illnesses associated with Vibrio infection can be minor, such as gastroenteritis or serious and result in death.   Moreover, several studies indicate that the majority of these potentially pathogenic Vibrios are of non-fecal origin and naturally occur in most of the shellfish harvesting areas of the United States. Therefore in order for these shellfish to be free of pathogenic Vibrios and other particulate materials, processing such as depuration may be required.

Depuration is a process of purification in which filter-feeding shellfish are placed in an environment in which the water has been disinfected by some means. The shellfish are allowed to pump, thus ridding themselves of accumulated bacteria and other pathogens (Blogoslawski and Stewart, 1983). This concept of cleansing pollutants from comtaminated shellfish was addressed as early as 1911 (Phelps, 1911).   Johnstone (1914) first used chlorine in the treatment of mussels and in 1929, Violle used ozone to depurate shellfish By far the most popular method to depurate shellfish in the US is the use of ultraviolet (UV) light depuration   The UV light possess bactericidal effects associated with its ability to disrupt unsaturated bonds in molecules such as purines and pyrimidines   These compounds which make up nucleic acids absorb UV light which can cause progressive and lethal biochemical changes in the bacterial cell (Huff et al., 1965).   For most bacteria studied, the greatest germicidal effects occur at a wavelength between 250 and 260 nm (Harm, 1980)

It has been known for some time that enteric bacteria such as E coli can be successfully depurated (Arcisz and Kelly, 1955).   Although a great deal is known about shellfish depuration varying environmental circumstances and with various species of bacteria, not much work has been done on the depuration of Vibrios   Greenberg et al. (1984) preformed a study on the persistence of several species of vibrio in the hardshell clam   Their study revealed that under certain conditions V. harveyi and V parahaemo-lyticus survived while E. coli was completely removed   Thus, investigations were deemed appropriate and necessary to determine use of both ultraviolet (UV) light and ozone assisted depuration of E. coli, V. vulnificus and non-O1 V. cholerae from both oysters and clams

MATERIALS AND METHODS

Approximately 600 specimens of oysters (Crassostrea virginica) were collected at Apalachicola Bay, Florida, from both approved and unapproved waters. Oysters were marked appropriately so mixing did not occur. More than five hundred specimens of clams (Mercenaria campechiensis) of approximately the same size (shell length 8-10cm) were collected off the south end of the Sunshine Skyway Bridge (Terra Ceia, FL). The clams were transported in coolers containing water from the collection site and placed in holding tanks (55 gal. aquaria) containing water of the same salinity. Both the clams and oysters were maintained in the holding tanks for 2-3 days to allow for adequate acclimatization to laboratory conditions. During this period the clams were fed a daily ratio of phytoplankton culture consisting mainly of Chlorella and Chlamydomonas. The culture had an average density of 1×10^6 cells/ml of which, 100 ml per clam was added daily by a slow drip (1 drop/sec) infusion into the holding tanks.

Phytoplankton consisting of Chlorella and Chlamydomonas was maintained on agar plates containing sea salts, F/2 algal culture media (Guillard, 1983), antibiotics and fungicide. The algae was cultured in two liter flask using F/2 media with mild aeration and 24 hour fluorescent lighting. Every 48 hours aliquots were drawn from the culture flasks and replaced with fresh F/2 media. Algae counts were determined by using a Neubauer cell counting chamber

Non-01 Vibrio cholerae and V. vulnificus were grown on brain-heart infusion (BHI) agar overnight at 37^oC. Isolated colonies were scrapped off the plate and resuspended in phosphate buffered saline (PBS). The optical density of the suspension was read spectrophotometrically and adjusted to 10^6cells/ml of Vibrios Viable counts were confirmed by calculating the number of colony forming units on thiosulfate citrate bile salts (TCBS) agar.

Three to five specimens (100 grams of meat) of C. virginica or M. campechiensis in each experiment were used for background controls. The oysters and clams were scrubbed under running tap water, then opened with an alcohol flamed oyster knife. The weighed oysters were placed in preweighed Waring blender jars. Equal weight/volume of PBS was added (pH 7 4) The resulting crude homogenate was then diluted with peptone water from 10^{-1} to 10^{-8} These dilutions were inoculated in a 3-tube MPN series containing alkaline peptone and incubated for 8 to 10 hours at 37^oC. Positive cloudy tubes were recorded, then streaked on to TCBS agar plates for confirmation of either non-01 V. cholerae or V. vulnificus.

Seawater samples were diluted in peptone water or PBS Dilutions were inoculated into alkaline peptone (1 ml in 9 ml) 3-tube series. Incubation and processing of positive tubes followed established techniques as described above The experiments dealing with the uptake, retention and fate of non-01 V. cholerae and V. vulnificus included numerous groups of 20 to 25 clams and oysters collected from Tampa Bay and Apalachicola Bay, respectively.

At the beginning of each experiment, the UV light system was deactivated One liter of algae and bacteria (non-01 V. cholerae, or V. vulnificus) was added to each aquarium, respectively After adequate mixing in each tank, a seawater sample was taken and processed bacteriologically as previously described for the determination of V. cholerae and V vulnificus. At predetermined time intervals, oysters and clams were removed, rinsed with 70% ethanol, scrubbed with a brush under tap water and processed bacteriologically as previously described Seawater samples were taken from each tank at the same time as the experimental clams, oysters and controls. As a result of these experiments, the kinetics of the uptake, and the fate of both non-01 V. cholerae and V. vulnificus were determined.

The experiments dealing with the effects of UV light irradiation involved separate groups (20 to 25) of oysters and clams inoculated in a glass aquarium as described in the uptake, retention and fate experiments. The initial seawater Vibrio concentration was determined using established techniques The groups of inoculated oysters and clams were then placed into 2 separate tanks. Seawater in tank number one was recirculating without UV light treatment. Seawater in the number two tank was recirculating with UV light treatment. A control tank without a UV light sterilizer, contained only seawater and bacteria. After 24, 48 and 72 hours, seawater, oysters and clams were examined bacteriologically for the presence of Vibrios, total Vibrio-like and fecal coliform.

Artificial seawater was prepared at a salinity of 23 parts per thousand (ppt) in a 40 liter aquarium. The seawater temperature was maintained at $19 \pm 1^{\circ}C$ by controlling the ambient air temperature in the experimental area. These conditions were maintained during disinfection studies and in the pilot-scale depuration studies. Artificial seawater was ozonated at a gas flow rate of 1 2 liters per minute for 60 minutes Ten 505 ml samples were drawn over a 60 minute period at 6 minute intervals. Five ml of each sample were added to a neutral potassium iodide (KI) solution at room temperature and allowed to react for 30 minutes in the dark (Shechter, 1973). The absorbance of triiodide produced was measured using a 1 cm path length at a wavelength of 352 nm The remaining 500 ml sample was titrated using the iodometric method number 408A (APHA, 1985) for determining residual chlorine. The method was modified for bromine residuals that are present in ozonated seawater. Values were expressed in parts per million (ppm) of bromine residual. A standard curve was prepared relating triiodide absorbance at 352 nm to oxidant concentration as measured by the iodometric titration method. The absorbance of triiodide produced in ozonated seawater was linear over the range from 0 to 2 0 absorbance units

Ozone gas was generated with an Annox OPT Portable Ozone Generator Model HFC-1000 supplied with medical grade oxygen (99.5% O_2 with a dew point of $-120^{\circ}C$). The ozone generator was set for high output. The gas flow was controlled by a Teflon flowmeter set at 1.2 liters per minute. According to the machine specifications, a maximum concentration of 34 mg/hour of ozone was produced. All gas lines from the ozone generator leading to the depuration tank were made of Teflon to ensure the least amount of reactivity and loss of ozone. Ozone gas was introduced to the system via a crystalline alumina gas diffusing stone located within the vertical mixing column. Oxidant concentrations were monitored using the modified neutral KI test (Shechter, 1973) and controlled manually by periodically operating the ozone generator

The experimental tank used in pilot-scale disinfection and depuration studies was of wooden construction. All seams were sealed with silicon, and the entire tank was painted with two coats of coal tar epoxy paint An epoxy paint was used to ensure a water tight system and to resist oxidation from ozonated seawater. The tank was divided into two identical but separate sections The dimensions of each section was 240 x 45 x 25 cm and held approximately 300 liters of seawater. Each section was equipped with a 1710 liter per hour (LPH) impeller type pump and was equipped with an in-line spiral wound cellulose 20 micrometer pore size cartridge filter to remove particulate matter. Flow rates were controlled by a PVC and stainless steel valve, and measured with a 0-38 LPH float-type flowmeter.

Oxidant demand free glassware was used in the disinfection and depuration experiments and in the quantitation of residual oxidant levels. The glassware was placed in a tank containing 20 liters of artificial seawater which was subjected to ozonation for a period of 1 hour. Glassware remained in the ozonated seawater for 24 hours, at which time it was removed and placed in a 103°C drying oven. The glassware was covered with aluminum foil until needed.

Pilot-scale disinfection experiments were performed with ozonated and aerated seawater. V. vulnificus grown in batch culture was centrifuged at 5875 x g for 10 minutes, washed and resuspended in sterile saline. Two hundred and fifty ml of bacterial suspension was added to each section of the depuration tank to provide a maximum bacterial density of 2.4 x 10^7/100 ml. Sample aliquots were removed for Vibrio enumeration to determine the effectiveness of natural die-away as compared to the ozonated seawater.

Depuration trials consisted of recirculating ozonated or aerated seawater through the pilot-scale system. Each section contained approximately 40 clams. Four specimens (100 g or greater of shellfish meat) were collected at timed intervals for enumeration of Vibrio organisms. The microbial levels in control specimens allowed for an evaluation of any natural microbial purging. Oxidant levels were monitored throughout the trials in the experimental section. Depuration trials were comprised of five sampling periods at 0, 2, 6, 12, and 24 hours. At each sampling period four specimens of M. campechiensis were collected. Seawater samples were also collected to monitor changes in Vibrio levels using the MPN technique as previously described

Water quality parameters for shellfish depuration were kept within the State of Florida, Department of Natural Resources guidelines (Table 1). The water temperature was held at 19±1°C, and salinity was maintained at 23 ppt. The experimental flow rate for all tests was 8 LPM. Seawater for the control section was aerated with ambient air. The gas was introduced in the same manner as the experimental group. Air flow was set to the same flow rate (1.2 liters per minute) as the ozonated section.

TABLE 1: Recommended and experimental environmental
parameters for depuration systems.

Parameter	Recommended	Experimental
Flow Rate	1 gal/min/bushel	1-3 gal/min
UV Wattage	Bactericidal	15 mwatts-sec/cm2
pH	7.0 - 8.4	8 2
Salinity	suitable	32 ppt
Temperature	suitable	25°C
Turbidity	0 - 20 JTU	2.0 JTU

To date, no guidelines for the use of ozone in depuration processes have been proposed in the United States.

RESULTS AND DISCUSSION

FATE, UPTAKE AND RETENTION

Tables 2 and 3 show a rapid uptake of non-01 V. cholerae and V. vulnificus from the seawater environment respectively. In the case of non-01 V. cholerae, several location within the animal achieved bacterial concentrations higher than those found in the surrounding seawater after 90 minutes (Table 2). At both the 60 and 90 minute sampling times, the digestive glands showed the highest accumulation of bacteria. The results for digestive gland tissue may be explained if one considers the path particles take. The distribution of non-01 V. cholerae throughout the rest of the clam seemed to be relatively even. The high counts associated with gonadal tissue appear to be somewhat anomalous.

TABLE 2: The distribution of non-01 <u>Vibrio cholerae</u> in selected tissues from the southern quahog, <u>Mercenaria campechiensis</u>. Specifically, clams were exposed to 4.3×10^7 non-01 <u>Vibrio cholerae</u> for 60 and 90 minutes in 5 liters of $30^\circ/oo$ seawater at $25^\circ C$ with 50 ml $(10^6/ml)$ algae.

Exposure Time (Minutes)	MPN per Gram of Clam Tissue				
	Digestive Gland	Gonadal Area	Mantle	Gill	Hemolymph
0	<3 0	<3.0	<3.0	<3.0	<3.0
60	5.0×10^6	4.0×10^5	$3 0 \times 10^4$	$4 0 \times 10^5$	2.0×10^4
90	2.1×10^7	2.0×10^6	4.0×10^5	1.0×10^5	$7 0 \times 10^4$

Table 3 displays results for V. vulnificus similar to those found for non-01 V. cholerae. The digestive glands had the highest concentration of associated bacteria at both the 60 and 90 minute sampling time periods. As with the non-01 V. cholerae trial, the gill tissue also displayed high counts of bacteria, which is to be expected. It is interesting to note that none of the tissue sampled for V. vulnificus had counts higher than the surrounding seawater. This may be due to the lower initial inoculum delivered to the clams and/or a reduced survival of V. vulnificus as compared to non-01 V. cholerae.

TABLE 3: The distribution of <u>Vibrio vulnificus</u> in selected tissues from the southern quahog, <u>Mercenaria campechiensis</u> exposed to $4 3 \times 10^5$ V. vulnificus for 60 and 90 minutes in a 1 gallon tank containing 2 liters of $32^\circ/oo$ seawater at $25^\circ C$ in the presence of algae

Exposure Time (Minutes)	CFU's per Gram of Clam Tissue				
	Digestive Gland	Gonadal Area	Mantle	Gill	Hemolymph
0	<3 0	<3 0	<3 0	<3.0	<3 0
60	8.4×10^4	7.6×10^3	$4 2 \times 10^3$	$8 0 \times 10^3$	3.0×10^3
90	$6 2 \times 10^4$	3.4×10^3	$2 2 \times 10^3$	1.4×10^4	2.5×10^3

These results were similar to those in the literature (Blake et al., 1985) which used a "piggy-back" method of using algae to introduced bacterium It should be noted that these results differ with other studies (Kelly and Dinuzzo, 1985) which did not use algae association to inoculate shellfish.

UV LIGHT IRRADIATION

To determine if UV light depuration is an effective method of depurating shellfish of bioaccumulated bacterium, it was first necessary to establish if the treatment would kill the organisms in the seawater. This form of depuration relies on the purging or release of the contaminating bacteria into the water column. Once in the seawater environment, the flow and the water will eventually draw the bacteria throught the UV light source, thus killing it and preventing reaccumulation. A flow-through system differs from the aforementioned recirculating system in that the UV light source only has to reduce (or eliminate) the bacteria from the water supply. Reduction is achieved by a slow release of bacteria with no recontamination.

Table 4 shows the results of UV light irradiation of non-01 $\underline{V}$. $\underline{cholerae}$ in seawater and depuration in clams. The UV treated seawater displayed a marked decrease in bacterial numbers No organisms were detected by MPN after 24 hours. Conversely, non-01 $\underline{V}$. $\underline{cholerae}$ was present even after 48 hours Results for the depuration trial were similar. No bacteria were recovered after 24 hours in UV light treatment shellfish. Control clams had a fairly constant level of bacteria throughout the 72 hour experiment

TABLE 4: The effects of ultraviolet light irradiation on the depuration of non-01 Vibrio cholerae in Mercenaria campechiensis in recirculating seawater.

| | MPN for Duration of UV Light Exposure | | |
Treatment	24hrs	48hrs	72hrs
Seawater without UV	9.3×10^2	2.3×10^3	< 3.0
Seawater with UV	< 3.0	< 3.0	< 3.0
Clams without UV	2.3×10^4	9.3×10^2	$< 2.3 \times 10^3$
Clams with UV	< 3.0	< 3.0	< 3.0

The results of UV light irradiation and depuration of $\underline{V}$. $\underline{vulnificus}$ are seen in Table 5. The results for seawater showed no $\underline{V}$. $\underline{vulnificus}$ recovery in the seawater at any time period. Organisms were found in the non-treated seawater at the 24 hour sampling time, but not at 48 or 72 hours. The UV light depuration of $\underline{V}$. $\underline{vulnificus}$ did not show any enhanced reduction over the control group. Both recorded 2 log cycle reductions.

TABLE 5: The effects of recirculating ultraviolet light irradiated seawater on the elimination of <u>Vibrio vulnificus</u> in the southern quahog, <u>Mercenaria campechiensis</u>.

	MPN per Duration of UV Light Exposure		
Treatment	24hrs	48hrs	72hrs
Seawater without UV	2.3×10^3	< 3.0	< 3.0
Seawater with UV	< 3.0	< 3.0	< 3.0
Clams without UV	1.5×10^3	9.1×10^2	< 3.0
Clams with UV	4.0×10^4	1.7×10^3	1.4×10^2

The effect of UV light irradiation on total Vibrio-like organisms in seawater and depuration from oysters can be seen in Table 6. The UV light treated seawater had lower numbers of recoverable Vibrio-like bacteria than the control. It is interesting to note that bacteria was undetectable at time zero and fluctuated throughout the experiment. Bacterial numbers never reached zero as was the case in Tables 4 and 5. The UV light depurated oysters had a two log cycle reduction, while the controls exhibited no decrease in bacterial numbers.

Table 6: The effects of recirculating ultraviolet light irradiated seawater on depuration of total Vibrio-like organisms form naturally infected oysters.

	MPN per Duration of UV Light Exposure		
Treatment	0hrs	24hrs	96hrs
Oysters without UV	9.3×10^4	4.3×10^7	2.3×10^4
Oysters with UV	9.3×10^4	4.3×10^5	9.3×10^2
Seawater without UV	< 30	2.3×10^4	4.3×10^4
Seawater with UV	< 3.0	2.3×10^3	9.0×10^1

This same experiment repeated with fecal coliforms (Table 7) exhibited different results from those of the total Vibrio-like organisms. In both the UV light treated and control seawater samples, no fecal coliforms were recovered. This might be due to low survivability in the seawater when not in association with an animal host. Depuration results show an equal elimination of bacteria in both the experimental and control tanks.

TABLE 7: The effects of recirculating ultraviolet light irradiated seawater on depuration of fecal coliforms acquired by oysters.

	MPN per Duration of UV Light Exposure		
Treatment	0hrs	48hrs	96hrs
Oysters without UV	7.5×10^3	4.3×10^4	< 3.0
Oysters with UV	75×10^3	2.3×10^3	< 3.0
Seawater without UV	< 3.0	< 23	< 3.0
Seawater with UV	< 3.0	< 40	< 3.0

OZONE

Blogoslawski et al (1976) analyzed the production of oxidants in ozonated seawater. They reported that the residual oxidant concentration formed in ozonated seawater are bromine species. It was noted the ozone does not exist in seawater but reacts with bromide. For this reason the procedure of Shecther (1973) was modified and a standard curve was generated for I_3^- production versus oxidant concentration as measured by an iodometric titration, 408A (APHA, 1985). The titration method for residual chlorine was used to determine oxidant concentration. Based on the knowledge that bromine is a halogen with similar reaction properties to that of chlorine and is a thermodynamically favored reaction when seawater is reacted with ozone (Pichet and Hurtubise, 1976), a titration method for residual chlorine was used (APHA, 1985). The equation was modified by substituting the molecular weight of bromine for that of chlorine.

The method of Shecther (1973) was used to determine the concentration of ozone in aqueous solution. The method involves mixing 5 ml of sample with 5 ml of a neutral KI solution. The resulting interaction between oxidant and reagent produces a proportional amount of I_3^-. This reaction product was measured spectrophotometrically, and correlated with the oxidant concentration measured by iodometric titration 408A (APHA, 1985).

Blogoslawski et al. (1979) reported that in depuration experiments with the softshell clam, pumping activity was severely reduced, or in some cases halted by oxidant concentrations in excess of 4 5 mg/L. Therefore it was necessary to establish the rates of oxidant formation and dissipation to ensure a bactericidal environment that is conducive to pumping.

The process of inactivation of V vulnificus in a pilot-scale system had problems that were not observed in the laboratory setting. Blogoslawski et al. (1975) used ozone to disinfect seawater in a large-scale experiment and found that increasing the oxidant demand in the system increased the rate at which oxidant was consumed. Thus in the pilot-scale system the amount of demand (in the form of algal growth, bacterial lysis products, oxidation of construction materials, etc.) caused a variation in the oxidant consumption and the inactivation of V. vulnificus.

Figures 1 illustrate the bactericidal action of oxidant against V. vulnificus action in a large scale system. The effects of mass transfer of ozone to seawater and the direct inactivation of bacterium in the mixing columns must also be considered as not to confuse ozone disinfection with bromine species inactivation.

In the disinfection trial (Figure 1) the inoculum was introduced to a pre-ozonated environment. The initial oxidant concentration was 0.9 mg/L. After 60 minutes, no bacteria were recovered. The control section exhibited a slow die-off of 1.1 log units after 240 minutes. A constant equilibrium of 200 CFU's per 100 ml of seawater was observed.

Success in the depuration of Vibrios seems to depend on the ability to maintain an oxidant residual and pumping shellfish concurrently. Depuration trials, the time of pumping coincided with the major reduction in bacterium recovered in shellfish meats.

Figure 2 illustrates the first depuration trial with artificially infected Mercenaria campechiensis. Clams were first observed pumping at the 12 hours into the experiment. At that time the oxidant demand increased as indicated by the decrease in measured concentration from approximately 3.0 to 1.0 mg/L. A 2 log unit increase in V. vulnificus

Pilot-Scale Disinfection

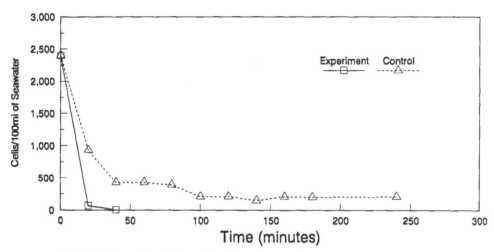

FIGURE 1 Pilot-scale inactivation of V. vulnificus in ozonated seawater at a salinity of 23 ppt, pH 8.2, and 19 C. Initial oxidant concetration was 0.0 mg/L.

Pilot-Scale Depuration

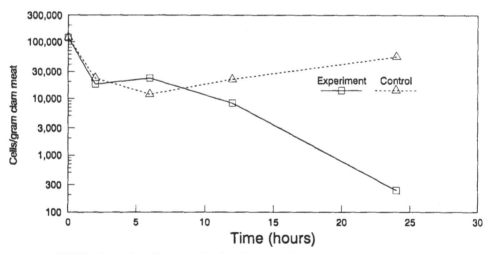

FIGURE 2 Depuration of M. campechiensis at 3.14 mg/L initial oxidant concentration. Clams were artificially dosed with V. vulnificus. Pumping was observed at 12 hours.

was noted in the overlying water in the experimental section during the time period when pumping began. In the first 12 hours the experimental section had only a 0.5 log unit difference in bacteria recovered in clam meat as opposed to the control section In the second 12 hours after pumping was initiated, the difference in bacteria recovered in clam meat increased to 2.5 log units. These results support similar finding of ozone assisted depuration of V. vulnificus done by Schneider et al. (1991).

REFERENCES

APHA, 1985. Standard Methods· For the examination of water and wastewater. 16th edition. American Public Health Association, Washington, DC.

Blake, N J., G E. Rodrick, M. Tamplin, and P. Luth 1985. Uptake and fate of bacteria in a laboratory depuration system. In: Proc. of the Tenth Annual Tropical and Subrtopical Fisheries Conference of the Americas. 10:227-248.

Blake, P A , R.E. Weaver, and D.G. Hollis. 1980. Diseases of humans (other than cholera) caused by vibrios. Ann. Rev. Microbiol. 34:341-367.

Blake, P.A., M.H. Merson, R.E. Weaver, D.G. Hollis, and P.C. Heublein. 1979. Disease caused by a marine vibrio: clinical characteristics and epidemiology. N. Engl J Med. 300:1-5.

Blogoslawski, W.J. and M.E. Stewart 1983. Depuration and public health. In: Proc 14Th Annual Meeting World Mariculture Society. 14:535-545.

Blogoslawski, W.J., M.E. Stewart, J.W. Hurst, J R., and F.G. Kern III. 1979.   Ozone detoxification of paralytic shellfish poison in the softshell clam (Mya _ arenaria). Toxicon 17:650-654.

Blogoslawski, W.J., C. Brown, E.W. Rhodes, and M. Broadhurst. 1975. Ozone disinfection of a seawater supply system. In: Proc. 1st Int. Sym. Ozone for Water and Wastewater, pp.·674-687. Washington, D.C.

Blogoslawski, W.J., L Farrell, R. Garceau, and P. Derrig. 1976 Production of oxidants in ozonized seawater In: Proc 2nd Int Sym. on Ozone Tech. International Ozone Institute, pp.:671-681. Jamesville, NY.

Greenberg, E.P., H.B. Kaplan, M. Duboise, and B. Palhof. 1983. Persistence   and distribution of marine vibrios in the hardshell clam. In: Vibrios in the Environment. pp. 479-493.

Guillard, R.R.L. 1983. Culture of phytoplankton for feeding marine invertebrates. In·Cultures of Marine Invertebrates (C.J. Berg, Jr. ed.) Hutchinson Ross Publishing Comp. Stroudsburg, PA. pp.108-132.

Harm, W. 1980. Biological effects of ultraviolet radiation. Cambridge University Press, Cambridge England.

Huff, C.B., H F. Smith, W.P. Boring, and N.A. Clark. 1985. Study of ultraviolet disinfection of water and factors in treatment efficiency. Pub. Health Rep. 80.695-704.

Hood, M.A., G.E Ness, G.E. Rodrick, and N.J. Blake. 1983. The ecology of <u>Vibrio cholerae</u> in two Florida estuaries <u>In</u>. Vibrios in the Environment pp. 399-409. John Wiley & Sons, NY.

Johnston, J. 1914. The methods of cleansing living mussels from ingested sewage bacteria. Report for 1914 on the Lancashire Sea Fisheries Laboratory. No. 23.

Kelly, M.T. and A. Dinuzzo. 1985 Uptake and clearance of <u>Vibrio vulnificus</u> from Gulf Coast oysters, <u>Crassostrea virginica</u>. Appl. Environ. Microbiol. 50:1548-1549.

Schneider, K.R., F.S. Steslow, F.S. Sierra, G.E. Rodrick, and C.E. Noss. 1991 Ozone Depuration of <u>Vibrio vulnificus</u> from the Southern Quahog Clam, <u>Mercnearia campechiensis</u>. J. Invert. Path 57:(In press).

Shechter, H. 1973. Spectrophotometric method for determination of ozone in aqueous solution. Water Res. 7:729-739

Violle, H. 1929. De la sterilisation de l'eau de mer par ozone: applications de cette methode pour le purification des coquillages contamines. Revue d'Hygiene et de Medecine Preventive 51:42-46.

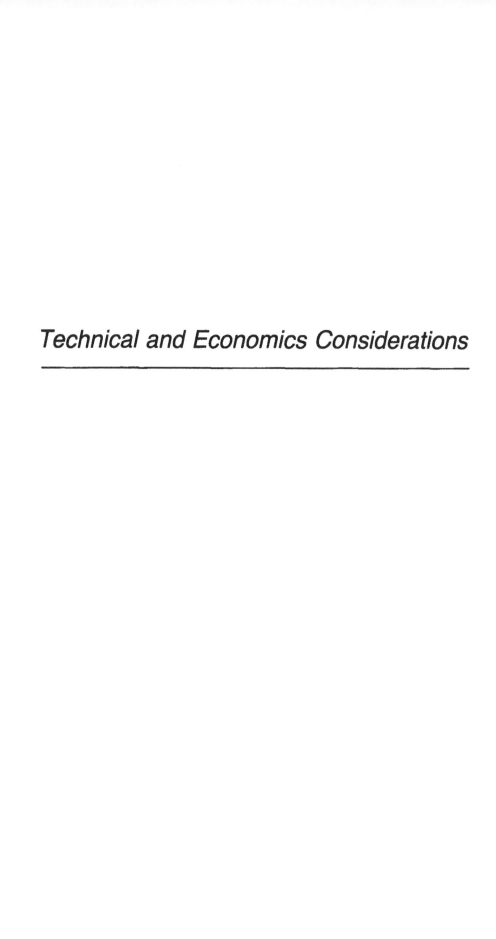

Technical and Economics Considerations

DESIGN OF DEPURATION SYSTEMS

Santo A. Furfari

INTRODUCTION

The purpose of depuration is to reduce microorganisms in molluscan shellfish so that they will be safe for human consumption (NSSP, 1988). Live shellfish are depurated by being held in a controlled clean water environment where they can naturally purge their digestive systems of harmful microorganisms. The National Shellfish Sanitation Program (NSSP) Manual of Operations, Part II (1990) sets forth the criteria for the depuration of shellfish. The sources of the shellfish can be only moderately polluted, with the median total coliform count not to exceed an MPN of 700/100 ml of water. This classification restricts areas applicable to depuration.

Depuration has recognized limitations. It is not intended to reduce organic or inorganic chemicals since the requisite purification time is too long to be economically feasible. A period of 2 to 3 days is considered economically feasible. It is also not intended to reduce heavy microbiological pollution or high virus levels which also require long purification times (Richards, 1988). The process relies totally on the shellfish themselves. Because of these environmental and biological constraints, the design and operation of depuration facilities must fulfill specific requirements for each species of shellfish, as well as for public health, sanitation, and economic aspects (Furfari, 1976).

This paper describes the minimum requirements for facilities and operations only as they affect generic design.

FACILITY - MINIMUM REQUIREMENTS

BIOLOGICAL

The biological ramifications of the depuration process with respect to design and construction are quite extensive. Shellfish are handled many times from harvest to process completion. The plant layout must move, handle, and store shellfish efficiently without affecting their ability to purify. No plant component should allow the shellfish to become contaminated or prevent their purification. The biology of the depuration process requires that shellfish be washed, cleaned, and stored in suitable containers which encourage purification. Since shellfish purify themselves by natural activity, seawater, whether natural or artifical, must be provided. The waste products generated during purification must not be allowed to recontaminate the shellfish.

STRUCTURAL

A full discussion and presentation of plant design as a whole or of its components can be found in several reports (Ayres, 1978; Furfari, 1966, 1976, 1982; Nielson et al., 1976; Richards, 1988). This paper lists each critical plant feature and component with appropriate design and construction criteria. A set of generic plant design calculations is appended (Appendix A)

PLANT AND LAYOUT

The basic plant requirements include loading/unloading areas, washing and culling areas for incoming and outgoing shellfish, depuration tanks, seawater systems (source, pumps, piping, treatment), and some storage space. Depending on a state's requirement, office, laboratory, and toilet space may be needed. Depending on local circumstances, the utilities may include water supply, lighting, heating, and ventilation. As with any food facility, the plant must be constructed to maintain sanitation. This includes not only floors and walls, but also anything that might contact the shellfish product. For outdoor plants a roof is required for protection against birds. The plant layout should prevent the inadvertent comingling of different harvest lots or shellfish from different phases of purification.

SEA WATER SYSTEM

The sea water system of a depuration plant encompasses pumps, piping, and treatment units. All parts of the system should be easily cleanable to prevent contamination of the water and the shellfish. The materials should not impart chemicals which might be toxic to humans nor retard the depuration rate of shellfish. Pumps should be large enough to provide the minimum flow requirements that are effective for the species. Compliance requirements are one gallon of seawater per minute per bushel unless studies show otherwise (NSSP, 1988; Appendix A).

Treatment may include units to reduce microbiological levels, add oxygen, raise or lower temperature and/or salinity, and reduce turbidity. All recirculating seawater systems require units to reduce coliform bacteria to less than detectable levels. In the U.S., only ultraviolet (UV) light treatment has been approved without reservation. Chemicals such as chlorine are not allowed. Ozone may be used provided that no ozone contacts the shellfish and the water is discarded at the end of the cycle for recirculated systems. Flow-through seawater systems with deep salt water wells have operated without the need for microbial treatment.

Aeration has been needed to increase oxygen for most recirculation systems. The minimum level of oxygen is 50 percent of saturation. Aeration should be done outside the tanks containing the shellfish to avoid causing turbulence in the tanks. Geographical differences affect the need to raise or lower seawater temperature, which is dependent on species and season The minimum and maximum temperature requirements, unless found otherwise by studies, are oysters, 10 to 25°C; hard clams, 10 to 20°C; and soft clams, 2 to 20°C.

Salinity may need to be adjusted if the processed seawater is outside the range of ± 20% of that of the harvest area. The pH should range from 7.0 to 8.4. Turbidity should not interfere with shellfish activity or any of the other treatment units. Thus filtration of incoming water may be needed in some situations for efficacy of UV light treatment. Maintenance of these seawater constituents requires reliable equipment that is easily maintained and/or repaired. Commercially available equipment is recommended, but approved "home-made" devices have been used successfully.

TANKS

Although seawater quality is the most important aspect of a depuration plant, without adequate tank design shellfish may not reliably purify. Criteria for good tank designs are enumerated in the shellfish program manual (NSSP, 1988); however, the details and specifics for tanks are lacking. Although the U.S. does not have standardized tank designs, England's are somewhat standardized (Ayres, 1978). Different designers using the same set of criteria will derive different tank configurations, which, in general, all work within some limits Some of the more important tank design criteria are as follows:

Dimensions. The total volume of the tanks is based on expected production of the plant. Usually with a 48-hour expected holding time, the tanks should be able to hold twice the expected daily production. For example, if 200 bushels per day or per batch is the desired production, then total tank volume should be 400 bushels. The NSSP manual (1988) recommendations are 5 cubic feet per bushel of soft clams, and 8 cubic feet per bushel of hard clams or oysters. With these factors the total volume of the tanks can be calculated. If operations are to be flexible, a number of smaller tanks would be more useful than a few large tanks. During certain seasons production may be low and small tanks that are full of shellfish are more economical than large tanks that contain relatively few shellfish. After the number of tanks needed is ascertained, the volume of each tank can be calculated.

If the volume of a tank and its shellfish loading are known, the actual dimensions can be calculated. The configuration of shellfish containers and the hydraulic features will determine the depth. Shallow tanks are easier to load, unload, and clean. Water is also more evenly distributed with shallow tanks (2 to 3 feet deep).

Hydraulic flow. The NSSP Manual (1988) requires uniform hydraulic flow with a minimum of turbulence. Experience has shown that shallow tanks are better for horizontal flow Upward or downward flow has been used with deeper tanks (4 feet deep or more).

Turbulence usually does not occur in the main tank portion with the recommended flow of 1 gallon per minute per bushel. However, the flow may be turbulent at entrance and exit points. Turbulence may occur if aeration by compressed air takes place in the tank. This should be avoided.

Shellfish loading. Shellfish containers should not be allowed to disrupt the uniform flow features. This means that containers need space between each other as well as along the walls and bottom. There should be a few inches of water above the shellfish in the top containers Although no clearance dimension is specified by the NSSP manual (1988), each plant design should specify a clearance dimension.

Materials and sanitation. Tank surfaces should be fabricated of nontoxic, corrosion-resistant materials. Their cleanability is important. Surfaces must be smooth (nonporous) for effective sanitizing and cleaning. Plastic based tanks and epoxy coatings on wood and concrete have been used. The tanks and foundation must be of sturdy construction to support the weight of water.

Slope and drainage. For easy draining and flushing out shellfish waste, the tank floor should slope toward a large drain. Quick tank draining requires a large drain opening. The size will depend on the volume of water in the tank and the time desired to empty the tank If more than one tank will be drained at the same time, the overall tank drainage system must have increased capacity.

Container placement. Although no set pattern is recommended, container placement and the number of layers are very important in purification operations. The spacing between containers was mentioned earlier. Containers must also be placed so that all shellfish are exposed to the flow of the water. The configuration of sea water input and output and the number of layers must be carefully designed. The species of shellfish also is a factor since different maximum container depths are recommended for different shellfish. Generally two or four layers of containers are recommended to fulfill the above criteria.

Containers. The purpose of containers is to hold shellfish in a manageable way for washing, moving, and purifying in the tanks. For biological reasons the maximum depth within the container is 3 inches for hard clams and oysters, and 8 inches for soft clam Since water flow is important, containers should be made of meshed material. The mesh size should not interfere with normal shellfish functioning. For example, oysters need to open easily, and thus the mesh should not wedge the shells. The siphons of clams must not be pinched by a mesh that is too small. Container materials should be durable because of the rough handling, and should allow easy cleaning and sanitizing. Storage space for possibly hundreds of containers must be provided. If containers are to be handled they should be light in weight. Plastic coated or plastic mesh containers have been used successfully.

Storage. Plant design must include adequate and segregated storage for incoming (untreated) and outgoing (treated) shellfish. The storage environment for untreated shellfish should not interfere with purification. To avoid interference with purification rates and to prevent microbial multiplication in shellfish. a storage time-temperature must be resolved. The season of the year, the geographical location, and the species of shellfish will determine this biological requirement Shellfish that are stored too cold may
be shocked if placed in much warmer purification water. They may even spawn. Bacteria may multiply in shellfish that are not stored in cool environment. It is better to avoid prolonged storage before purification. After purification the NSSP requires the usual refrigeration to minimize the potential for microbiological growth (NSSP, 1988).

Washing and culling. The washing of detritus from the shellfish is an essential component of depuration. Shells must be clean and free of excess mud, sediment, or other material which may contain microbial contaminants. Disinfection of shellstock is not allowed; however, washing either manually or mechanically is acceptable. In mechanical washing, shellfish are placed on continuous belts or in shallow trays. Designs are not standardized. Soft shell clams are difficult to handle because of the fragile shells. Oysters are difficult to clean because their shells lack smooth surfaces. Hard clams are easiest to wash because they have rugged shells. Commercially available machine washer-graders have been used in hard clam depuration plants. The size and number of these machines depends on general operations of the enterprise.

Manual washing with spray from hoses or submersion of containers in tanks has been used. With either method, potable water or water from an approved source should be used. The end result is a clean shell that will not impart bacteria or turbidity to the depuration water.

Dead or broken shellfish are removed by culling. These shellfish probably would not purify and should not be placed in depuration tanks. Culling must be repeated after purification. Washing equipment may be designed to carry out this process simultaneously.

Meters and measuring devices. Because certain features of seawater systems are regulated to some extent, measuring or recording devices may be needed. The NSSP (1988) does not explicitly require installed devices. Regulations of some states are stricter than others. Devices typically installed are temperature probes with recorders and flow meters or flow indicators. Measurements of salinity, turbidity, and dissolved oxygen may be needed periodically but continuous readings are seldom required. Many models of UV light units can be purchased with built in meter(s) for UV light intensity and timer(s) to record the hours on the lamps.

PLANT SIZING AND SCALING

A difficulty in depuration plant design is sizing the plant. The scale up factors are not straight forward and there is little experience to draw on. Usually production requirements are estimated through the experience of the local shellfish industry. The plant design is based on a broad range of physiological requirements reflecting seasonal fluctuations and on marketing considerations of supply and demand for the particular shellfish species. At times, a plant may have to operate at a production rate of one-tenth maximum capacity. Although intensity of operations may vary with this range of production, the plant design must accommodate these fluctuations.

To accommodate a highly fluctuating operation, a modular design is desirable (Nielson et al., 1976). The seawater/tank system should be easily adjusted to handle full plant capacity as well as a much lower production without loss of efficiency. The extreme case is a plant with only two large tanks: one for the new lot of shellfish and the other for the treating lot which was begun the day before. To minimize sea water use and treatment a tank should be capable of adjusting the water level. Even better would be smaller tanks with a fixed design of unchanging hydraulics that would be more easily loaded with fewer containers of shellfish.

Current flexible plant designs have a single water treatment unit which handles several tanks with a recirculating sea water system. This is termed a depuration unit (NSSP, 1988). The advantage is that only the necessary tanks are filled with shellfish and seawater. Although the treatment unit can accommodate the entire depuration unit, dual small pumps and dual small water treatment units installed in parallel provide an economical cutback in flow.

The design and sizing of a depuration plant is a challenge often met by trial and error. Table 1 shows how design technology must change with scaling up. Advanced technology is applicable to larger plants. The figures are for demonstration only. The final plant design and operational acceptability must be confirmed by a process verifications study (NSSP 1988).

Table 1.   Scaling factors for depuration plants

Operating range bushels/day	Plant capacity bushel[a]	Shellfish handling methods	Number of tanks[b]	Tanks sizes[c]	Water treatment units[d]	Container sizing[e]
<50	<50	manual	4-10	small	2	1/2 bushel
50-100 bushel	100-200	manual to machine	>10	small	2	1/2 - 1
100-300	200-600	machine	4-10	large	---[d]	---[e]
300-500	600-1000	machine	>10	large	---[d]	---[e]
>500	>1000	machine	>10	large	---[d]	---[e]

Notes
[a] Assume 48-hour purification
[b] Species dependent
[c] Small: typical range 4 x 8 x 2 feet
   Large: typical range 10 x 15 x 3 feet
[d] Depends on whether plant is flow through or recirculation
[e] Large trays holding several bushels are more appropriate

DEPURATION UNIT

The depuration unit is the heart of the overall plant scheme layout and operation. According to Part II of the NSSP Manual (1988) a depuration unit is a tank or series of tanks supplied by a single seawater system. The reason for this designation is that quality assurance requires a certain sampling protocol based on a harvest lot and a process batch in a depuration unit. Sampling can be a major cost item for a depuration plant. The selection of the size (tank size and number) of the depuration unit with respect to overall plant capacity and typical routine operations is very important.

Because of the 48-hour minimum time and the desire for daily output, the minimum number of depuration units will be two. Since on a given day, some shellfish are through treating and can be removed from a unit, this unit may be filled with a fresh lot of shellfish. With low capacity (less than 100 bushels) the number of depuration units will probably be two. With larger plant capacity more depuration units are recommended to facilitate a flexible operation. Small treatment units are less expensive but more individual units will be needed. The treatment capacity will govern the flow rate, which in turn governs the quantity of shellfish (1 gallon per minute per bushel).

RELIABILITY OF TECHNOLOGY AND PROCESS

Because depuration deals with public health, plant reliability must be high. Although some operational features, such as sanitation, upkeep, and flow rates affect reliability, the design and construction of the plant are major factors in ensuring high reliability. Without standardized designs it may be difficult to determine if a plant will be reliable even though the intent is to fulfill all necessary NSSP manual (1988) requirements.

The plant design features that inherently affect reliability are given in Table 2. Current technology cannot provide a relative critical scoring of several features. Two critical criteria assure adequate depuration: shellfish activity, to ensure depuration of the pollution, and prevention of recontamination. Shellfish activity is slowed or stopped because of inadequate seawater, but design of seawater system components and tanks may cause decreased activity even with a good seawater source. Recontamination may occur because of poor design of components.

TABLE 2. Depuration plant design criteria and reliability technology design features

Plant Redundancy components	Shellfish depuration criteria	Reliability features	required
Pre-storage	Shellfish activity	Temperature control	No
Seawater system	Shellfish activity recontamination	Treatment control	Yes
Washing	Shellfish activity recontamination	Wash water	No
Tanks	Shellfish activity recontamination	Loading, cleanability hydraulics	Yes
Containers	Shellfish activity	Depth, mesh size	NA

Certain plant components should be designed with redundancy in mind. Seawater disinfection systems, pumps, tanks, and the like should have spare parts and/or duplicate units, which should be carefully considered during the design and purchasing phases.

Process verification (NSSP, 1988), although an operational requirement, is intimately linked to plant design. The NSSP requires that plant design and operation be verified by studies of tank - loading, sea water flow rate, water quality parameters, and purification rate. Container designs and arrangements in tanks also require verification. Verification studies involve sampling programs and usually are conducted when a plant begins operation. The studies may continue for a year to cover seasonal variations in water and shellfish quality. The results of the study may indicate the need for design changes.

ITEMS FOR TECHNOLOGICAL DEVELOPMENT IN THE USA

The technology available to other fishery enterprises would probably be applied to depuration plants if depuration activity would expand in the United States. The following is a list of major concerns in the development of depuration technology.

*Disinfection of large volumes of seawater.
*Repeated reuse of artificial seawater:

*Large plant tank designs for flexible operations relative to volume and flow.
*Better shellfish handling methods.
*Improving washing and culling methods.
*Use of two-stage process for shellfish depuration in areas exceeding the restricted criteria.
*Humidity and corrosion control in depuration plants.
*Standardized plant and tank design for each species.
*Improving reliability throughtout the plant design and facilities.

REFERENCES

Ayres, P.A. 1978. Shellfish purification in installations using ultraviolet light. Laboratory leaflet No. 43. Ministry of Agriculture, Fisheries, and Food, Burnham-on-Crouch, England. 20 pp.

Furfari, S.A. 1966. Depuration plant design, Publication No. 999-FP-7. U.S. Public Health Service, Washington, DC. 119 pp.

Furfari, S.A. 1976. Shellfish purification: a review of current technology. In "Advances in Aquaculture" FAO Technical Conference on Aquaculture, Kyoto, Japan Fishing News Books Ltd, Farmham, England, p. 385-394.

Furfari, S.A. 1982. Purification of shellfish: Current practices. Administrative document. U.S. Food and Drug Administration, Washington, D.C. 104 pp.

National Shellfish Sanitation Program (NSSP). 1990b. Manual of Operations, Part II: Sanitation of the Harvesting, Processing and Distribution of Shellfish, Public Health Service, U.S. Food and Drug Administration, Washington, DC.

Nielson, B.J., Haven, D.S. and Perkins, F.O. 1976. Practical consideration for bacterial depuration of oysters in the Chesapeake Bay region, Volume I (p.69). FDA contract report. Virginia Institute of Marine Sciences, Gloucester Point, VA. Vol. I-III.

Richards, G P. 1988. Microbial purification of shellfish: a review of depuration and relaying. Journal of Food Protection 51:218-251.

APPENDIX A
GENERIC PLANT DESIGN

Given:

Species: Oysters   Maximum daily production: 100 bushels (BU)

Find:

Total plant capacity with 48 hour purification (2 days)
100 bushels/day x 2 days = 200 BU

Total plant pumping capacity (flow-through)
200 BU x 1 gpm/BU = 200 gpm

Tanks

Total tank volume (water)
200 BU x 8 cubic feet/BU = 1600 cubic feet

Desire 4 tanks available each day

4 tanks/day x 2 days = 2 tanks

Individual tank volume (200 BU/8 tanks = 25 BU/tank)

1600 cubic feet/8 tanks = 200 cubic feet/tank

Maximum tank depth = 3 feet

Area of each tank

200 cubic feet/3 feet = 67 square feet

Assume length to width ratio of 4:1

width = approx. 4 feet

length = approx. 16.75

Assume 3-inch freeboard

Tank dimensions 4 feet x 16.75 feet x 3.25 feet deep

Recirculation pump capacity of each tank = 25 gpm

Drainage capacity of 4 tanks x 200 cubic feet/tank = 800 cubic feet (5980 gallons)

Time desired 20 min. Flow = 300 gpm

Container design 3-inch depth, for 25 bushels

Approx. area = 3.5 feet x 16 feet = 56 square feet = 8064 square inches

Volume 1 layer = 8064 x 3 = 24192 cubic inches (2150 cubic/BU) = 11.25 bushels

2 layers needed = 24.5 bushels

USE OF ULTRAVIOLET LIGHT IN DEPURATION

Thomas L. Herrington

INTRODUCTION

All white light contains ultraviolet (UV) light. The UV light for use in water disinfection in depuration plants is generated by low pressure mercury vapor bulbs, creating short-wave radiation at a wavelength of 254 nm (2537 Angstrom units) In a UV treatment system, water is disinfected by exposure to this radiation at a specific dose rate or intensity, i.e., milliwatts sec/cm^2. No microorganisms have been found to survive UV treatment when irradiated at a sufficient intensity and proper wavelength. The UV light disinfects by damaging the DNA and RNA of pathogenic organisms (Shechmeister, 1977). In practice, water flows through an irradiation chamber known as a photoreactor. The selection of flow rate for an effective resident time for the area under the bulb is a function of the intensity of the UV tube.

The first use of UV light was in Marseilles, France, in 1910 to disinfect the water supply (Somerset, 1990). It is currently used to disinfect treated sewage effluent, water supplies, and equipment surfaces Use of UV water disinfection for shellfish depuration in the United States was introduced in the early 1960's as an alternative to chlorination. About 20 UV depuration plants were certified in the U.S in the early and mid 1980's. Currently, there are fewer than 10 certified depuration plants in the U S , all of which use UV disinfection Several more plants are being designed Depuration using water treated by UV rather than other types of disinfection methods is compared in Table 1.

UV REACTOR DESIGN

Three general UV unit designs are commonly used for water treatment in the U S The oldest design, which uses UV bulbs installed over a relatively large, shallow, free-flowing water surface, is commonly referred to as a Kelly-Purdy unit (Figure 1). A second type of unit consists of a UV tube encased in a quartz tube. Water is circulated parallel to the tube and within a jacket of stainless steel or polyvinyl chloride (PVC) (Figure 2) The third type encloses the water in fluorocarbon polymer (Teflon) tubes, which are adjacent to the UV bulbs (Figure 3). Each of these types of reactor units has specific maintenance requirements for maximum UV transmission, because the ability of UV to penetrate into the water is directly related to its ability to disinfect These designs may also require specific operation temperatures and warm-up times.

Fluorocarbon polymer and quartz encasements are used because of their high transmittance properties. Quartz sleeves allow 90-95% UV transmittance, whereas fluorocarbon polymer tubes allow 70-85% transmittance of the available UV (Harris et al., 1987) Most commercial UV units provide the minimum required disinfection at 60% operating efficiency of a UV bulb. The National Shellfish Sanitation Program (NSSP) suggests discarding bulbs at about 7500 hours of usage A policy statement issued April 1, 1966, by the Division of Environmental Engineering and Food Protection, Public Health Service states that UV radiation at a level of 2537 Angstrom units must be applied at a minimum dosage of 16.0 milliwatt sec/cm^2 at all points throughout the water disinfection chamber. This policy was issued for drinking water treatment aboard ships. Higher dosage rates may be required for disinfection of waters used in depuration plants.

TABLE 1. Comparison of three water disinfection systems

Operation/condition	Ultraviolet Light	Chlorine/ chlorine compound	Ozone
Capital costs	Low	Medium	High
Operating costs	Lowest	Low	High
Installation	Simple	Complex	Complex
Ease of maintenance	Easy	Moderate	Difficult
Cost of maintenance	Low	Medium	High
Performance	Excellent	Possible growth	Unreliable
Source water clarity	High	Low	Medium
Virucidal effect	Good	Poor	Good
Personnel hazards	Medium (eyes, skin)	High	Medium (oxidant)
Toxic chemical	No	Yes	Yes
Residual effect	No	Yes	Some
Effect on water	None	Trialomethanes	Toxic by-products
Operating problems	Low	Medium	High
Contact time (min.)	1-5 sec.	30-60 min.	10-20 min.
Effect on shellfish	None	Irritant	Oxidant

From Zinnbauer, Pharmaceutical Engineering, March-April, 1985.

Water characteristics that affect UV systems are bacteriological load, UV absorbency, and suspended solids. The quality of the source water used for depuration must meet or exceed the bacteriological requirements of a restricted area. That is, a median of samples collected cannot exceed 88 fecal coliforms (FC)/100 ml and not more than 10% of the samples may exceed 260/100ml by the 5-tube most probable number (MPN) test. The water must also be free of direct discharges of pollution.

All waters absorb different amounts of UV light. Transmission of UV through water is an inverse function of its mineral and organic content. Dissolved materials such as manganese, organic compounds, and iron cause most of the variation. Block (1977) reported that, in practice, the variation of degree of absorption was due almost entirely to dissolved iron salts. Most UV equipment specifications require the water to absorb no more than 50% of UV at a 1 cm depth

Common UV systems are not effective for treating raw oyster supplies such as rivers or estuarine waters that contain high levels of physical and organic compounds usually in the form of suspended solids. Total suspended solids should not exceed 20mg/L (turbidity units) Not only

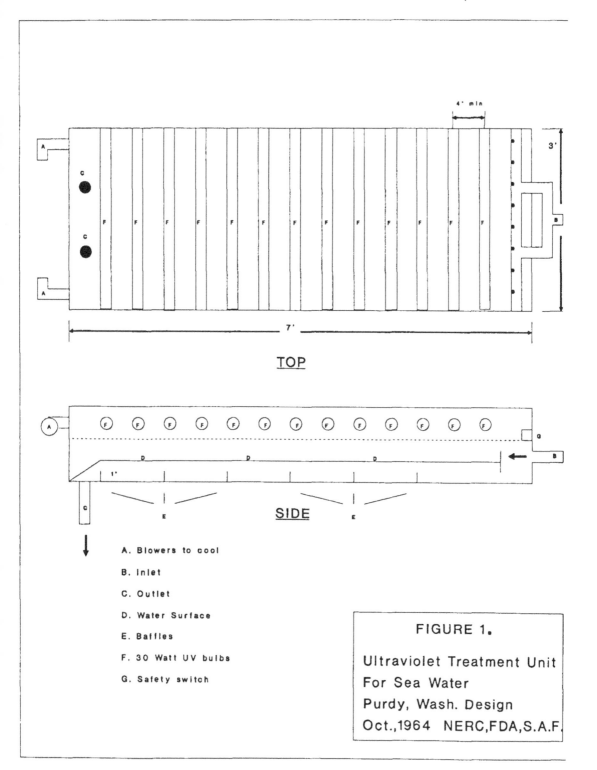

TOP

SIDE

A. Blowers to cool

B. Inlet

C. Outlet

D. Water Surface

E. Baffles

F. 30 Watt UV bulbs

G. Safety switch

FIGURE 1.

Ultraviolet Treatment Unit
For Sea Water
Purdy, Wash. Design
Oct.,1964 NERC,FDA,S.A.F.

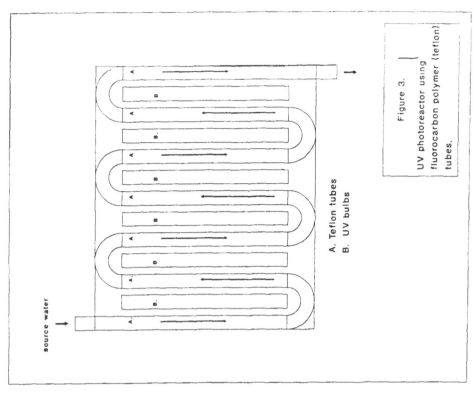

A. Teflon tubes
B. UV bulbs

Figure 3.

UV photoreactor using
fluorocarbon polymer (teflon)
tubes.

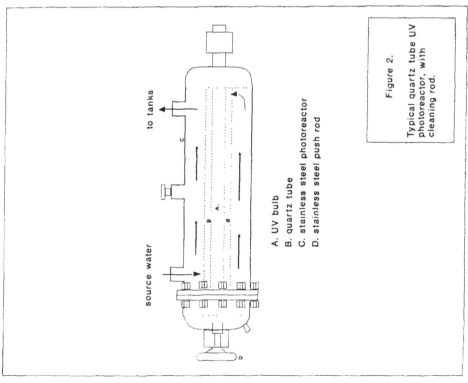

A. UV bulb
B. quartz tube
C. stainless steel photoreactor
D. stainless steel push rod

Figure 2.

Typical quartz tube UV
photoreactor, with
cleaning rod.

do these suspended solids and dissolved materials directly inhibit the effectiveness of UV, their subsequent deposition on the water conveyance systems, quartz sleeves, and Teflon tubes physically prevents UV transmittance.

The NSSP Manual of Operations (NSSP, 1990a,b) sets forth the following operational requirements:

- There are no detectable coliform organisms as measured by the standard 5-tube MPN test for drinking water or a test of equivalent sensitivity in the tank effluent ("influent").
- The installed water treatment system shall provide adequate quantity and quality of water for operating the controlled purification process.
- The treatment does not leave residues that may interfere with the process.
- The quality of the water prior to final disinfection shall meet the requirements of restricted or approved growing areas.
- Turbidity does not exceed a value that will inhibit normal physiological activity and/or would interfere with process water disinfection. UV units must be cleaned and serviced frequently to ensure effective treatment of process water.

As an operational concern, the need to monitor the dose rate of the UV is paramount. However, because of the unavailability of accurate in-line monitors and questionable results from improper placement of monitors, the dose rate of the treatment unit cannot always be effectively monitored. Effective in-place recording UV monitors would be beneficial to the depuration plant. Operators and manufacturers should be urged to design and fabricate adequate monitors

OPERATION AND MAINTENANCE CONSIDERATIONS

General considerations of UV disinfection unit installation and operation in a shellfish depuration plant should include the following:

- Signs should be posted in conspicuous locations, stating "ULTRAVIOLET LIGHT - DANGER TO EYES - DO NOT LOOK AT BULBS WITHOUT EYE PROTECTION." Exposure of skin and eyes to UV radiation can cause damage that may require hospitalization.
- An automatic shut-off switch must be installed on all units to disconnect the circuitry when reactors are opened. There is danger not only of UV radiation exposure but also of electrical shock. All units must be properly grounded, and extreme care must be taken when these reactors are opened or serviced.
- All parts of the UV reactor should be corrosion-resistant and able to withstand a salt water environment.
- To measure bulb life, in-line metering clocks should be installed and records kept of each bulb's hour usage. Bulbs should be replaced after at least every 7500 hours of usage. However, these bulbs do not necessarily last 7500 hours. In fact, some bulbs have blown within one day of installation. Each bulb, therefore, should be checked for intensity at least monthly. When the intensity falls below 60% of initial intensity, the bulb should be replaced.
- Routine cleaning and sanitizing must be conducted on a predetermined schedule, and this must be specified in the scheduled control purification process In addition to detergents and sanitizers, descaling solutions, citric acid, or sodium hydrosulfite should be used to remove any build-up of inorganic salts.
- The photoreactor must be designed to ensure that each part of fluid passing through the unit will receive the required UV dosage.

Recent work with the three basic types of systems used in the U.S. offer specific operational and maintenance problems and advantages. The Kelly-Purdy (KP) unit is easy to maintain because the top portion of the reactor is hinged, allowing easy access for cleaning UV bulbs and washing and sanitizing the unit. Baffles on the water carriage side along the bottom provide turbulence and thus prevent any possible short-circuiting or bypassing of UV radiation. No back pressure is created because of the gravity flow of water through the unit. If, however, the water contact region is not cleaned and sanitized at least weekly, algae and inorganic substances may build up, decreasing water volume and increasing flow, thereby preventing adequate resident time within the photoreactor. The UV bulbs in a KP unit are not protected against salt water intrusion as in the other units and must be removed from the unit for proper cleaning. An automatic electrical shut-off is needed when the top portion of the unit is raised to prevent eye and skin damage by the UV radiation.

The commercial quartz unit provides residence time of the process water within a cylinder adjacent to the quartz sleeve which houses the UV bulb. The water within the outer tube is mixed radially or axially to ensure homogeneous radiation. Removal of the UV bulb and the quartz sleeve should be easy and should not require any special tools. These units usually contain a cleaning system consisting of gaskets that are around the quartz tube and are connected by a stainless steel rod which can be pushed back and forth from the outside of the reactor. Because gaskets wear out and most have design flaws, the gasket rings should be designed as a split-ring to eliminate buckling and deformation. A deformed gasket can cause streaks of inorganic salts to remain on the sleeve that prevent UV transmittance.

Some stainless steel quartz units have been prone to rust, causing iron salts to dissolve into the process water and absorb UV energy. To eliminate this problem the units were returned to the manufacturer for coating with Teflon. In subsequent use the Teflon coating flaked off and the units continued to rust. Another problem with the quartz units was that incomplete drainage of the reactor tube allowed growth of algae which was subsequently resuspended upon restart. To solve this problem, the drain cock was relocated and the unit was slightly tilted for complete drainage.

Removal of commercial Teflon units for cleaning can be difficult. The tubes can be accidentally broken while being removed for cleaning and should be replaced with plastic tubes. However, plastic can decrease UV transmittance to less than design specifications. The housing covers do not have automatic cut-offs and the cover is not easily removed; thus cleaning and maintenance of bulbs is more troublesome.

Kreft, et al. (1986) demonstrated that cleaning of polymer tubes dramatically increases UV transmittance. Tube transmittance as low as 5% increased to 72% after cleaning. Similarly, after overnight cleaning with 15% sodium hydrosulfite, the original UV transmittance by quartz tubes (roughly 92%) was restored.

SUMMARY

Estuarine water may be effectively disinfected with UV lighting, provided that the water has been treated to reduce suspended solids and dissolved organic and inorganic compounds. However, regular routine maintenance of UV units, proper design of reactors, and observance of safety precautions are essential for disinfecting and producing a depurated, cleansed product and for maintaining the health of employees. With proper sanitary design and monitoring of UV water treatment units, commercially sterile water may be provided for shellfish.

REFERENCES

Harris, G.D., Adams, V.D., Sorensen, D.L., and Dupont, R.R. 1987. The influence of photoreactivation and water quality on ultraviolet disinfection of secondary municipal wastewater. J. Water Pollut. Control Fed. 59:781-787.

Kreft, P., Scheible, O.K., and Venosa, A. 1986. Hydraulic studies and cleansing evaluations of disinfection units. J. Water Pollut. Control Fed. 58:1129-1137.

National Shellfish Sanitation Program (NSSP). 1990a. Manual of Operations, Part I: Sanitation of Shellfish Growing Areas. Public Health Service, U.S. Food and Drug Administration, Washington, DC.

National Shellfish Sanitation Program (NSSP). 1990b. Manual of Operations, Part II: Sanitation of the Harvesting, Processing and Distribution of Shellfish, Public Health Service, U.S. Food and Drug Administration, Washington, DC.

Shechmeister, I.L. 1977. Sterilization by ultraviolet radiation.   In: Disinfection, sterilization, and preservation. Seymour S. Block, (ed.), pp. 522-541, Lee and Febiger, London.

Somerset, I. 1990. Personal communication. U.S. Food and Drug Administration, Washington, DC

Zinnbauer, Frank E. 1985. UV water disinfection comes of age. Pharm. Eng , March-April, 36-37, 43. Reprinted from Ultrapure Water.

ENHANCING SHELLFISH DEPURATION

Walter J. Blogoslawski

The elimination of potential human pathogenic microbes from the tissues of bivalve mollusks is termed depuration. The process can occur naturally when contaminated shellfish are exposed to purified seawater. Their tissues are cleansed by the passage of water through the filter-feeding mechanism. This process of depuration can also occur through human manipulation of the shellfish environment. The earliest method used to enhance depuration consisted of relaying contaminated bivalves to clean waters and harvesting the cleansed shellfish after an appropriate time interval.

As naturally available shellfish stocks declined due to over-harvest, and areas suitable for natural depuration diminished due to increased coastal pollution, it became expedient to examine alternate methods of depuration. Specialized facilities were constructed which employed methods to ensure the delivery of disinfected seawater to depurated mollusks taken from potentially contaminating waters. These methods of water purification included the use of ultraviolet light, chlorine, and/or ozone. This paper explores each of the various methods of water disinfection for depuration purposes.

RELAYING

Relaying consists of transporting shellfish harvested from contaminated areas to areas, which according to standard tests, are free of microbes (or have acceptable levels of microbes) considered pathogenic to humans. After an appropriate time interval during which the animals are allowed to depurate, usually a minimum of 14 days, they are then harvested for market Though this method is inexpensive and does not require sophisticated equipment or technology, the yield of saleable shellfish is reduced due to breakage and loss incurred during repeated movement and harvesting. A further constraint is weather. Harvesting of shellfish may be delayed by excessive rainfall which may cause closure of approved relaying areas due to potential contamination of the shellfish from the nearshore discharge of untreated sewage from plants having combined storm sewers or plants which are improperly treating or discharging domestic wastes.

Temperature and oxygen are also limiting factors in any depuration system. Minimal temperatures and minimal oxygen content must be maintained which are specific to a species to insure the continuation of pumping for effective depuration (Fitzpatrick and Eble, 1973, VanWinkle et al., 1976; Canzonier, 1971). For example, in relaying in Connecticut State waters, the ambient water temperature must remain at 50°F or greater for the period of the relay or the shellfish cannot be harvested until spring or when water temperatures remain at 50°F.

While relaying has a low initial cost, its drawbacks are a lower yield of marketable product and a less steady supply of shellfish due to the vicissitudes of the environment used for depuration. Moreover, as the world energy shortage grows such treatments are very energy expensive.

ULTRAVIOLET LIGHT

Ultraviolet (UV) light can eliminate microbial pathogens from water by causing irreversible damage to the DNA of the pathogen. The principles of UV sterilization of seawater and the use of UV to depurate bivalves have been described by several authors (i.e., Wood, 196la; Ayres, 1978). Briefly, seawater is exposed to UV light in either a single pass, flow-through system or a recirculating system before being passed through shellfish to be cleansed.

Ultraviolet light was first used in Japan (Tohyama and Yasukawa, 1936; Shelton and Green, 1954) and later in Spain (Hurlburt and Hurlburt, 1975) to sterilize seawater for the depuration of bivalve shellfish. Presently, six depuration plants in the United States are operating which use UV-light treated seawater. Several plants in England and Wales use UV light depuration (Wood, 196lb). Approximately, 360 plants in Australia employ UV disinfection because depuration of shellfish to be used for human consumption is mandatory (Quadri et al., 1976; Souness et al., 1979, Souness, 1989).

Ultraviolet light provides a practical method for fast and efficient seawater disinfection if certain requirements are followed. Ultraviolet light penetration is related directly to water depth and turbidity. Commercial UV apparatus manufacturers limit depth in UV contactors to 1-3 mm of pre-filtered water to achieve 99.9% disinfection (Aquafine Operating Manual, 1979). High flow rates are limited, making large-volume UV contactors very bulky and expensive. Mechanical wipers are also usually installed in continuous treatment systems to keep quartz tubes free of bacterial slime which absorbs the disinfective radiation (Oda and Eng, 1969; Huff et al., 1965). Since UV treatment leaves no residual or antimicrobial compounds, UV-treated seawater cannot retain sterility if stored in large quantities. The size of a depuration facility then becomes limited by the amount of water which can be moved through the UV-treatment system. Investigators have used differential equations to demonstrate that in a closed water system the bacterial count never reaches zero even if the UV unit is operating at 100% efficiency (Spotte and Adams, 1981) This suggests that it would be possible for pathogens to become established in a recirculating system using UV light as the sole means of disinfection. Thus, UV light is a proven method of seawater disinfection for depuration purposes, but should probably be used only in smaller facilities employing single-pass, UV application.

CHLORINE

Chlorine has been the disinfection method of choice for many depuration facilities around the world because of its reliable cleansing action and ease of residual measurement (Carmelia, 1921; Wood, 196lb).

Chlorinated seawater is used in depuration facilities to cleanse the outer shells of bivalves by removing the adhering dirt and fecal material containing microbes pathogenic to humans. An extremely powerful oxidant, chlorine is never used in direct contact with the tissues of shellfish since it would cause the animals to stop pumping (Galtsoff, 1946). The tissues are cleansed of microbial contaminants with seawater that has been treated by filtration or by seawater that has been chlorinated to achieve disinfection and then dechlorinated prior to shellfish tissue contact.

Chlorination of seawater is a successful method to achieve disinfection, but it does have some drawbacks. Dechlorination of seawater depletes the oxygen content of the water, placing an increased stress on the bivalves undergoing the cleansing process and necessitates the aeration of the holding water. Chlorine treatment is costlier in terms of equipment and manpower than some other disinfection methods (Fauvel, 1963). Perhaps one of the most severe restrictions in chlorine use arises from the formation of chemical compounds that result from the addition of this oxidant to marine waters. Most of the halogenated organic compounds formed upon chlorination display complex chemistry (Block et al , 1977). Studies show that chlorine-produced compounds in seawater appear to have no fatal effect on adult bivalves, but elevated concentrations of chlorine in seawater can cause a cessation of pumping activity, thus preventing the complete cleansing of the contaminated shellfish (Hedstrom and Lycke, 1964).

In summary, chlorine is an effective means for disinfection of bivalves shell surfaces and for disinfection of contaminated waters used for depuration facilities. This oxidant does not interfere with the cleansing process if chlorine levels do not exceed the tolerance levels of the species. Depuration stations that use chlorinated seawater must first establish the dose level of chlorine necessary to insure seawater disinfection, and then establish adequate dechlorination procedures prior to exposing the shellfish to the water so that the metabolic activity of the mollusk remains normal.

OZONE

Several studies have shown that ozone gas, the triatomic form of oxygen, is able to disinfect seawater and bivalves exposed to ozonized seawater (Voille, 1929; Salmon et al. 1937). A commercial plant using ozonized seawater was established in France in 1963 (Fauvel, 1963). It was so successful that ozone-treated seawater is now the depuration method employed in all the major shellfish cleansing stations in France (Fauvel et al., 1982).

Ozone increases the dissolved oxygen content of seawater so that the oxygen content of the holding water for the depurating shellfish does not become a limiting factor in the cleansing process. The initial cost of equipment for ozone disinfection of seawater and/or bivalves is usually higher than a comparable station using chlorine. Over time, however, ozone is less expensive because the method does not require the additional purchase of any other chemicals. This is in contrast to chlorine which must be purchased on a continual basis to replenish the supply of that oxidant in a chlorine system (Blogoslawski, 1989) Ozone treatment does not produce a food product that is chewy in consistency nor does it induce a chlorine-like taste or odor; these undesirable characteristics have been noted in chlorine-treated shellfish (Fauvel, 1963).

As for chlorine, ozone is introduced in a contacting chamber separate from the shellfish to be depurated. Removal of chemical residuals, however, is different in the two systems. Residual ozone is removed by degassing, using compressed air which results in a beneficial enrichment in oxygen content of the seawater. Residual chlorine is removed by dechlorination which reduces the oxygen content of the seawater.

Ozone disinfection should not exceed a dissolved concentration of more than 0.5 ppm ozone because hypobromous acid and its by-products may form at that concentration in seawater of 35 ppt (Blogoslawski, et al 1976). If residual ozone or its oxidant by-products (i.e., hypobromous acid) should remain in disinfected seawater during shellfish depuration, the shellfish may cease to pump or could accumulate the chemical by-products. The technology for monitoring oxidants formed above 0.5 ppm dissolved ozone is made difficult by of the presence of interfering substances such as bromide, manganese, and ferric ions.

SUMMARY

Increasing pollution of natural shellfish beds dictates an increasing need for the use of depuration to ensure a product that is safe for the health of human consumers. Facilities that employ a depuration system which guarantees the disinfection of the water and, thus, of the bivalves within an economically acceptable time frame, are increasing in number. This increase will continue as the necessity for, and profitability of, such facilities expands. A depurated shellfish product should allay the fears concerning public health hazards that many potential shellfish consumers presently harbor by offering them a product that is certified to be free of potential human pathogens.

REFERENCES

Aquafine Operating Manual. l979. Model AL-SL-l, Ultraviolet Sterilizer. Aquafine Corporation, Burbank California 9l504.

Ayres, P.A. l978. Shellfish purification in installations using ultraviolet light. Lab-leaflet, MAFF Direct. Fish. Res., Lowestoft, 43.

Block, R.M., Helz, G.R., and Davis, W.P. l977. The fate and effects of chlorine in coastal waters. Ches. Sci., 18:97-101.

Blogoslawski, W.J. l989. Depuration and clam culture. In: Clam Mariculture in North America. ed by J.J. Manzi and M. Castagna,pp. 415-426, Elsevier Science Publishers, Netherlands.

Blogoslawski, W.J., Farrell, L., Garleau, R. and Derrig, P. 1976. Production of oxidants in ozonized seawater. In "Proceedings of Second International Symposium on Ozone Technology," ed. R. Rice, P. Pichet and M. Viscent, pp. 671-679. Ozone Press, Norwalk, CT.

Canzonier, W.J. 1971. Measurement of the oxygen requirements of a large population of Mercenaria. Proceedings of the National Shellfisheries Association, Volume 61, p. 2 (Abstract).

Carmelia, F.A. 1921. Hypochlorite process of oyster purification. Public Health Reports 30(16): 876-883.

Fauvel, Y. 1963. Utilisation de l'ozone comme agent stérilisateur de l'eau de mer pour l'epuration des coquillages. Commission Internationale pour l'Exploration Scientifique de la Mer Mediterranée. Procés-verbaux des Réunions, 17: 701-706.

Fauvel, Y., Pons, G., and Legeron, J.P. 1982. Ozonation de l'eau de mer et epuration des coquillages. Science et peche, Nantes, 320: 1-16.

Fitzpatrick, G. and Eble A.F. 1973. Preliminary investigations of the ecological aspects of re-laying the hard shell clams Mercenaria. Bull. N.J. Acad. Sci., 18(1): 20 (Abstract).

Galtsoff, P.S. 1946. Reaction of oysters to chlorination. U.S. Department of the Interior, Fish and Wildlife Service, Research Report ll, pp. 1-28. U.S. Government Printing Office, Washington, D.C.

Hedstrom, C.E., and E. Lycke. An experimental study on oysters as virus carriers American Journal of Hygiene 79: 134-142.

Huff, C.B., Smith, H.F., Boring, W.D. and Clarke, N.A. 1965. Study of ultraviolet disinfection of water and factors in treatment efficiency. Public Health Reports 80: 695-705.

Hurlburt, C.G., and Hurlburt, S.W. l975. Blue Gold: Mariculture of the edible blue mussel (Mytilus edulis). Mar. Fish. Rev., 37(10): 10-17.

Oda, A., and Eng, P. 1969. Ultraviolet disinfection of potable water supplies. Ontario Water Resources Commission Division of Research, Paper No. 2012. 47 pp

Quadri, R.B., Buckle, K.A., and Edwards, R.A. 1976. Reduction in sewage contamination in Sydney rock oysters. Food Tech. Australia, 28: 4ll-4l6.

Salmon, A., Salmon, J., LeGall, J., and Loir, A. 1937. Projet type d'une station d'epuration des coquillages par l'eau de mer ozonee. Ann. d'Hygiene, 15: 581-584.

Shelton, L.R., and Green, R.S. 1954. Sanitary Aspects of the Shellfish Industry of Japan, U.S. Dept. Health, Education, and Welfare, Washington, DC.

Souness, R. 1989. Personal commuication. Dept. Food Science and Technology, Univ. of New South Wales, Kensington, Australia.

Souness, R., Bowrey, R.G., and Fleet, G.H. 1979. Commercial depuration of the Sydney rock oyster, Crassostrea commercialis. Food Tech. Australia, 31:531-537.

Spotte, S., and Adams, G. 1981. Pathogen reduction in closed aquaculture systems by UV radiation: Fact or artifact? Mar. Ecol. Prog. Ser., 6: 295-298.

Tohyama, Y. and Yasukawa, Y. 1936. Purification of Polluted Oysters Japan Dept of Food Control, Government Inst. for Infectious Diseases.

Van Winkle, W., Feng, S.Y., and Haskin, H.H. 1976. Effect of temperature and salinity on extension of siphon by Mercenaria. J. Fish. Res. Bd. Canada., 33(7): 1540-1546.

Voille, H. 1929. De la sterilisation de l'eau de mer par ozone: applications de cette methode pour le purification des coquillages contamines. Revue d'Hygiene et de Medecine Preventive, 5l: 42-46.

Wood, P.C. 1961a. The Principles of Water Sterilization by Ultra-Violet Light. Fish. Invest., Series II, 23 London.

Wood, P.C. 1961b. The production of clean shellfish. Proc. Royal Soc. Health, Chelmsford, England. pp. 1-11.

ROUTINE TESTS TO MONITOR DEPURATION

John J. Miescier

INTRODUCTION

The National Shellfish Sanitation Program (NSSP) recognizes that controlled purification or depuration of molluscan shellfish is a complex biological process. It also recognizes that effective shellfish purification is dependent upon the control of a wide range of interrelated variables, which include but are not limited to shellfish species, seasonal effects, water temperature, salinity, dissolved oxygen, turbidity, and the bacteriological and chemical quality of the process water. Consequently, the NSSP requires each depuration plant approved by a State Shellfish Control Authority (SSCA) to develop a scheduled controlled purification process (SCPP) to deal with these variables. The SCPP must include criteria for variables which are the result of process verification studies. The operator must meet the limits of these variables while processing shellfish in accordance with good manufacturing practices (FDA, 1986). This chapter discusses NSSP required quality assurance practices, including routine physical, chemical, and microbiological tests, for monitoring critical variables associated with depuration process water and shellfish under operational conditions. Most of this information is directly referenced from the current NSSP Manuals of Operations (NSSP, 1990a, b).

QUALITY ASSURANCE REQUIREMENTS

The NSSP requires the development and maintenance of an adequate routine sampling program to ensure that the depuration plant continues to operate pursuant to the SCPP, particularly to determine if critical environmental variables are suitable for depuration; and that the shellfish produced by the process meet the current end-point criteria set for the shellfish species being depurated. Part II of the NSSP Manual specifies a minimum sampling plan. Unfortunately, because of variations in plant size and design, it may be necessary to increase the frequency of sampling. As a result, each plant may have to develop a sampling program. At a minimum, the sampling and examination program should include the following:

- total coliform densities in the raw source water and in the treated tank water;
- fecal coliform densities in incoming (zero-hour) and final product (end-product) shellfish;
- temperature, turbidity, and salinity values for the depuration process water; and
- dissolved oxygen levels in process water as measured at the shellfish holding tank outlet.

LABORATORY ANALYSES

With respect to certain critical physical, chemical, and microbiological parameters for process water and shellfish, the NSSP requires that samples be "analyzed in a laboratory which has been evaluated and approved pursuant to the requirements of Manual Part I, Section B. Laboratory analyses shall be performed by a state laboratory or a laboratory approved by the SSCA and shall use American Public Health Association (APHA), Association of Official Analytical Chemists (AOAC, 1984), or Interstate Shellfish Sanitation Conference (ISSC) approved methods to assure uniformity and acceptance of results. The laboratory shall be inspected as part of the annual plant certification and reinspected as necessary to assure that standard analytical methods are being applied. Any major changes in the laboratory personnel or facilities shall be followed by a reinspection. Laboratory inspections shall be conducted by state laboratory evaluation officers because of the special technical expertise required.

The Food and Drug Administration (FDA) does not certify private laboratories to conduct NSSP required analyses. Therefore, the State must have an FDA-approved laboratory evaluation officer to certify private laboratories used to support commercial depuration operations. One significant requirement is that "all quality assurance records shall be maintained for at least the preceding two years at both the depuration plants and at the laboratory" (NSSP, 1990b).

DEPURATION PROCESS WATER

To ensure effective water treatment unit operation (e.g. ultraviolet light), safety, and normal physiological activity of each shellfish species, the NSSP requires that untreated and treated process waters be sampled daily and subjected to certain routine physical, chemical, and microbiological tests. The results of such tests are in satisfactory compliance with Manual Section I, Part II when values are as follows (the original source (NSSP, 1990a, b), should be referred for exact wording and interpretation):

TEMPERATURE

Water "temperature is to be a minimum of 10°C (50°F) for oysters and hard clams and 2°C (35°F) for soft clams. Water temperature is to be a maximum of 20°C (68°F) for hard and soft clams and 25°C (77°F) for oysters. Water temperatures outside of these specified ranges can be used where shellfish are adapted to higher or lower temperatures. In such cases, maximum and minimum water temperatures different from the values specified above must be determined during process verification studies. These studies must establish that the alternative temperatures promote normal physiological activity and efficient purification".

Several types of temperature-measuring devices are used in estuarine water (APHA, 1970). The conventional laboratory thermometer may be used for surface sample temperature determinations. Subsurface water temperature may be measured by a conventional thermometer in an insulated sample bottle, a reversing thermometer, or a continuous recorder. Regardless of the type, the reading must be within ± 0.05°C of a National Institute of Standards and Technology (formerly National Bureau of Standards) thermometer (APHA, 1970). A metal dial-type thermometer may also be used.

TURBIDITY

"A water treatment system is to be installed, where necessary, to provide an adequate quantity and quality of water for operating the controlled purification process. The treatment does not leave residues that may interfere with the process. The treatment quality of the water prior to final disinfection shall meet the requirements of restricted or approved growing areas. Turbidity does not exceed a value which will inhibit normal physiological activity and/or would interfere with process water disinfection. The maximum allowable level of turbidity must be established for each plant during process verification studies."

For process water receiving UV disinfection, turbidity must not exceed 20 nephelometric turbidity units (NTU's) measured as described in Standard Methods for the Examination of Water and Wastewater (APHA, 1985; NSSP, 1990b).

FLOW RATE

"The minimum flow rate of process water in each depuration tank shall be one gallon per minute per standard U.S. bushel (bushel) of shellfish. The minimum volume of process water in depuration tanks shall be 8 cubic feet of water per bushel for hard clams and oysters and 5 cubic feet per bushel for soft clams based upon the effective empty tank volume. Deviations from these criteria can be allowed only if process verification studies show that efficient depuration and end product bacteriological criteria are consistently obtained." A flow meter should be used to measure the flow rate of process water in each depuration tank.

CHEMICAL TESTS

Dissolved oxygen (D.O.). "Minimum D.O. at any point in the tank must be capable of sustaining normal physiological activity by the shellfish. The D.O. levels are to be specified as percent saturation for a given salinity and temperature of depuration process water. In no case shall the D.O. be less than 50% of saturation." D.O. levels may be measured either by the titration method or by using an oxygen analyzer (APHA, 1970). The oxygen analyzer involves the combined use of a membrane-surfaced probe and a galvanic cell.

Hydrogen ion concentration (pH). "The pH is to be 7.0-8.4." Determinations of pH may be made by a pH meter based on criteria specified by APHA (1970).

Salinity. "Salinity is to be ± 20% of the median salinity regimes of the harvest area as determined by the SSCA. Salinities outside this range may be used if justified by data obtained during process verification studies." Salinity determinations are based on any of four methods: electrical conductivity, density (hydrometric), silver nitrate titration, and the measurement of refractivity (APHA, 1970).

Ozone. The FDA has recently stated that it has no objection to using ozone in wet storage or depuration as long as certain conditions are met. One of the conditions states that when ozone is used, the process water must receive special handling to limit the formation of oxidation products that are potentially toxic and that oyster-contact waters be closely monitored to ensure that no free ozone is in contact with the oysters (L. R. Lake, FDA Office of Compliance, personal communication to C. Conrad, Louisiana Department of Health and Hospitals, 1989). The basis of the statement is that ozone will not come in direct contact with shellfish and thus there is no chance that any of the ozone will migrate into the shellfish or affect the shellfish in any way. In this situation, the ozone is not a food additive.

The same conclusion cannot be reached, however, about the use of free ozone in direct contact with shellfish. In this situation, the ozone may well become a component, or affect the characteristics of the shellfish and thus be a food additive. Because the use of ozone in contact with shellfish is not presently approved by FDA, the use of free ozone in shellfish-contact waters would render any shellfish held in the waters adulterated. For this use to be approved, a food additive petition or a "generally recognized as safe" (GRAS) affirmation petition, in accordance with the Code of Federal Regulations, title 21 Sections 171.1 or 170.35, respectively, would need to be submitted to FDA.

Methods for determining and measuring levels of free ozone in seawater are currently being evaluated.

Microbiological Tests

Coliform most probable number (MPN) test. "No detectable coliform organisms as measured by the standard five-tube MPN test for drinking water or a test of equivalent sensitivity in the tank effluent." The MPN method is fully described in Standard Methods for the Examination of Water and Wastewater (APHA, 1985).

SHELLFISH SAMPLING

The NSSP Manual requires that "A routine sampling program is developed and applied to each process batch of shellfish to document that a harvest lot has been adequately cleansed and that the end product criteria are met. The sampling program provides for an adequate number of shellfish samples collected from representative locations within the tank and analyzed prior to and after depuration to assess the effectiveness of the process. The minimum sampling scheduled shall be in accordance with [Table 1]."

TABLE 1. Minimum shellfish sampling schedule for bacteriological analysis

No. of harvest areas	Pollution variability[1]	Minimum number of samples	
		Incoming shellfish	Final Product[2]
Single area	Low	Periodic[3] single sample	Single samples of each harvest lot[4] in process batch [5]
Single area	High	One sample from each harvest lot	Duplicate sample of each harvest lot in each process batch.
Multiple areas	Variable	Periodic single sample from each area	At least one duplicate sample from each highly variable area each week in each processing batch. Single samples of each process batch at other times during the same week.

[1]Pollution variability is determined by sanitary survey and water quality data evaluation as part of the SCPP (based on low, high, or variable pollution variability as evidenced by statistical analyses of bacterial indicator levels).
[2]If final product criteria are met before 48 hours, the product may be released after 48 hours of depuration within a finished product sample.
[3]The frequency of sampling is determined by the SSCA based upon the information from the SCPP.
[4]A harvest lot is defined as all shellfish harvested from a particular area at a particular time and delivered to one depuration plant. The designation of areas is left to the SSCA.
[5]A process batch is a quantity of shellfish used to fill each separate depuration unit. A depuration unit is a tank or series of tanks supplied by a single process water system.

The NSSP Manual, Part II, provides that "At the discretion of the SSCA, zero hour samples of unprocessed shellfish and samples of partially processed shellfish may be taken midway through the depuration cycle and analyzed for the end-point indicator. Such sample results may be used to predict whether the shellfish are acceptable for release. This approach can be used if an appropriate statistical analysis, preformed or approved by the SSCA, of the samples collected during the process verification study and other historical sample results demonstrates this is feasible. Appropriate criteria and procedures are established to assure that the shellfish will be adequately cleansed when released. These beginning and mid-process samples must be supplemented with routine end-product samples to enable continuous verification of the reliability of the criteria "

The NSSP Manual, Part II, states that "The continuing evaluation of depuration plant performance and process efficiency is measured on the basis of end-point bacteriological assays of the depurated product using the geometric mean and upper 10% level (i.e., no more than 10% of the samples analyzed can exceed this value) for the species listed in [Table 2]. In determining these values all of the final product data from every group of 10 consecutive process batches is used. For plants processing less than 10 batches in a three-month period, if any one sample exceeds the upper 10% level or if the geometric mean is in excess of the value specified, the plant is not in compliance."

TABLE 2. End-product standards for overall depuration plant performance evaluation

| Species | Fecal coliforms per 100 grams | |
	Geometric mean[1]	Upper 10%
Soft Clam (Mya arenaria)	50	130
Hard Clam (Mercenaria mercenaria)	20	70
Oyster (Crassostrea gigas)	20	70

[1]Each sample having an indeterminate low value will be assigned a value of 10 for computational purposes.

The NSSP Manual, Part II (1988a) also states that "More intensive sampling of specific harvest lots is required by the SSCA when the shellfish are from a new harvest area and the effectiveness of depuration of those shellfish has not been established; and when the shellfish are from a harvest area which has historically exhibited highly variable water and/or shellfish quality, and end-product results are at times marginal or unsatisfactory."

With regard to compliance with end-product microbiological standards for each process batch of shellfish, Part II of the NSSP Manual (1988a) requires that "Shellfish from single process batches are not released to market unless laboratory results confirm that the end-point fecal coliform (F.C.) criteria established by the SSCA and included in the SCPP are met at 24 or 48 hours for each process batch. The number of samples to be analyzed is based upon the information from the SCPP. The criteria [in Table 3] have been established."

FECAL COLIFORM TEST

The fecal coliform content of shellfish may be determined in two ways: the 48-hour standard methods MPN procedure of the APHA (1970) and the 24-hour Elevated Temperature Coliform Plate Count (ETCPC) method of Heffernan and Cabelli (1967). Alternative test procedures such as the 24-hour mFC agar pour plate method and the 48-hour MUG (4-methylumbelliferyl-ß-D-glucuronide) multitube MPN method may be used after AOAC and ISSC approval and acceptance as official NSSP procedures are received (Andrews, 1987; Rippey et al., 1987). Approval is contingent upon the successful performance of an AOAC interlaboratory collaborative study which could take up to 3 years to complete.

INTERPRETATION OF TEST RESULTS

The NSSP recognizes that shellfish associated with the depuration process require an increased level of control relative to shellfish from approved areas because of the increased potential for contamination. Such control can be exerted only through strict adherence to the SCCP, including the routine tests for monitoring process water and shellfish for certain critical physical, chemical, and microbiological parameters. The results of these tests are valuable public health tools for judging the effectiveness of the purification process and the safety of the shellfish which have undergone depuration.

TABLE 3. End-product standards for each process batch of shellfish

No. of samples	Shellfish species	Fecal coliforms per 100 grams		
		Geo. mean not to exceed	One sample may exceed	No sample to exceed
1	SC	--	--	170
	O, HC	--	--	100
2	SC	125	--	170
	O, HC	75	--	100
3	SC	110	--	170
	O, HC	45	--	100
5	SC	50	100	170
	O, HC	20	45	100
10	SC	50	130	170
	O, HC	20	70	100

SC = Soft clam; O = oyster; HC = Hard clam; -, no value

REFERENCES

American Public Health Association (APHA). 1970. Recommended procedures for the examination of sea water and shellfish, 4th ed APHA, Washington, DC.

American Public Health Association (APHA). 1985. Standard methods for the examination of water and wastewater, 16th ed. APHA, Washington, DC. in conjunction with the American Water Works Association and Water Pollution Control Federation.

Andrews, W.H. 1987. Recommendations for preparing test samples for AOAC collaborative studies of microbiological procedures for foods. Journal of the Association of Official Analytical Chemists 70.931-936

Association of Official Analytical Chemists. 1984. Official methods of analysis, 14th ed AOAC, Arlington, VA

FDA. 1986. Current good manufacturing practice in manufacturing, packing, or holding human food. U.S. Food and Drug Administration. Federal Register 51(118):22475-22483, June 19.

Heffernan, W.P. and Cabelli, V.J. 1967. Modified MacConkey agar plate technique for fecal coliform determinations. Shellfish Sanitation Research Center, U.S. Public Health Service, Department of Health, Education, and Welfare, Washington, DC

National Shellfish Sanitation Program (NSSP). 1990a. Manual of Operations, Part I: Sanitation of Shellfish Growing Areas. Public Health Service, U.S. Food and Drug Administration, Washington, DC.

National Shellfish Sanitation Program (NSSP). 1990b. Manual of Operations, Part II: Sanitation of the Harvesting, Processing and Distribution of Shellfish, Public Health Service, U.S. Food and Drug Administration, Washington, DC.

Rippey, S.R., Chandler, L.A., and Watkins, W.D. 1987. Flourometric method for enumeration of Escherichia coli in molluscan shellfish. Journal of Food Protection 40:685-690.

ECONOMIC CONSIDERATIONS FOR CLAM DEPURATION

Raymond J. Rhodes and Kenneth L. Kasweck

INTRODUCTION

This econimic assessment is based on shoreside facilities used for depuration of the hard clams, <u>Mercenaria</u> sp. in Florida during the 1980's in accord with the Comprehensive Shellfish Control Code of the state of Florida (16R-7), Florida Administrative Code. Although depuration techniques may be applicable to oysters, clams and mussels, Florida rules only apply to hard clams.

In addition to summarizing the "typical" costs and returns of Florida's clam depuration plants, this assessment also considers the impact of regulatory costs on the profitability of clam depuration. The intent is not to advocate or criticize current regulatory policies and rules impacting depuration, but to objectively explore private sector costs and returns most pertinent to the United States' heterogeneous shellfish industry.

FLORIDA HARD CLAM DEPURATION PLANTS

The first depuration facility in Florida became operational at Grant in 1982, just prior to the discovery of highly productive clam sets in the Indian River Lagoon. After the high level of harvest reduced hard clam stocks in open waters, five additional depuration facilities were constructed to augment market demand by depurating clams harvested from restricted shellfish harvesting areas The harvest of hard clams has declined substantially since 1985, and only two Florida clam depuration plants remained operational in 1989.

Depuration facilities in Florida use prefiltered recirculated water, disinfected with ultraviolet (UV) lights. Samples of water and shellfish are taken at 0, 24 and 48-hour periods for microbial testing at two private laboratories in the neighboring region. Prior to its sale, depurated product must meet the Interstate Shellfish Sanitation Conference (ISSC) bacteriological standards. Fecal coliforms in water must be less than 2 MPN/100ml; the median fecal coliforms in shellfish meats can not exceed 20 MPN/100 grams; and no sample can be greater than 70 MPN/100 gram.

The depuration facilities that came into existence on the east coast of Florida represent several distinct approaches to plant design. This presented unique problems to the regulatory authorities as each plant had a distinct Scheduled Controlled Purification Process (SCPP). The plants differed in the size, shape and number of tanks. The tank construction materials varied fiberglass, to plywood or cement blocks. All tanks were required to have a non-toxic inner surface. All plants employed stackable plastic trays for use in the depuration tanks.

Florida plants typically started the process with the pumping of a saltwater source through a sand filter and into the plant. The water passed through a bank of UV lights to a mother tank which was connected to a chiller that lead into the processing units by way of PVC piping laid out to provide free circulation as well as easy disassembly for cleaning. In some cases the processing unit was a single tank. In other cases there were two, four, seven or eight interconnected tanks. All tanks were arranged with a slope to aid in efficient draining onto a concrete slab floor. Currently, all depuration facilities have a roof, and one is in an enclosed structure. This enclosure may become a requirement for Florida depuration plants.

ASSESSMENT METHODS

Data used in this analysis was collected from various sources The major sources of information were 1989 interviews of Florida depuration plant operators and/or owners, various reports and documents related to the production economics of depuration, price information provided by equipment vendors, data provided by the Florida Department of Natural Resources, and the authors' direct observations of depuration plant operations during the 1980's

Based upon the above information and professional judgement of the authors, the operating costs and returns for a small (160 bushel) and a medium sized (345 bushels) plant were simulated using a simple, deterministic financial model. This model was developed on a microcomputer spreadsheet software, Lotus 1-2-3 (Version 2.01). Plants with capacities over 400 bushels have operated in Florida, but the analysis of these plants has not been included due to the lack of data.

When performing this economic analysis, several major financial and operating assumptions were made by the authors based on professional judgement and direct observation. The major assumptions included plants only processing once a week over 26 weeks at about 50% of their NSSP rated capacity This reflects the reduction in shellstock available for relaying to depuration facilities in the late 1980's. Other assumptions included a gross margin of about 6 cents per clam between the price paid to fishermen and the plant's wholesale price for depurated clams The small plant would process approximately 2,100 bushels (an estimated 250 clams/bushel) per year The medium sized plant would produce about 4,500 bushels per year. No debt capital (i e loans) were considered to be available to finance construction of plants The cost of land is not included in the analysis.

RESULTS

The total investment costs excluding land for the "small" and "medium" sized plants are about $86,000 and $135,000, respectively (Table 1). In addition, industry respondents tended to underestimate labor costs and/or rely upon recall of early 1980's construction costs. Consequently, these investment costs may significantly underestimate the costs of constructing a modern depuration plant in 1989

TABLE 1. Estimated capital investment (excluding land) in two sizes of Florida in two sizes of Florida hard clam ultraviolet depuration plants, 1989

Plant size in bushels	160	345
Building	$37,590	$ 60,480
Tanks and Piping	16,450	24,000
Coolers and Water Chillers	14,350	23,540
UV System	8,000	12,390
Tank Trays	3,600	7,200
Other Equipment	6,210	8,130
Total Investment·	$86,200	$135,740

For the quantities processed, the estimated total cost per clam processed for a 160 bushel and 345 bushel plant was 5 5 cents/clam and 4.4 cents/clam, respectively (Table 2) Wages and

Table 2.   Estimated Costs and Returns for
Florida Hard Clam Depuration Plants, 1989.

ASSUMPTIONS	"SMALL"	"MEDIUM"	AVERAGE
CAPACITY IN BUSHELS:	160	345	253
TOTAL OPERATING WEEKS:	26	26	26
CAPACITY USED/WEEK:	50%	50%	50%
TOTAL BUSHELS PROCESSED:	2,080	4,485	3,283
Price per Clam:	$0.25	$0.25	$0.25
Wholesale Price/Bushel:	$62.50	$62.50	$62.50
Total Bushels Processed:	2,080	4,485	3,283
NET SALES:	$127,400	$274,706	$201,053
COST OF SALES:			
Hard Clams Purchased	$95,510	$205,944	$150,727
GROSS PROFITS (MARGIN):	$31,890	$68,762	$50,326
VARIABLE COSTS:			
Processing Labor	$4,056	$9,672	$6,864
U.V. Lights	960	2,400	1,680
Utilities	2,100	3,050	2,575
Laboratory Fees @ $125/lot	3,250	6,500	4,875
Handling & Other Expenses	2,600	5,606	4,103
TOTAL VARIABLE COSTS:	$12,966	$27,228	$20,097
FIXED COSTS:			
Salaries	$8,320	$10,400	$9,360
Depreciation	5,285	8,382	6,834
General Maintenance	1,000	1,500	1,250
Miscellaneous Supplies	900	1,800	1,350
TOTAL FIXED COSTS:	$15,505	$22,082	$18,794
TOTAL OPERATING COSTS:	$28,471	$49,310	$38,891
NET INCOME BEFORE TAXES:	$3,419	$19,452	$11,435
ESTIMATED INCOME TAX:	$684	$3,890	$2,287
NET INCOME AFTER TAXES:	$2,735	$15,562	$9,148
NET ANNUAL CASH FLOW:	$8,704	$27,834	$18,269
5-YR. NET PRESENT VALUE*	-$49,576	-$36,899	#N.A.
5-YR. INTERNAL RATE OF RETURN	-19.2%	0.8%	#N.A.

COSTS & RETURNS PER CLAM	"SMALL"	"LARGE"	AVERAGE
GROSS PROFITS (MARGIN)	$0.061	$0.061	$0.061
LABOR COST/CLAM	$0.008	$0.009	$0.008
TOTAL VARIABLE COST/CLAM	$0.025	$0.024	$0.024
TOTAL FIXED COST/CLAM	$0.030	$0.020	$0.023
TOTAL COST/CLAM	$0.055	$0.044	$0.047
NET INCOME AFTER TAXES	$0.005	$0.014	$0.011
NET CASH FLOW/CLAM	$0.017	$0.025	$0.022

* The after tax discount rate is 15%.
# N.A. = Not Applicable

162

salaries represented about 40% of the total costs for both plant groups   Required laboratory services constituted over 10% of the total costs based upon a laboratory fee that is currently about $125 per lot. The private benefits and costs of maintaining and operating a laboratory were not examined in this study. The net income after estimated taxes is only about 0 5 cents/clam for the smaller plant and 1 4 cents/clam for the medium sized plant (Table 2).

IMPACT OF DEPURATION REGULATIONS

The possible effects of regulatory costs on Florida style clam depuration plants were simulated by using the laboratory testing fees as a proxy for regulatory costs (Table 3). If depuration plants in Florida were not required to incur the cost of testing clam lots, as was the case in South Carolina depuration facilities during the 1980's, the total cost of depuration would decline to about 3.8 cents/clam for a 345 bushel plant. In addition, the projected 5-year internal rate of return (IRR) would increase to about 8%. In contrast, a doubling of testing fees would increase processing costs to 5 cents per clam and decrease the IRR to -7.5%, a negative return on the plant owners' investment over a five year period.

TABLE 3.   Sensitivity analysis of lab testing fees on the cost/clam and interal rate of return (5-year) of a Florida hard clam depuration plant (345 bushel capacity)

Lab testing costs	5-yr. irr	cost/clam
$ 0 (No Fee)	8.4%	$0.038
$125 (Base Case)	0.8%	$0.044
$200	-4 1%	$0 048
$250	-7.5%	$0.050

Current laboratory fees are a fixed fee per lot regardless of the size of the lot Consequently, the financial impact of these fees might be mitigated by increasing the volume of clams processed per lot. This fee structure could favor a plant with larger processing units (e.g 800-bushel capacity), assuming other factors are the same. Whether current or future market conditions will allow depuration plant owners to "pass on" incremental increases in regulatory related depuration costs to harvesters and/or buyers is unknown.

CONCLUSIONS

Based upon the hard clam depuration experience in Florida, the estimated initial investment cost in a plant (excluding land cost) ranges from $86,200 to $135,740 for a "small" (160-bushel) to "medium" (345-bushel) sized facility. Based upon selected assumptions, the estimated total cost per clam processed for a 160 bushel and a 345 bushel rated plant are 5 5 cents/clam and 4.4 cents/clam, respectively. In addition, the profitability of these clam depuration plants displays a significant sensitivity to changes in regulatory costs (i.e laboratory testing fees) If future regulations increase variable costs, the financial survival of U.S. hard clam depuration plants is questionable based upon assumptions used in this paper

ECONOMIC CONSIDERATIONS FOR OYSTER DEPURATION

Kenneth J. Roberts, John E. Supan and Charles Adams

INTRODUCTION

The cleansing of oysters via relaying and depuration procedures is a costly means of marketing oysters in the United States. In the absence of broodstock maintenance, feeding, disease control, etc., the procedures could be viewed as processing. Mariculture would also be a defensible viewpoint. Depuration involves securing, maintaining, and producing living organisms for market. The apparent semantic distinction can be significant beyond academic discussions. For example, financing through public programs and access to permits may be involved. Regarding the latter, the state of Louisiana currently has a limit on the number of mariculture permits. Therefore, matter of terminology and classification serves as the initial indication that oyster depuration from an economic perspective may defy generalization. With this caution clearly noted, the following discussion was organized to include both fundamental information and a detailed analysis. The former was represented by a listing and discussion of general characteristics of importance to the economic performance of depuration systems. The latter, detailed analysis, aspect describes a specific system and related proforma financial statements.

Suggesting that depuration is a costly means of marketing oysters implies that depuration business planning must include a market focus. This matter merits presentation prior to the detail of financial analysis.

MARKET CONSIDERATIONS

The oyster market situation and condition in the United States indicates a general investment opportunity. Domestic production of eastern oyster, Crassostrea virginica, decreased dramatically between 1985 and 1988 (NMFS, 1989). The harvest decrease was concentrated in the Chesapeake Bay production area. The 1985-88 average Chesapeake oyster landings was half of the 1975-79 average. Production in Gulf of Mexico states and Pacific states did not offset the Chesapeake supply shortages (Table 1). The supply shortfall of eastern oysters could not be relieved by imports or domestic production from other areas. Per capita oyster consumption, therefore, decreased substantially during a period of unprecedented seafood price and consumption increases. A missed market opportunity was evident to those knowledgeable about the magnitude of consumer price increases. The consumer price index for seafood increased to 137 in 1988 from a 1982-84 base equal to 100. This was triple the price index increase of beef and beef and 50 percent larger than the index for poultry, (USDA, 1989.) Per capita consumption of seafood exhibited a total increase of 22 percent from 1982 to 1988. The scenario of supply induced oyster consumption decreases when total seafood consumption gains occurred under increasing prices makes a business plan preparer's task easy. It is, in fact, too simplistic of an analysis. The favorable relationships relate to the demand aspect only.

The feasibility of marketing at a profit depends on both the market conditions and cost of supply at market. Regardless of the depuration business location, a supply shortfall seldom produces a strong enough market to cash flow a business in any producing situation. Production efficiency emerging from wise initial investment in highly regulated facilities and well trained management remains a key element. Depuration of oysters is a significantly more costly means of getting a product to sell. If it were not, there would be numerous oyster depuration businesses making use of oysters from restricted areas.

The domestic supply situation of a given area may be one of a shortage. This must be reviewed from both the implications to market needs not being met and the fact that a chronic supply shortage of oysters from local approved grounds could indicate a continuously sparse supply of oysters in general. A continuous shortage will have implications to cost of raw material supply for a depuration facility. It is this aspect of supply considerations which seldom receives appropriate attention in what can often be depuration euphoria.

CHARACTERISTICS OF ECONOMIC IMPORTANCE

A desirable means of approaching economic considerations is to analyze and report results from a survey of oyster depuration businesses. The current United States oyster depuration industry is small with no prospect that analysts can develop data representing a general situation. The more desirable survey approach was not an alternative for the authors. Thus, the necessary approach is that of focusing on the inherent unique conceptual aspects of depuration which translate to significant cost and revenue considerations. Five categories of relevance were established to include: 1) regulations, 2) raw material supply, 3) operation efficiency, 4) marketing success, and 5) capital acquisition.

REGULATIONS

The impacts of federal and state regulations are comprehensive. System design specifications, raw material procurement procedures, and operating requirements translate to cost impacts at facility construction and during production. Investors should analyze with engineers the federal (NSSP, 1990a,b) and state (LSC, 1989) regulations in order to identify minimal capital expenditures. Will companies be allowed to develop systems which are efficient and reliable from engineering and health standpoints? Or will the approach be one of regulations stipulating a design with no regard for operating cost and technical improvements by the firm? Technical freedom may prevail placing the burden on a firm to prove that the system is capable of repeatedly yielding reliable product. Freedom to demonstrate an alternative approach will require additional capital and result in more uncertainty over initial cash flow.

Including an additional regulatory related cost to the standard harvesting cost is essential. Oyster harvest from restricted growing waters must be strictly controlled to protect public health and encourage business investment. Higher harvesting costs can arise via regulations concerning the presence of supervisory personnel, bonds, licensing and limitation on number of operating days. A depuration firm must also determine how operating regulations could affect efficiency A few noteworthy matters are: allowable depth of shellfish in containers, salinity variation, depuration water temperature, turbidity, conditions under which the system must be shut down temporarily, received/storage temperatures, and aeration.

There are few opportunities for regulations to restrict sales. However, the impacts of the regulations which could be implemented are large. Prominent among the shipping, marketing, and sales regulations from federal or local levels is the prospect of limited sales days, licensing, additional market records, and potential for recall of product.

RAW MATERIAL

The cost of raw material, i.e., oysters, to put into the system will be a major expense The average price paid may reflect a mix of oysters from both approved and restricted grounds. Use of approved oysters may be beneficial for a few years. Such initial use of approved oysters would permit management to get experience with the system and gain the added confidence of buyers later attempting to purchase shellstock from restricted grounds. Use of shellstock from approved grounds may also be justified on the basis of improving the acceptability of oysters to occasional consumers, provision of oysters with cleaner shells, and perhaps more uniform saltiness. The input cost differential between oysters from approved or restricted grounds can be large, up to 50 percent lower without inclusion of regulatory costs of harvesting. However, the cost difference may narrow as use of oysters from restricted grounds gains acceptance and the resource becomes depleted. The prospects of oyster leaseholders obtaining oysters at a lower cost from firm-owned grounds may initially appear to assure a least-cost operation. It must be emphasized that in such a situation the oyster input cost should reflect more than the cost of company harvest because the firm could have sold the oysters on the open market for shellstock prices. Profitability must be determined on the basis of the facility making oysters from approved waters a more valuable product and/or making something salable which normally is not. In fact, firms that have oyster leases do not necessarily have a competitive advantage in the depuration business. Since depuration systems are basically businesses requiring skills in the fields of hydraulics and microbiology, input cost advantages of existing oyster processing firms can be unimportant if superior depuration engineering and biological talent is not available.

OPERATION EFFICIENCY

The choice between closed recirculating and flow through systems is one which may be site specific. Water supply and circulation systems which repeatedly circulate water may be more costly in terms of initial investment but cheaper on an operating cost basis. Salt costs, aeration and temperature requirements will likely result in closed (recirculating) systems using well water. Facilities on natural waterways may be more costly to operate due to variation in water quality, which may limit the number of operating days.

Lower operating costs and the spreading of overhead costs can be expected up to a point as business scale increases. There is not enough available information or experienced management to identify at which production level per unit costs begin to increase as scale increases. The possible scale savings per unit would likely be in the area of plant management, well water costs, laboratory expense, and marketing cost. It is recommended that an investor plan on developing the facility over a few operating years. Begin the business in a manner which allows for some expansion and retrofitting of original space used. This will allow management to benefit from production and market experience, technological changes common to such a new industry, and evolving regulations.

The mortality rate in the system is obviously important to the performance and eventual economic survival of the company. Experience may be the only teacher on this topic. Prospective investors must, however, incorporate some mortality estimate into the business plan. Mortality in pond aquaculture of catfish is generally below 10 percent. Prospective oyster depuration should plan on a lower mortality in a system with a holding duration of only 48 to 72 hours.

Availability of oysters throughout the year may not be equal. This characteristic of a depuration facility can be a significant problem. The system may have to be designed for a peak period. Operation during off-peak periods may be well below efficient levels. Thus, poor business performance can be expected when near full capacity use can not be maintained year round. Each facility's management will be faced with achieving a duration time in the system equal to the regulatory minimum of 48 hours on a reliable basis. Success in this regard means a firm will have the number of batches or repetitions approximately equal to a maximum number possible for the time period operated each year. This factor is perhaps the key element to minimizing per unit overhead cost due to mortgage debt, insurance, management, etc. Attention to developing a system and management skills required to achieve a maximum number of repetitions may be more important to profitability than trying to identify the optimum scale of a facility.

MARKETING SUCCESS

When little competition exists for a market, premium prices may be achieved At the current time, the prospects are good that a reliable quality boxed oyster (i.e. washed, single oysters

for the half-shell trade sold in a waxed seafood box) will retain a market advantage. As depuration firms increase in number and expand individual production, premium market prices may not be maintained. This is the point at which establishing a marketing program, rather than simply selling (i.e., filling orders), will pay dividends. Maintenance of a premium price over sack shellstock may require expansion of sales to new buyers who value reliability and safety. The basic marketing objective with depuration in the United States is to take a sacked shellstock, improve the oyster in the mind of buyers, and sell it for boxed oyster price high enough to exceed depuration costs. If the authors' premise that depuration is a
costly means of producing a marketable oyster has merit, then marketing plans must be as detailed as production plans.

CAPITAL ACQUISITION

Lenders will have no performance data with which to evaluate funding proposals submitted by prospective depuration companies. The experience and character of the principals organizing the venture and their business plan must serve as the basis for evaluation. The prospective borrowers must expect to provide a large percentage of facility construction and operating capital costs from equity sources. That is, owners should anticipate lenders viewing the loan as especially risky and one requiring down payment at a higher percentage of total capitalization than is the norm. In the United States, various public agency direct loan and loan guarantee programs may keep the owner's equity to as low as 10 percent initially. A realistic prospect is for owners to anticipate committing 25 percent of facility capital needs. In addition, it is evident that depuration businesses, due to uniqueness, will require prolonged planning which results in organizational capital costs prior to loan application.

The exclusive use of company generated funds through stock sales to speculative investors, however, can have two critical impacts. First, the expected rate of return on these funds is 50 to 100 percent higher than from commercial sources. Second, the anticipated pay back period is shorter for venture capitalists than through financial institutions. Investor actions based on 20 percent return and payback in five years can have a substantial impact on cash flow. The speculative capitalization method places an additional $15,000 demand on annual cash flow per $100,000 of capital than a conventional loan of 10 years at 12.5 percent   The source of financing can clearly impact financial feasibility.

OYSTER DEPURATION SYSTEM SPECIFICATION: UNITED STATES

Ultimately, financial feasibility can only be analyzed through specification of a system to handle a certain capacity. The current absence of oyster depuration companies to survey necessitated the development of of a pro forma approach. This was accomplished by proposing a specific system and estimating its financial feasibility. The financial analysis is based on a depuration plant of the following description:

total system capacity: 1,000 bushels (34,020 kg)/week; (75 lbs (34 kg)/bu)

depuration rate: 500 bushels (17,010 kg)/48 hours; 1,000 bu (34,020 kg)/wk; capacity/tank: 50 bushels (1,700 kg)

tank size and number: 10 tanks-75'x 3' x 2' (22.9m x 0.9m x 0.6m) inside dimensions

tank circulation rate: 1.5 gal/min/bu (0.11 liters/min/kg)

system/site specifics: flow-through design utilizing ambient water; saltwater well assisted, daily/morning
                              harvesting schedule

influent treatment: pressure sand filtration; ultraviolet light

effluent treatment: gravity sedimentation; sanitary sewerage

pumps(s): long-column immersed type (low-lift)

water distribution/measurement: gravity-fed subfloor channels with weirs

System Design Criteria. The system was designed based on current federal and state regulations (NSSP, 1990a,b; LSC, 1989), literature reviews, specifically Canzonier (1982, 1984); Furfari (1966); Bond, et al (1979); and Wheaton (1977) and personal interviews and experiences

System Capacity and Depuration Rate. The system capacity and bushel weight were determined by individual consultations with Louisiana oyster industry members. We are using a weight of 75 lbs/bu (34 kg/bu), which is 3.3 lbs (1.49 kg) greater than that found by Bond, et al., (1979). The capacity was based on the needs of Louisiana oyster processors to supply count (box) oysters to interstate half-shell markets.

Only two 48-hour depuration runs per week were planned to allow time for longer depuration periods, if found necessary, based on bacterial and shellfish response parameters This schedule also allows time for required maintenance procedures (i.e., tank cleaning after each 24-hour depuration interval, water distribution system cleansing, etc.), based on current regulations within the National Shellfish Sanitation Program (NSSP, 1988). The most disastrous economic consequence to a depuration plant is a regulatory closure, especially if illness is caused by the consumption of its product. Plant managers may discover that the operational requirements are so onerous that they are either circumvented or the process is abandoned altogether (Canzonier, 1988). Business planners should be very aware of the required and recommended operational procedures and their time management prior to finalizing financial strategies.

Tank Capacity and Size. A one-tenth capacity scale was chosen for each tank unit. Much of the literature recommends a number of small tanks used within a single unit, adding flexibility to production. Eight cubic feet of water per bushel is required by federal regulations (NSSP, 1990a,b), which is achieved with a tank size of 75' x 3' x 2' (22.9 m x 0.9m x 0.6m) providing 450 cubic feet (12.4 cubic meters) of water for a 50 bushel capacity.

The size of the tank is based primarily on publications by W. J. Canzonier (1982, 1984), conformed to a 3' x 2' x 1' container stacked two high and 25 long. The 75' x 3' x 2' tank dimension exceeds the recommended 2:1 minimum length to width ratio and utilizes a most efficient water depth of 2 feet to accommodate two baskets (Furfari, 1966). The tank bottom is sloped one-half inch per foot from inlet to outlet, as recommended for tank drainage (Furfari, 1966). The water flow is directed through the width of the tank (3 feet) rather than the length (75 feet) to insure consistent water exchange rate throughout the tank (Figure 1).

Economic failure will result from inconsistent cleansing, often a result of improper tank design and hydraulics. Such improprieties will reduce water quality parameters in the tank from the influent end to the effluent end, particularly with regards to dissolved oxygen. Such reduced water quality will stress the shellfish, resulting in reduced pumping rates and inconsistent cleansing. Tank hydraulics are important, therefore, not to simply achieve required minimum flow rate of one gal/min/bu, but to provide good water quality (Furfari, 1966) equally throughout the tank of containerized shellfish. Consistency can be achieved by utilizing a short hydraulic axis, directing the water flow through a short geometric axis of the tank and container array (Canzonier, 1982).

Circulation Rate. A circulation rate of 1.5 gal/min/bu is utilized, based on recommendation by Canzonier (1982), which exceeds the required minimum of 1 gal/min/bu (NSSP, 1990a,b). This requirement was derived from previous shellfish oxygen consumption data (Furfari, 1966) An extra

0.5 gal/min/bu will provide a slight increase in this important parameter at a reasonably low additional cost without creating turbulence that would re-suspend shellfish fecal material deposited on the tank bottom.

Site/System Specifics. Since the proposed design utilizes ambient waters, the plant design is very site specific. Economic viability begins with proper site selection for all depuration plant designs. Since sustainable water quality must be available for a flow-through system, a detailed watershed investigation must be conducted prior to site selection. Prior property ownership provides an economic advantage, however, it should be one of the last criteria used for site selection and is not assumed in this analysis. Site selection recommendations and water quality requirements can be found in federal regulations (NSSP, 1990a,b) and other recommended references (Furfari, 1966; Canzonier 1982, 1984: Wheaton, 1977). The state shellfish sanitation program, is another important source of information to evaluate site suitability, particularly growing water classification data, shoreline and sanitary survey results.

Ground water resources can provide a source of suitable saline water for depuration purposes, especially to be used as a backup source when ambient waters are impacted by various environmental changes. Water with a constant salinity, temperature, clarity and low microbial content may be available, depending on the well location. Iron and dissolved organics can be a problem, however, depending on the substrata. The Water Resources Division of the United States Geological Survey can provide "electric logs" for the nearest potential well location (Cake, et al., 1986).

Good oyster physiology is another important factor specific to site selection, and successful cleansing in general. Close proximity to the harvesting area(s) not only helps in obtaining oysters for plant operation, it will reduce the time the oysters are out of the water, resulting in an improved oyster response to the depuration process. Time-temperature abuse (i.e., the exposure of shellfish to high temperatures after harvesting, particularly during the summer months) is not only a common cause of oyster stress and mortality during depuration, but is often the reason for rising microbial values in shellfish (Cook and Ruple, 1989).

Oyster depuration mortality averaged approximately one percent after a three day depuration period during the summer months along the Mississippi Gulf Coast (Bond, et al., 1979). The oysters were harvested generally within 24 hours prior to being placed in the depuration system. Typically, the oysters were placed in the depuration system between 12 and 3 p.m. of each harvest day. Although the oysters were not always collected in a commercial manner, the results show that excessive mortality, recently encountered during commercial depuration attempts in Louisiana, can be avoided by reducing time-temperature abuse. High shellfish mortalities and economic failure will surely result when time-temperature abuse occurs.

Oyster physiology is also affected by spawning and the resulting poor conditioning, predation, disease, high temperatures in shallow water, improper dredging and culling, exposure to urban/industrial pollutants, etc. Most of these effects can be minimized or eliminated with proper site location and planning.

Influent Treatment. The chosen design treats the influent through a commercial pressure sand filtration and ultraviolet light system designed for a minimum flow rate of 750 gpm. A 96 inch sand filter capable of receiving 755 gpm at 15 gpm/sq. ft of media was recommended (personal communication: John Hans), A submerged-type UV unit is capable of treating a flow rate of 1,250 gpm, based on restricted growing water fecal coliform values (88 MPN/100 ml) with a 16 mWs/sq. cm dosage, allowing a theoretical reduction of 99.9 percent for $E.$ coli. (personal communication. Fred Zinnbauer).

Furfari (1966) does not highly recommend influent filtration since microbial life used as food by shellfish also may be filtered out. Bacterial reduction, however, can be accomplished with sand filtration, but sole reliance on filtration for bacterial removal is not recommended. Filtration with UV treatment is acceptable provided this method is shown not to affect shellfish activity and feeding.

The importance of turbidity reduction prior to UV treatment necessitates the utilization of sand filtration. Increased turbidity, whether from plankton, silt or other material, will reduce transmission of UV radiation. Large fluctuations in turbidity can be expected using ambient coastal waters. Sand particle size combined with suitable backwashing of the filter bed can be designed to allow specific size particles to pass through (Wheaton, 1977). The use of sand filtration will minimize the loss in UV effectiveness due to turbidity.

Effluent Treatment. Effluent treatment may not only be a regulatory necessity in some states, but an economic necessity as well. One should be aware of the consequences of contaminating the intake water of the plant itself. Thus, provisions should be taken to minimize such problems with appropriate location, engineering and treatment of effluents (Canzonier, 1984). Since suspended solids are not as critical with effluent as with influent, more economical solids removal techniques can be utilized, such as gravity sedimentation. Designs for settlement basins (Wheaton, 1977) can utilize the solids-generation information by Bond et. al. (1979). The effluent system in the paper is gravity sedimentations.

Pumps. Low-lift axial type pumps were chosen based on experiences with other aquaculture uses. Correct pump selection is important since pumping costs may be a major expense item. Poor pump selection can double or triple pumping cost and can significantly increase maintenance cost and/or risk of pump failure at critical times. Sizing and selection should be based on analysis of the total head encountered in the system and performance curves (Wheaton, 1977).

Water Distribution/Measurement. Water distribution in this design is based on gravity flow within subfloor channels. Utilization of elevated troughs is suggested by Canzonier (1982, 1984), similar to plants in Spain. Water flow is easily monitored. The troughs are easily cleaned and allow precise water flow control utilizing weirs and variable height standpipes. Such a system greatly eliminates the cost of piping, valves and their maintenance. The troughs can be covered with fiberglass grating in high traffic areas at a cost of $10-12/sq. ft. (Lusk, 1989)

The troughs or subfloor channels can be built, along with the tanks, into the concrete floor, their construction costs incorporated into the cost of the plant structure. Wheaton (1977) states that concrete channels usually are rectangular or trapezoidal in cross section, since construction cost for other shapes are high. Impervious linings, such as concrete, increase allowable velocities, which allow smaller channel cross sections to carry the required flow rate. This reduces construction costs and can be economical if sufficient slope is available. Since the tanks, influent and effluent distribution channels are incorporated into the concrete floor of this analysis, the engineering of these must be exact.

Laboratory/Personnel. Onsite laboratory costs for a 500 bushel depuration plant can range from $25,000-$40,000, utilizing 600 sq. ft. of floor space (Kasweck, 1989). Personnel needs would include two laboratory employees if the plant manager is experienced in bacteriological testing. Three laborers would also be needed to help with plant operations and maintenance. Furfari (1966) recommends a minimum labor force of four personnel for a 400-800 bushel plant capacity.

Off-site laboratory costs can range from $7.00-$11.50/coliform analysis, depending on the laboratory and sample delivery.

Sampling costs are higher during initial start-up, due to the establishment of a required scheduled controlled purification process (SCPP), under the supervision of the state shellfish control agency (NSSP, 1990a,b). The SCPP takes into account the critical process variables that may be encountered and requires they be supported by studies showing that the process will consistently produce shellfish meeting end-point criteria. Every process variable is considered, including seasonal effects on the shellfish (i.e., physiological and microbiological), water temperature, salinity, dissolved oxygen, turbidity, sources of shellfish and process water, treatment of the process water, tank design and construction, hydraulics, clearance between shellfish containers, clearance between containers and tank walls, processing time, raw product quality, end-point criteria, process monitoring and general plant sanitation (NSSP, 1990a,b).

Although there are few oyster depuration firms, depuration regulations exist as part of a state's shellfish sanitation program. In Louisiana, the required water quality parameters are similar to federal regulations. However, sampling effort is more specific. The regulations, which are supervised by the Louisiana Department of Health and Hospitals, require three times the bacterial sampling effort on oyster meats from the tanks during the SCPP (LSC, 1989). After the department has reviewed the SCPP results and verified the process is successfully reducing bacterial levels, a routine sampling procedure shall be followed that conforms with the requirements of the latest edition of the NSSP. Louisiana also requires a $5,000 cash performance bond and immediate supervision of harvesting activities by a commissioned municipal, county, state police officer or a bonded security guard.

Shellstock Storage. Treated and untreated shellstock must be stored in a manner which does not compromise sanitary quality. Treated shellfish must be placed in cold storage no higher than 10 C (50 F). Untreated shellfish must be kept separate from treated shellfish to prevent accidental commingling and should be stored so that additional contamination does not occur and physiological activity is not so adversely affected that purification might not occur (NSSP, 1990a,b).

Better cooling methods have been demonstrated (Moody and Bankston, LSU unpublished data), utilizing a fan-induced air circulation system within a walk-in cooler. Containerized shellfish are stacked in a radiator-like fashion and a large fan draws the refrigerated air through the shellstock. With properly sized refrigeration units, this method can quickly reduce shellstock temperature, controlling the bacterial quality and increasing shelf-life and marketability. This method is economically suitable for both treated and untreated shellstock storage, since containerization is already utilized.

Containers. The system design is based on utilizing a 3' x 2' x 1' depuration container made of polyethylene structural foam (Brooks chicken coop, Brooks Products, Inc., Flora, MS). The container, with mesh-type construction, holds a commercial quantity of oysters 3" deep (one 75 lb (34 kg) bushel). The modular design allows for easy stacking. Other possible container types are documented by Furfari (1966)

Plant Construction. Regulations regarding the building's construction affect economics as well. The building's grounds, floors, walls, ceilings, lighting, heating and ventilation, water supply, pest control, plumbing, and sewage disposal must comply with regulations (NSSP, 1990a,b). Such building requirements increase the cost over cheaper, unregulated structures, but provide the necessary sanitary controls for processing molluscan shellfish.

A FINANCIAL EVALUATION

There was no population of depuration companies from which to solicit a sample of relevant business performance. The general economic considerations previously delineated when combined with the detail of a proposed facility served to facilitate a financial evaluation. Economic considerations of a regulatory nature, those associated with raw material input to the system, facility design operating efficiency, marketing components, and capital acquisition were all quantified. Financial evaluations can produce a wide range of results as on levels of key input and output variables are changed. The current status of the of United States oyster depuration technology dictates only general agreement on plausible input and output unit costs and prices. In this situation, the need for explicit delineation of major system costs and prices chosen to initiate the analysis is clear.

The evaluation focused on a system using flow-through technology. AQUADEC (Adams, 1989) was the software used in the evaluation. A waterfront location (0.8 ha) needed to secure water may result in higher initial investment due to land cost. Property cost was estimated at $70,000. An additional higher cost component of locating on a watercourse is the pier and dock. These are needed for water intake and offloading of oyster boats. The alternative was to rely on a deep saltwater well at more available sites away from the water. Non-waterfront land may be less expensive but saltwater wells in Louisiana recently drilled for aquaculture purposes are in the $25,000 to $30,000 range. Dock and pier capital investment used in the evaluation amounted to $28,000. Additional capital investment was grouped in facility and a machinery/equipment categories. The facility inclusive of the main structure, depuration tanks, piping, wiring, grounds preparation, and office represented an estimated investment of $418,330.

Machinery and equipment requirements included two coolers, a lift truck, a microbiological lab, pumps, engines, filter system, containers, grader/washer, and office operations as major items. The capital for machinery and equipment was estimated to be $290,058. Total capital was estimated to be $708,388. Twenty-five percent of capital needs were to be met by non-financial institution sources. That is, owner equity and/or investor contributions would amount to $177,097 of the total investment capital. Additional investment to meet possible site specific regulatory requirements are possible. Failure of the system to operate within regulatory requirements may necessitate costly system retrofitting. Cost recovery of the fixed investment was arranged on depreciable schedules derived from United States tax regulations. The cost recovery schedule reflecting the first full calendar year of operation is presented in Table 1.

TABLE 1. Cost recovery of capital investment in hypothetical 500 bushel oyster depuration system

	PURCHASE PRICE	LIFE (YRS)	COST RECOVERY [1]
Building	319,464	20	15,973
Tanks & piping[2]	79,866	15	7,987
Filter system	7,000	10	1,000
Land improvements	12,000	20	600
Machinery & equipment			
Coolers	105,000	7	15,005
UV system	32,750	10	4,680
Dock & pier	28,000	20	1,400
Truck	26,000	7	3,715
Lab equipment	25,000	10	3,573
Generator	15,000	7	2,144
Lift truck	14,000	10	2,001
Pump & engine	12,000	10	1,715
Office equipment	12,000	7	1,715
Containers	10,308	5	2,062
Washer/culler	10,000	5	2,000

1 The accelerated cost recovery method was not selected for any accounting purposes such as tax minimization.

2 Inclusive of shallow well to assist salt water needs which may occasionally not be met from surface sources.

Raw material considerations of the proposed system to be analyzed are complex and significant. A study of available oyster resources on restricted grounds must be a part of business planning. The essential information includes quantity, oyster size, likely cull rate, proximity to facility and traditional fishing grounds, and stability of supply. Stability of oyster populations could be less than open grounds because of proximity to sources of freshwater runoff. Population variation would result in financial payback delays unless specifically included in planning. Supplementing raw material supply input to the facility could be required in certain situations. A number of factors should be considered regarding source of raw input. Oysters from approved grounds at higher input costs may extract revenue from the business to the point of a financial loss. A supply of oysters from grounds that have not been fished could produce sizes outside of normal market experience. Culling of such oysters results in extra labor expense and higher effective raw material costs. The financial evaluation of the 500 bushel system included a 15 percent cull rate (Nick Vinterella, personal communication). The result is that each dollar of raw material purchase converts to $1.18 when compared to salable product. The cost of oysters to fill the tanks was initially based on $15.53 per sack (1.5 bu per sack). An effective post-cull rate of $18.27 actually resulted. There was no allowance for recovery of revenue from the 15 percent culled. The oysters from restricted waters can not be marketed if culled because of size, broken shell, or other reasons.

The system evaluated was a flow through system using ambient water from an approved water source. This makes the number of available sites small when combined with raw material proximity priorities. The operating efficiency of the flow through system is perhaps as much a function of management as engineering. It is only possible to differentiate between the two via a specific investigation of an operating facility. The system evaluated herein, because of the flow through technology, was limited to non-summer month operations in the Gulf of Mexico region due to high ambient water temperatures. Management in the September-May operating period has a critical role to play in keeping the system operating. That is, interruptions in supply, machinery and equipment malfunction, and retrofitting are among capacity use determinants. The financial evaluation was based on a 90 percent capacity use estimate during the September-May period of each producing year. Other important system efficiency levels which were assumed include 5 percent shell breakage, 5 percent mortality, and two 48 hour repetitions. Shell breakage and mortality during the depuration process when added to the 15 percent cull allowance amount to 75 percent of purchased raw material ultimately generating sales revenue. The limitation of the system to two 48 hour depuration repetitions represents a cautious management approach. Shellstock to be depurated will come from varied grounds under fluctuating inter and intraseasonal conditions. This coupled with the evolving technology nature of the business suggests business planning can not include reliance on tightly scheduled operations Salaries, hourly employee wages, and raw material costs comprise 65 percent of total cash outflow. Oyster cost alone is 44 percent ($407,818) of total cash outflow. Management's skillful purchase of shellstock that keeps the cull rate shellbreakage, and mortality below the assumed levels will obviously yield significant impacts. When these rates are lower and management can operate the system near or above the 90 percent capacity rate, net income increases are likely.

The financial evaluation included use of a 10 percent price premium over prevailing prices. The reference price was a 1989 Louisiana wholesale price for single boxed half-shell oysters. A financial evaluation price of $.205 per oyster is inclusive of the 10 percent premium. Maintenance of some premium is necessary because oyster depuration becomes an expensive means of marketing oysters. It may also be necessary at some point in the operation to utilize shellstock from approved grounds. This means input price of shellstock would be identical to the prices wholesalers currently pay for their product. Depuration in this case will add costs above those experienced by the conventional wholesaler. Consumers must, through market intermediaries, indicate a preference for the assured quality (safety) by supporting a premium price Sales from the facility at the prevailing market price of $.185 per oyster results in a gross revenue decrease and net income becoming negative (Table 2). Tables 2-4 concisely portray the evaluation in the first and second years of operation. The tables are income statements and balance sheets. The assumptions regarding the facility, raw material (shellstock), operating efficiency, marketing, and capital acquisition have been clearly stated. The first year of operation includes facility construction and trial runs with depurated oysters produced from mid-September through December. This was depicted as 1990 with 1991 being the first full year of operation. The impact of a $3 per sack decrease to $12.50 raw material cost is indicated in the net income entry of the income statement. A failure to attain the 10 percent price premium resulted in a major decrease in net income.

TABLE 2. Income statement for a hypothetical, 500 bushel oyster depuration statement.

	1990	1991
Gross Revenue	$454,134	$1,036,832
Operating Expense	387,130	831,588
Interest Expense	60,717	62,099
Gross Margin	6,286	143,145
Cost Recovery	32,772	65,568
Net Income	(26,486)	77,577
* $12.50/sk.	9,061	157,145
** $.185 per oyster	(71,897)	(26,106)

*   Effect on net income of oysters costing $12.50 per sack instead of $15.53 per sack.
**   Effect on net income of oyster sales price not achieving 10 percent premium.

TABLE 3. Balance sheet for a hypothetical 500 bushel oyster depuration facility, December 31, 1990

Currect Assets		Current Liabilities	
Cash	$5,000	Operating Ins.	$ 13,906
Total current assets	5,000	accrd. int.	162
		interm. Ins.(due 12 mths)	21,056
Interm. & fixed assets		long-term Ins.(due 12 mths)	8,104
Mach. & equipment	270,066	total	43,228
Land	70,000	interm. & long-term	
Bldg. & improv.	405,550	interm. loans	180,850
Total	745,616	long-term loans	306,484
All assets	750,616	all liabilities	530,562
		net worth	220,054
		net worth + liab.	750,616

TABLE 4. Balance Sheet for a Hypothetical 500 Bushel Oyster Depuration Facility, December 31, 1990

Current Assets		Current Liabilities	
Cash	$104,916	interm Ins.(due 12 mths)	$ 25,758
Total	104,916	long-term Ins.(due 12 mths)	9,914
		Total	35,672
Interm. & fixed assets		interm & long-term	
Mach. & equipment	230,059	interm Ins.	155,092
Land	70,000	long-term Ins.	288,465
Bldg. & improv.	379,990	All liabilities	479,229
Total	680,049	Net worth	305,736
All assets	784,965	Net worth + liab.	784,965

An overlooked aspect in most financial evaluations is the origin of the capital investment. The pertinent point was stated previously as one of borrowed funds from financial institutions versus funds from equity investors. A financial institution was the assumed source of 75 percent of capital investment in the evaluation. Use of private investor capital generally involves rates 50 to 100 percent higher and payback periods one-half of that used by financial institutions. Net income would be reduced approximately $80,000 if private investor terms had to be met in this evaluation.

State and municipal business incentives may exist for new investment that results in employment. There was no means by which to include specific information on this matter. Property tax exemptions, job training programs, low interest loans, and tax credits for employment are among the possibilities. The exemptions can apply for several years while the job tax credits may expire within a year. Each program influences the cash flows which are critical in the initial year of a business.

SUMMMARY

A detailed discussion of the evaluation was not warranted. The authors' focus was on targeted economic considerations and their impacts. The evaluation was prepared to indicate the need to carefully quantify the various considerations.

It must be noted that the cull, mortality, and breakage rates experienced by depuration companies can effectively increase raw material costs to undesirable levels. Planning and management must focus on this matter because raw material is 44 percent of total cash costs. Uncertainty over the sustainable supply of oysters on restricted grounds and the efficient clustering of grounds will preclude the oyster depuration industry from including a large number of companies. It must be recalled that restricted grounds are generally not the recipient of public agency management and husbandry. Yields may in fact be less reliable with concomitant problems created for companies.

The general economic characteristics were proposed conceptually and then quantified in a financial evaluation. Monthly cash flows for 1991 (Table 5) indicated a noteworthy characteristic of an oyster depuration business. The first complete calendar year of operation was 1991. Recalling the documented prices, costs, and efficiency factors of the financial evaluation, it should be evident that rapid positive cash flow can occur quickly. The convention is an expectation of positive year-end flows occurring in years three of four of aquaculture operations. Depuration businesses do not involve time delays associated with broodstock establishment, hatchery operation, feed expense, and prolonged periods of fish growth. Quick turnover of the major cash expense, oysters to be cleansed, offers prospective owners an enticing business opportunity. Whether or not potentials are realized depends on owner and management familiarity with the key economic considerations.

Table 5. Monthly cash flow for a 500 bushel oyster depuration facility, 1991

	TOTAL	JAN	FEB	MAR	APR	MAY	JUN
				1000'S			
Beginning cash	5	5	24	39	81	125	98
Sale	1,039	129	129	129	129	68	0
Cash inflow	1,044	134	153	168	210	193	98
Operating expenses							
Oysters	411	51	51	51	51	27	0
Salaries	108	9	9	9	9	9	9
Hired labor	72	8	8	8	8	8	0
All other	243	20	38	11	9	43	4
Total	834	88	106	79	77	87	13
Loan repayment	96	8	8	8	8	8	
Total cash outflow	930	96	114	87	85	95	21
Operating loan							
Payments	14	14	0	0	0	0	0
Ending cash							
Balance	100	24	39	81	125	98	77

TABLE 5 (Con't). Monthly cash flow for a 500 bushel oyster depuration facility, 1991

	JUL	AUG	SEP	OCT	NOV	DEC
			1000's			
Beginning cash	77	56	9	16	45	60
Sales	0	0	68	129	129	129
Cash inflow	77	56	77	145	174	189
Operating expenses						
Oysters	0	0	27	51	51	51
Salaries	9	9	9	9	9	9
Hired labor	0	0	8	8	8	8
All other	4	30	9	24	38	13
Total	13	39	53	92	106	81
Loan repayment	8	8	8	8	8	8
Total cash outflow	21	47	61	100	114	89
Operating loan						
Payments	0	0	0	0	0	0
Ending cash						
Balance	56	9	16	45	60	100

REFERENCES

Adams, C. M., Alderman, R. 1989. "AQUADEC" Aquacultural Decision Support, Budgeting, Financial Analysis Tools. Florida Cooperative Extension Service, Institute of Food and Agricultural Sciences, University of Florida, Gainesville, Florida, Circular 843.

Banoub, S., Blake, N.J. and Rodrick, G.E. 1986. An approach to a cost-analysis of shellfish depuration. Excerpted from a draft final report to DOC/NOAA, award no. NA 83-GA-H-00007. University of South Florida. pp. 34.

Bond, M. T., Truax, D.D. Cake, E.W. and Cook, D.W. 1979. Oyster depuration facility - Engineering assessments. Miss-Ala. Sea Grant Consort. Publ. No. MASGP-78-038 pp. 85.

Cake, E.W., Demoran, W.J., and Cook, D.W. 1986. Environmental, legal and management aspects of proposed oyster depuration facility. Miss-Ala. Sea Grant Consort. Publ No. MASGP-85-023 pp. 66.

Canzonier, W.J. 1982. Depuration of bivalve molluscs- What it can and cannot accomplish and some practical aspects of plant design and operation. Intl. Seminar Management of Shellfish Resources. Tralee, Ireland, pp. 27.

Canzonier, W.J. 1984. Technical aspects of bivalve depuration plant operation: pipes, pumps and petri plates. In: Mussel Bound, A. J. O'Sullivan (ed.), Proc. Intl. Shellfish Seminar. Bantry, Ireland. pp. 68-96.

Canzonier, W.J. 1988. Public health component of bivalve shellfish production and marketing. J. Shellfish Research 7(2) 261-266.

Cook, D W and Ruple, A.D 1989. Indicator bacteria and Vibrionacea multiplication in post-harvest shellstock oysters. J. of Food Protection, 52(5) 343-349.

Furfari, S.A. 1966. Depuration plant design. U. S. Public Health Serv. Publ. No. 999-FP-6 pp. 119.
Hans, J. 1989. Personal communication. Gulf Pool Equipment Co., San Antonio, TX.

Kasweck, K. 1989. Personal communication. Shellfish Testing Service, Grant, FL.

Louisiana Sanitary Code, (LSU). 1989. Chapter IX. Proposed rule for shellfish depuration and/or onshore wet storage. Dept. of Health and Hospitals. Seafood Sanitation Branch, New Orleans, La

Lusk, T 1989 personal communication. Chemgate Corp., Woodinville, WA.

National Marine Fisheries Service. 1989. Fisheries of the United States, Washington, D. C., pp. 116.

Venterella, N. 1989. Personal communication. Venti Oyster Company, Amite, Louisiana.

National Shellfish Sanitation Program (NSSP). 1990a. Manual of Operations, Part I: Sanitation of Shellfish Growing Areas. Public Health Service, U.S. Food and Drug Administration, Washington, DC.

National Shellfish Sanitation Program (NSSP). 1990b. Manual of Operations, Part II· Sanitation of the Harvesting, Processing and Distribution of Shellfish, Public Health Service, U.S. Food and Drug Administration, Washington, DC.

Wheaton, F. W., 1977. Aquacultural engineering. John Wiley & Sons, New York, NY. pp. 708.

Williams, D.C., Jr., Etzold, O.J. and Nissan, E. 1980. Oyster depuration facility - economic assessment. Mississippi - Alabama Sea Grant Program, MASGP-79-011, pp. 28.

Zinnbauer, F. 1989 Personal communication. Aquionics, Inc., Erlanger, KY.

Related Issues and Developments

BACTERIAL EVALUATION OF A COMMERCIAL CONTROLLED PURIFICATION PLANT IN MAINE

Stephen H. Jones, Thomas L. Howell and Kathleen O'Neill

INTRODUCTION

Discharge of inadequately treated sewage from wastewater treatment plants and other sources to coastal U.S. waters has resulted in the contamination of vast areas of shellfish beds with potentially hazardous microorganisms. Consumption of raw or undercooked shellfish that are contaminated with microbial pathogens is a relatively common source of food poisoning. Guidelines have been determined for contamination levels for the protection of public health, and regulation of shellfish harvesting is based in sanitary surveys and routine monitoring of overlying waters. This process has proved to be less than completely effective, as outbreaks of shellfish-related illnesses occur frequently and with regulatory (Richards, 1985).

Shellfish harvesting from moderately polluted areas is allowed only when shellfish undergo controlled purification (CP), or depuration, to remove potentially pathogenic microorganisms prior to marketing or consumption. The final product can be sealed and marketed to give distributors and ultimately the consumer confidence that the oysters have been certified as safe for consumption, similar to the pasteurization programs instituted for milk years ago. The effectiveness of CP facilities for removing a variety of microorganisms has been demonstrated in a number of studies (Kelly and Dinuzzo, 1985; Metcalf et al., 1979; Son and Fleet, 1989; Timoney and Abston, 1984). Other studies (Canzonier, 1971; Power and Collins, 1989; Rowse and Fleet, 1984) have shown that, under certain conditions, depuration does not effectively remove microbial contaminants. Differences in conclusions between studies could result from a number of factors, including differences in design of CP facilities; the scale of the test plant, i.e., ranging from benchtop, aquarium-size systems to operating commercial facilities; the microorganism(s) being studies; whether shellfish are seeded or naturally contaminated; the physiological condition of the shellfish; and the level of initial microbial contamination.

The purpose of this study was to evaluate the effectiveness of an operating, commercial CP facility in Maine for depurating coliform bacteria, vibrios, and specific bacterial pathogens from oysters (Crassostrea virginica), mussels (Mytilus edulis), quahogs (Mercenaria mercenaria), and soft-shell clams (Mya arenaria) harvested from restricted Maine waters.

MATERIALS AND METHODS

Oysters and mussels were harvested from conditionally-restricted waters of the Piscataqua River in southern Maine and quahogs and softshell clams were harvested from restricted areas in the Samariscotta River in mid-coastal Maine. All shellfish were processed at the Spinney Creek Oyster Company (SCOC) CP facility located on Spinney Creek in Eliot, Maine SCOC is an aquaculture business which farms oysters and littleneck clams, and processes all marketed shellfish through CP. The SCOC is an operating commercial facility consisting of a building that includes a FDA-approved microbiology laboratory, purification tanks and equipment, utilities, storage and cooler areas, and an office. Spinney Creek is a salt pond that provides the seawater used for CP process The seawater is pumped through a buried intake pipe to the CP facility and through a 30"x24" sand filter (Neptune Benson, Inc., W. Warwick, RI) filled with #18 sand to remove particulate matter of >20 μm diameter. Filtered water flows into a 135 gallon reservoir tank (Utilities

Supply, Medford, MA) and is pumped through a 100L UV disinfection unit with Teflon tubing an a 100 gallon per minute capacity. The disinfected seawater then flows through perforated T-shaped pipes that act as passive aeration devices and into six purification tanks (AP-CHEM, Inc., Prince Edwards Island, Canada). The purification tanks are 4'x8'x3' fiberglass tanks coated on inside surfaces with a smooth gel coat. Fitted baffles of PVC sheet stock are placed in the rear of tanks to minimize agitation in the water produced by the water jet aeration devices and to initiate water flow from the bottom of each tank through 6" air stones (Aquatic Ecosystems, Inc. Apopka, FL). The dissolved oxygen in process waters is thus maintained at >70% saturation. Process water drains through overflow pipes into underlying drain pipes pitched to the reservoir tanks where it is pumped backed through the UV disinfection unit and into the tanks.

The SCOC facility has a design capacity of 200 bushels of shellfish per week and a flow rate of 80+ per minute. Maintenance of the required 10 ft^3 of water per bushel and one gallon per minute per bushel flow rate requires a minimum flow rate through the system (six tanks, 10 bushels per tank) of 60 gallon per minute. Harvested shellfish are washed, culled, and placed in trays for purification. The trays are plastic Nestier Chill Trays (Nestier Corp., Cincinnati, OH) that allow adequate water flow through the trays and shellfish. Each tank can contain six trays stacked four high, or 24 trays per tank. Each tray will accommodate approximately 0.5 market bushels of shellfish of ≤3" depth. Purification trays are not used for any purpose other than purification; separate baskets are used for harvest. After each process period, the whole facility is washed down and the purification tanks disinfected and cleans with Chlorox and detergent.

During the purification, shellfish are inspected for verification of functional purification. Visual inspection should reveal siphons from clam species, slightly open shells of oysters and mussels, and accumulation of debris (pseudofeces, sand, mud, etc.) on the bottom tanks. If spawning occurs as it does on occasion in the summertime, the process is stopped, tanks drained, and restarted at 0 hour again. After a 24 hour purification period, the tanks are drained and the tanks and shellfish are sprayed with fresh water to remove foam and debris. End-point microbiological criteria are often met after this initial 24 hour process period. The sand filter is then backwashed and the process begun again for another 24 hour period. After the 48 hour processing time, shellfish are washed, culled, graded, and packaged. Each package is sealed, tagged with a health tag giving the destination, type of shellfish, lot number, plant identification, and date and placed in a cooler for storage prior to shipping. Samples of shellfish processed for 48 hours are tested for fecal coliforms. Lots that meet required standards are released for sale. A detailed description of the facility can be found in a manual produced in conjunction with the New England Fisheries Development Association (NEFDA, Inc., Boston, MA) that includes siting and species considerations, blueprints and specifications, operation manuals, laboratory manual, management considerations, and cost detailing (Howell and Howell, 1989)

Eight to twelve individual shellfish fro freshly-harvested and depurated smaples of each weekly harvested of shellfish were processed for bacterilogical analysis. Shellfish were aseptically shucked and the contents homogenized with equal parts of buffered peptone water Homogenates were then split into equal parts and analyzed at the SCOC laboratory and at the Jackson Estuarine Laboratory (JEL), Durham, New Hampshire.

The method (ETPC) for fecal coliform detection used at SCOC was a 24 h elevated temperature pour-plate method (Greenberg and Hunt, 1985) modified by using mFC agar, and the method (APHA) used for the detection of total and fecal coliforms at JEL was the standard multiple tube fermentation, 5 tube MPN assay (APHA, 1985), carried through confirmed or completed tests in accordance with recommended procedures. Positive EC tubes were considered positive for fecal coliforms, and a portion of these tubes were streaked for colony isolation onto EMB agar. Dark colonies with metallic sheens or other representative colony types were transferred to nutrient agar slants and further analyzed to confirm the presence of Escherichia coli using routine IMViCs procedures, oxidase, and Gram stain tests and fermentation of lactose at 44.5°C.

At JEL, samples were also analyzed for the presence of Salmonella by the method of Hussong et al. (1984), Shigella, by Standard Methods (APHA, 1985), and vibrios, with further analysis of suspect colonies for detecting V. vulnificus and V. parahaemolyticus. Vibrio analysis involved decimal dilution of samples in alkaline peptone water as a three-tube MPN assay. Turbid dilution broth tubes were streaked onto thiosulfate-citrate-bile-sucrose (TCBS) agar and all different resultant colony types transferred to peptone broth containing 0% and 3% NaCl. Isolates that grew in 0% NaCl were further characterized using growth in 6%, 8%,and 10% NaCl, ornithine and lysine decarboxylase, arginine dehydrogenase, and fermentation of cellobiose tests, as well as API 20E identification system.

Fecal coliform (FC) data for the two methods utilized were statistically analyzed using the t test analysis as recommended by the U.S.E.P.A. (Bornder and Winter, 1978) and a paired t-test. Data analyses included separate analysis of data that were less than 3000 FC per 100 g by the APHA method, as preliminary analysis showed a consistent bias in favor of the APHA method for data that were greater than 3000 FC per 100 g.

RESULTS

The results for all the fecal coliform analyses conducted at JEL are summarized in Table 1 Processing of the four different species of shellfish by controlled purification resulted in a consistent and marked decrease in fecal coliforms to levels near or below the detection limits of the APHA method. Figures 1 and 2 show that final FC levels in purified oysters and other shellfish were relatively independent of initial levels, which varied over a wide range during the course of the study. Controlled purification was much less effective in eliminating total coliforms (Figure 3), which in some samples were higher in purified than in freshly harvested oysters.

Table 1. The Effect of Controlled Purification on Fecal Coliform Levels in Maine Shellfish

SHELLFISH SPECIES	NUMBER OF SAMPLES	MEAN CONCENTRATION OF FECAL COLIFORMS* Before Purification	After Purification
All shellfish	27	476±9†	<16
Crassostrea virginica	18	886±7	<21
Mercenaria mercenaria	3	292±6	<7
Mytilus edulis	3	1905±4	11
Mya arenaria	3	188±13	<9

* 90.2% of + EC tubes tested contained *E. coli*
† Mean value and standard deviation of log-transformed density per 100 g.

All shellfish analyzed for coliform bacteria were simultaneously analyzed for Salmonella, Shigella, and vibrios, including V. vulnificus and V. parahaemolyticus. No Salmonella, Shigella, or V. parahaemolyticus were detected in any of the different shellfish (Table 2). No vibrios were detected until June 8, and V. vulnificus were not detected until July 8. Vibrio concentrations were relatively high and, except for the June 8 samples, were apparently unaffected by controlled purification. The evidence for elimination of V. vulnificus, based on one mussel and two oyster samples in July, indicates an inconsistent response to CP. A more extensive study of the response of vibrios and V. vulnificus to CP presented by O'Neill et al., these proceedings

Table 2 The Effect of Controlled Purification on Vibrios and Bacterial Pathogens in Shellfish

DATE	SHELLFISH	SALMONELLA AND SHIGELLA SP 0 Hour	48 Hour	VIBRIO SP.* 0 Hour	48 Hour
		per 100 g		per 100 g	
5/9-6/2	Oysters	<2	<2	<2	<2
6/8	Oysters	<2	<2	>2400000	2400
6/13	Oysters	<2	<2	>2400000	>2400000
6/22	Oysters	<2	<2	43000	150000
6/28	Oysters	<2	<2	460000	210000
7/8	Oysters	<2	<2	1100000	1,100,000/110,000†
	Mussels	<2	<2	1100000	43000
7/14	Oysters	<2	<2	>2,400,000/240,000†	>2,400,000
	Mussels	<2	<2	93,000/24,000†	>2,400,000
7/21	Oysters	<2	<2	150000	1100000
	Mussels	<2	<2	1100000	24000

* nonpathogenic, environmental isolates, no pathogenic species.
† *Vibrio vulnificus*

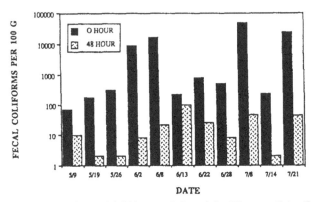

Figure 1. Fecal coliform levels in Maine oysters before and after 48 hour controlled purification.

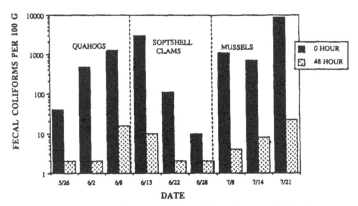

Figure 2. Fecal coliform levels in different shellfish before and after 48 hour controlled purification.

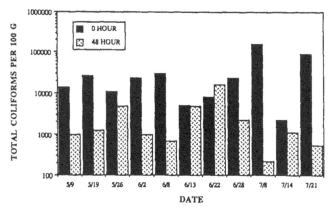

Figure 3. Total coliform levels in Maine oysters before and after 48 hour controlled purification.

Two methods for the detection of fecal coliforms were used for all shellfish samples tested. The APHA method conducted at JEL is the standard, traditional, and currently-accepted method, while the ETPC method conducted at SCOC offers several advantages over the APHA method, especially the shorter time (24 versus 72 hours) required for the assay. A summary of the comparative analyses is presented in Table 3. A t test analysis, as recommended by the U.S. E.P.A., conducted on replicated samples indicated no significant difference between the two methods. A paired t test analysis of all the data showed a significant difference between methods at a .05 significance level. The APHA method always gave significantly higher results than the ETPC method for shellfish samples having greater than 3000 per 100 g. Analysis of data that were less than 3000 FC per 100 g by the APHA method showed no significant difference between the two methods. Furthermore, there was little difference in sensitivities of the two methods: FC were detected in 39 and 38 of 49 samples for the APHA and ETPC methods, respectively. In five samples, FC were detected using the ETPC but not the APHA method, and in seven samples, FC were detected with the APHA but not the ETPC method. There was no significantly difference between the two methods for detection of FC in shellfish that were processed by CP for 48 h.

Table 3. Comparison of the ETPC and APHA Methods for Determining Fecal Coliforms (FC) in Shellfish

	FC DETECTION METHOD	
	APHA	ETPC
Mean density: All samples	476±9†*	313±5*
Mean density: APHA FC<3000/100 G	190±4	186±4
Total samples	49	49
FC positive	39	38
FC negative	10	11
Only method negative	5	7

† Mean value and standard deviation of log-transformed density per 100 g.
* Mean values significantly different at .05 significance level.

DISCUSSION

The process of controlled purification at the Spinney Creek Oyster Company in Eliot, Maine was effective in eliminating fecal coliforms bacteria from four different species of shellfish. An important factor in this study was the operation of the facility, where shellfish were maintained under conditions that minimized stress (low salinities; spawning; low dissolved oxygen) during the CP process. Stressful conditions can dramatically affect the elimination of microorganisms, especially $E.$ coli, from shellfish during controlled purification (Power and Collins, 1989). CP was effective in removing potentially pathogenic microorganism for initial contamination levels ranging from 20 to 50,000 FC per 100 g. However, shellfish with high initial levels of fecal coliforms, indicating elevated levels of fecal contamination, should probably not be processed in order to avoid possible health risks to the consumer.

Fecal coliforms are the indicator group used as the accepted standard of sanitary quality of shellfish. In this study, FC levels were dramatically reduced in shellfish processed by CP, often to below detection limits and always well below the 230 per 100 g retail limit. Total coliforms constitute a broad range of different bacteria, and are poor indicators of fecal contamination (Grimes, 1987). However, the comparative responses of the fecal and total coliforms to CP in this study illustrates the enhanced elimination response of the fecal coliforms in relation to the broad group of total coliform organisms.

The fecal coliform levels in shellfish and water samples from the Piscataqua River were never very high during the time period of this study. Therefore, the level of fecal contamination may not have been high enough to have detectable levels of Salmonella and Shigella present. Vibrios, however, are natural inhabitants of estuarine environment (Oliver et al., 1983). They were apparently either nonculturable or were present in the Piscataqua River at undetectable levels during the spring and summer. V. vulnificus has previously been reported only as far north on the east coast as Boston Harbor (Oliver et al., 1983), and was not detected in this study until July, after water temperatures had risen to relatively high ($\sim$20 C) levels in the river. The absence of V. vulnificus in the river during colder periods is consistent with the lethality of cold stress to V. vulnificus (Oliver, 1981). The lack of response to the process of CP by the vibrios, possibly including V. vulnificus, may be a natural response of autochthonous estuarine bacteria being able to remain associated with other estuarine organism such as bivalve mollusks under the conditions imposed by the CP process.

A rapid method for detecting fecal microbial contaminants in shellfish is of great interest to people in the shellfish industry and those that regulate it (Rippey et al., 1987). Such a method would be extremely helpful for CP facilities, where operators could decide sooner whether a given batch of shellfish are too contaminated to be depurated, minimize holding times for adequately purified shellfish prior to marketing, and decrease losses caused by extended storage. Even though the APHA method gave consistently and significantly higher results for the most highly contaminated shellfish samples in this study, the ETPC method also indicated high levels of contaminated in same samples; shellfish with >3000 FC per 100 g should probably be considered unsafe for consumption, even after CP. More importantly, the ETPC method was not significantly different from the APHA method for detecting FC in low to moderately contaminated shellfish, and its sensitivity was essentially equal to the APHA for detecting low levels of FC. Thus, the results of this study indicate that the ETPC method may be adequate for determining the safety of shellfish by establishing that little or not fecal contamination remains in shellfish processed in a CP facility, as well as being useful for indicating whether shellfish are too contaminated to be processed. More research is needed to determine the cause of the difference between the two methods for more highly contaminated shellfish. The shorter assay time for the ETPC method is an important advantage, and this method warrants consideration as an alternative accepted method for determining FC levels in shellfish, especially for CP facilities.

ACKNOWLEDGEMENT

This study was supported by grants from the New England Fisheries Development Foundation and the National Institute of Health.

REFERENCES

American Public Health Association (APHA). 1985. "Standard Methods for the Examination of Water adn Wastewater." 16th Edition. Amer. Publ. Health Assoc., Washington, DC.

Bordner, R. and Winter, J. (eds.). 1978. "Microbiological Methods for Monitoring the Environment." EPA-600/8-78-017, U.S. E.P.A., Cincinnati, OH.

Canzonier, W.J. 1971. Accumulation and elimination of coliphage S-13 by the hard clam, Mercenaria mercenaria. Appl. Microbiol. 21:1024-1031.

Greenberg, A.E. and Hunt, D.A. (eds.). 1985. "Laboratory Procedures for the Examination of Seawater and Shellfish." 5th Edition. Amer. Publ. Health Assoc., Washington, DC.

Grimes, D.J. 1987. Assessment of Ocean Waste Disposal: Pathogens and Antibiotic-and Heavy Metal-Resistant Bacteria. Final report. Ofice of Technology Assessment, Washington, DC.

Howell, T.L. and Howell, L.R. 1989. The Controlled Purification Manual. New England Fisheries Development Association, Inc., Boston, MA. 77 p.

Hussong, D., Enkiri, N.K., and Burge, W.D. 1984. Modified agar medium for detecting environmental salmonellae by the most-probable-number method. Appl. Envir. Microbiol. 48:1026-1030.

Kelly, M.T., and Dinuzzo, A. 1985. Uptake and clearance of Vibrio vulnificus from Gulf Coast oysters (Crassostrea virginica). Appl. Environ. Microbiol. 50:1548-1549.

Metcalf, T.G., Mullen, B., Eckerson, D., Moulton, E., and Larkin, E.P. 1979. Bioaccumlation and depuration of enteroviruses by the soft-shell clam, Mya arenaria. Appl. Environ. Microbiol. 38:275-282.

Oliver, J.D. 1983. Lethal cold stress of Vibrio vulnificus in oysters. Appl. Environ. Microbiol 41:710-717.

Oliver, J.D., Warner, R.A., and Cleland, D.R. 1983. Distribution of Vibrio vulnificus and other lactose-fermenting vibrios in the marine environment. Appl. Environ. Microbiol. 45:985-998.

Power, U F., and Collins, J.K. 1989. Differential depuration of poliovirus, Escherichia coli, and a coliphage by the common mussel, Mytilus edulis. Appl. Environ. Microbiol. 55:1386-1390.

Richards, G.P. 1985. Outbreaks of shellfish-associated enteric virus illness in the United States: Requiste for development of viral guidelines. J. Food Prot. 48:815-823.

Rippey, S.R., Chandler, L.A., and Watkins, W.D. 1987. Fluorometric method for enumeration of Escherichia coli in molluscan shellfish. J. Food Prot. 50:685-690.

Rowse, A.J. and Fleet, G.H. 1984. Effects of temperature and salinity on elimination of Salmonella charity and Escherichia coli from Sydney rock oysters (Crassostrea commercialis) Appl. Environ. Microbiol. 48:1061-1063.

Son, N.T., Fleet, G.H. 1980. Behavior of pathogenic bacteria in the oyster. Crassostrea commercialis, during depuration, relaying and storage. Appl. Environ. Microbiol. 40:994-1002.

Timoney, J.F. and Abston, A. 1984. Accumulation and elimination of Escherichia coli and Salmonella typhimurium by hard clams in an in vitro system. Appl. Environ. Microbiol 47. 986-988.

OCCURRENCE OF VIBRIO VULNIFICUS IN WATER AND SHELLFISH FROM MAINE AND NEW HAMPSHIRE

Kathleen O'Neill, Stephen H. Jones, Thomas L. Howell and D. Jay Grimes

INTRODUCTION

Vibrio vulnificus, a halophilic, lactose-fermenting vibrio, is the causative agent in two types of illness associated with shellfish or marine environments. Exposure to this organism via sea water or raw shellfish can cause a primary septicemia which can lead to secondary cutaneous lesions. Vibrio vulnificus can also infect a preexisting wound, causing swelling and erythema and sometimes leading to necrosis and septicemia. Increased severity of the disease and mortality is seen in patients with previous liver damage (Blake et al. 1980).

V vulinificus has been isolated from various estuarine environments in the United States It has been recovered from the water column, sediment, animals and plants along the Eastern seaboard from Florida to Massachusetts (Oliver et al. 1983); from the Gulf Coast region (Kelly 1982) and from water, shellfish and sediment along the West Coast (Kaysner et al. 1987).

Although Maine and New Hampshire recreational and commercial shellfishing are north of the northernmost site of isolation previously reported for V. vulnificus (Oliver et al 1983), prevailing concern necessitated a study to determine the incidence of V. vulinificus in depurated and undepurated shellfish from Maine waters, in shellfish from open and prohibited areas in the Great Bay Estuary of New Hampshire and in the water of the Great Bay Estuary.

MATERIAL AND METHODS

Oysters (Crassostrea virginica), quahogs (Mercenaria mercenaria) and mussels (Mytilus edulis) harvested from conditionally restricted Maine waters were obtained from a commercial depuration plant before and after 48 hours of depuration (plant description, Howell and Howell, 1989). Additional oysters were collected from 2 restricted sites and 2 approved sites in New Hampshire. Water samples were also collected from 7 different sites within the Great Bay Estuary of New Hampshire.

Samples were analyzed in the following manner to determine the presence of V. vulnificus. Shellfish were placed under running water, cleaned with a stiff brush and shucked Approximately 100 grams of shellfish meat and liquor from 8 to 12 shellfish were ground with an equal volume of buffered peptone water in a Waring blender for 90 seconds

Alkaline peptone water containing 1% NaCl was used in a most probable number (MPN) procedure for selective enrichment of vibrios in shellfish and water samples. Inoculated MPN tubes were incubated for 12 to 18 hours at 37°C, all tubes showing turbidity were streaked onto Thiosulfate-citrate-bile-sucrose (TCBS) agar plates and incubated at 37°C for 18 to 24 hours. Representative yellow and blue colonies on TCBS plates were inoculated into peptone water containing 0%, 3%, 6%, 8% and 10% NaCl; arginine dehydrogenase (ADH), ornithine decarboxylase (ODC) and lysine decarboxylase (LDC) broths; triple sugar iron agar, and phenol red broth containing cellobiose or salicin. Suspected isolates were also gram stained and tested for the ability to produce oxidase. Oxidase positive, gram negative rods which grew in only 3% and 6% peptone salt broths, which were negative for ADH, positive for LDC and fermented cellobiose or salicin were retained as presumptive <u>Vibrio vulnificus</u>. These isolates were confirmed by the API 20E Identification System and the MPN per 100 g or ml was calculated.

RESULTS

No vibrios were detected in the Maine oysters until early June, at which time they were detected at levels of greater than 10^6 (Figure 1). Total numbers of vibrios persisted at levels between greater than 10^4 to 10^6 through late September when the levels began to decrease. <u>Vibrio vulnificus</u> first appeared in depurated oysters at the end of July and was detected in both depurated and undepurated oysters through mid-September at levels of 10^4 to 10^6. This organism was also detected in undepurated Maine mussels and quahogs at similar levels (Table 1). <u>V. vulnificus</u> was cultured from oysters harvested from 2 closed oyster beds in New Hampshire, but it was not found in oysters harvested from two open beds (Table 2). Water samples taken from 4 out of 5 closed areas in New Hampshire were positive for <u>V. vulnificus</u>; no <u>V. vulnificus</u> was detected in either open area (Table 3).

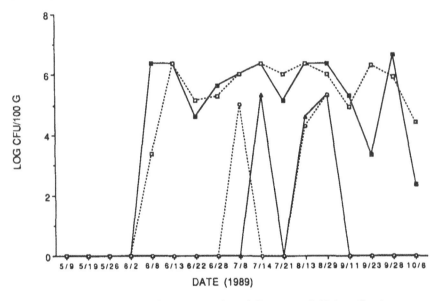

Figure 1. Vibrios in Undepurated and Depurated Maine Oysters

─■─ UNDEP VIBRIO	···□·· DEP VIBRIO	─▲─ UNDEP V VULNIF	···○·· DEP V VULNIF

DISCUSSION

Studies have indicated that <u>Vibrio parahemoylticus</u> and <u>Vibrio vulnificus</u> can be cleared from oysters during depuration (Son and Fleet 1980; Kelly and Dinuzzo 1985). These studies involved inoculating clean oysters in a laboratory setting. Kelly and Dinuzzo (1985) noted that <u>V. vulnificus</u> took longer than two weeks to be cleared from oysters infected by exposure to contaminated seawater while Son and Fleet (1980) found decreased levels of <u>Vibrio parahemolyticus</u> after 48 hours of depuration. The present study examined naturally colonized oysters being depurated in a commercial depuration plant for 48 hours and detected no overall decrease in the numbers of <u>V. vulnificus</u>; in the samples where <u>V. vulnificus</u> was deem in paired undepurated and depurated samples the levels were identical or within one log of each other (Figure 1).

Table 1. Vibrios in Undepurated and Depurated Maine Shellfish

DATE		0 HOUR		48 HOUR	
		TOTAL VIBRIOS /100 g	*VIBRIO VULNIFICUS* /100 g	TOTAL VIBRIOS /100 g	*VIBRIO VULNIFICUS* /100 g
6/2	QUAHOGS	15000	<2	ND	
7/8	MUSSELS	1100000	<2	43000	<2
7/14	MUSSELS	93000	24000	>2400000	<2
7/21	MUSSELS	1100000	<2	24000	<2
8/3	OYSTERS	240000	750	ND	
		2100000	43000	ND	
		240000	<2	ND	
8/17	OYSTERS	2400000	1100000	ND	

ND - 48 hour sample not cultured

Table 2. Vibrio vulnificus in New Hampshire Oysters

	TOTAL VIBRIOS /100 g	*VIBRIO VULNIFICUS* /100 g	TEMPERATURE °C	SALINITY °/oo
SITE 1				
8/18	>2400000	150000	23.1	18
8/30	460000	110000	20.1	-
9/18	240000	<2	20.5	-
10/6	400	<2	12.5	-
10/16	18600	<2	11.8	20
10/27	4300	<2	7.8	2
SITE 2				
8/30	20000	3000	20.1	27
9/18	460000	<2	19.1	-
10/6	4300	<2	12.7	-
10/16	48000	<2	12.2	25
10/27	9300	<2	8.3	14
SITE 3				
8/2	24000	<2	23.5	23
10/16	4300	<2	12.1	26
SITE 4				
10/16	9300	<2	12.1	25

Site 1 and 2 - closed to shellfishing
Site 3 and 4 - open to shellfishing

Table 3. Vibrio vulnificus in New Hampshire Waters

	TOTAL VIBRIO /100 g	*VIBRIO VULNIFICUS* /100 g	TEMPERATURE °C	SALINITY °/oo
SITE 1				
7/25	4600000	75000	-	-
8/10	2400000	90000	26.3	15
8/18	170000	12000	23.1	18
8/30	14000	<2	201	-
9/18	1100000	<2	20.5	-
10/6	80000	<2	12.5	-
10/16	2400	<2	11.8	20
10/27	430	<2	7.8	2
SITE 2				
8/30	13000	3000	20.1	27
9/18	75000	<2	19.1	-
10/6	22000	4300	12.7	-
10/16	930	<2	12.2	25
10/27	930	<2	8.3	14
SITE 3				
8/2	24000	<2	23.5	23
10/16	230	<2	12.1	26
SITE 4				
10/16	930	<2	12.1	25
SITE 5				
8/23	93000	<2	24.2	5
8/27	8000	<2	20.2	6
10/6	4600	700	13.1	5
10/11	20000	7000	11.1	6
SITE 6				
9/28	9300	<2	-	8
10/6	1500	<2	17.1	7
10/11	400	<2	12.1	8
SITE 7				
8/29	130000	<2	24.1	10
9/28	460000	400	12.8	5

Sites 1,2,5,6,7 - closed to shellfishing
Sites 3 and 4 - open to shellfishing

V. vulnificus was detected in New Hampshire water and oysters at temperatures ranging from 11.0 to 26.3°C and salinities between 5.0 to 23.0 parts per thousand. These results agree with the ranges for temperature and salinity seen by Tamplin (Tamplin et al. 1982) and Oliver (Oliver et al. 1983) for other areas of the United States. Statistical analyses has not been performed at this time to determine the correlation, if any, between temperature, salinity and the presence of Vibrio vulnificus from New Hampshire oysters or waters. However, there was an apparent decrease in the level of vibrio species in Maine oysters which may be attributed to a decrease in temperature with the change in seasons. Oysters are a natural ecological niche for estuarine bacteria, including vibrios This study indicated that such autochthonous bacteria appear to be recalcitrant to depuration.

REFERENCES

Blake, P.A., Weaver, R.E. and Hollis, D.G. 1980. Diseases of humans (other than cholera) caused by vibrios. Annual Review of Microbiology, 34:341-367.

Howell, T.L. and Howell, L.R. 1989. The Controlled Purification Manual. New England Fisheries Development Assoc , Inc., Boston, MA. 77 pp.

Kaysner, C.A. et al. 1987. Virulent strains of Vibrio vulnificus isolated from estuaries of the United States West Coast. Applied and Environmental Microbiology, 53.1349-1351.

Kelly, M.T. 1982 Effect of temperature and salinity on Vibrio vulnificus occurrence in a Gulf Coast environment. Applied and Environmental Microbiology, 44.820-824.

Kelly, M.T. and Dinuzzo, A. 1985 Uptake and clearance of Vibrio vulnificus from gulf coast oysters (Crassostrea virginica). Applied and Environmental Microbiology, 50:1548-1549.

Oliver, J.D., Warner, R.A. and Cleland, D.R. 1983. Distribution of Vibrio vulnificus and other lactose-fermenting vibrios in the marine environment. Applied and Environmental Microbiology, 45:985-998.

Son, N T. and Fleet, G.H. 1980. Behavior of pathogenic bacteria in the oyster, Crassostrea commercialis, during depuration, relaying, and storage. Applied and Environmental Microbiology, 40.994-1002

Tamplin, M., Rodrick, G.E., Blake, N.J. and Cuba, T. 1982. Isolation and characterization of Vibrio vulnificus from two Florida estuaries Applied and Environmental Microbiology, 44.1466-1470.

DESIGN OF A BENCH SCALE AUTOMATED RECIRCULATING SYSTEM FOR USE IN THE DEVELOPMENT OF PURGING/TASTE ENHANCEMENT CRITERIA FOR THE RANGIA CLAM (RANGIA CUNEATA)

Kelly A. Rusch, James E. Robin and Ronald F. Malone

INTRODUCTION

The successful marketing of Louisiana Rangia clams (Rangia cuneata) as a food product is hindered by two main obstacles, an off-flavor detectable upon steaming and bacterial contamination. Geosmin, a volatile compound produced by actinomycetes and blue-green algae, has been qualitatively identified as the off-flavor source in Rangia clams (Hsieh et. al., 1988). Relaying, a method which involves suspending contaminated shellfish in baskets or bags in unrestricted waters for a period of time, has been successfully implemented to reduce bacterial contaminants (Andrews, 1988), but off-flavor removal remains a problem. One potential method of resolving both problems is to use a closed, recirculating water system equipped with an automated algae chemostat and biological treatment unit to allow extended storage of the clams. By including a geosmin reduction unit and bacteria removal loop, the problems of off-flavor and bacterial contamination can be simultaneously eliminated. In addition, a recirculating system easily lends itself to computer automation, thereby reducing labor. This description of the bench scale automated system includes designs for the holding system and algae chemostat, the overall system operation, and the computer control/monitoring.

METHODS AND MATERIALS

The primary components of the recirculating system include the holding tank/trays, biological filter for water quality maintenance, sump, ozone generator and activated carbon column for geosmin removal, ultraviolet light (UV) unit for bacterial/viral eradication, constant head chamber, and algae chemostat for food production The schematic of the overall system is illustrated in Figure 1. The computer control block is not shown

HOLDING TANK AND TRAYS

The clam tank and trays were designed in accordance with the federal shellfish depuration guidelines (Furfari, 1966; NSSP, 1990a,b). These guidelines along with the design criteria are summarized in Table 1 and the holding tank is illustrated in Figure 2. Criteria for Rangia clams are not specifically covered in the guidelines; thus, the more conservative design requirements were used in the design calculations.

The overall dimensions of the tank are 6 00' X 2.50' X 1.25' with a 0 5-inch per longitudinal foot bottom slope for solids removal. Trays with the dimensions 2.50' X 2.00' X 0.50' provide a three inch clearance with the sides and bottom Two trays with these dimensions placed side-by-side on the support lip in the holding tank hold one bushel of Rangia clams.

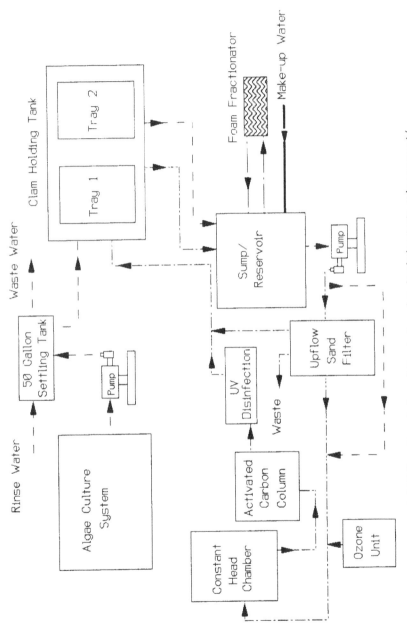

FIGURE 1. Overall view of the designed recirculating taste enhancement/depuration system for Rangia clams.

TABLE 1.   Depuration regulations and design criteria for recirculating rangia clam holding tank

Parameter	Regulation	Design Criteria
Tank Shape	length to width ratio of 2:1 to 4:1	6.0' X 2.5' rectangular tank
	4-foot maximum depth	1.25' depth
	0 25 to 0 50-inch per longitudinal foot bottom	0.50 inches per foot
Flow Rate	1 GPM per bushel	2 GPM
Basket Size	minimum bottom and sidewall clearance of 3 inches; maximum height of 6 inches	2.5' X 2.0' X 0.5'
	3-inch maximum stacking depth	2.8" (clams stacked 2-high)
Water Volume	8 ft^3 per bushel	13 ft^3 (98 gallons)

BIOLOGICAL FILTER

Proper water quality maintenance was accomplished using a pressurized upflow sand filter to entrap suspended solids and provide high nitrification capabilities at low substrate levels Design criteria for the filter followed those established for soft crawfish recirculating systems (Malone and Burden, 1988). Because of the low ammonia excretion rate of the clams (0 07 g/bushel-day), the filter was sized to meet 10 g/bushel-day solids removal requirements. Subsequently, a 6-inch diameter filter with a sand bed depth of 15 inches (0.25 ft^3) was determined to provide adequate water quality in the recirculating system.

Water, flowing upward through the filter, passes over a coarse sand bed (1.19 - 2.38 mm) supporting a bacterial biofilm composed partly of <u>Nitrosomonas</u> and <u>Nitrobacter</u> Twice daily, the flow rate to the filter is increased from 2 to 13 gallons per minute (gpm) for 30 seconds such that the sand bed is expanded to 50 percent, releasing captured solids. The waste water is discharged to a drain. Figure 3 shows the two operational modes for a typical pressurized upflow sand filter

SUMP

The sump provides water quality buffering capacity, collects the water draining from the holding tank, receives the excess overflow from algae input, and provides a water source for the main system pump. A constant sump volume of 20 gallons provides enough water for the pump to run at least five minutes without water returning to the sump and also provide the necessary volume to accommodate the water obtained from the chemostat. A sump of this size results in a total system volume of approximately 120 gallons or 14 gallons/lb wet clam meat

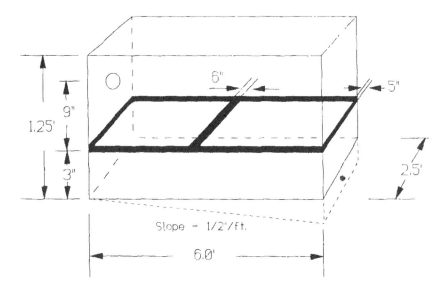

FIGURE 2. Holding tank for the Rangia clam recirculating purging/
taste enhancement system. This system was designed
to hold one bushel of clams (not drawn to scale).

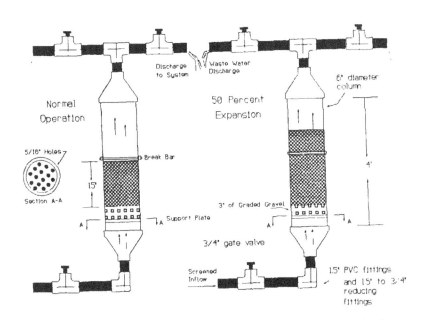

FIGURE 3. Normal and expanded operational modes for a pressurized
upflow sand filter used as the biological filtration
unit in the Rangia clam recirculating purging/taste
enhancement system (after Malone and Burden, 1988).

GEOSMIN REMOVAL UNIT

According to Richard and Fiessinger (1972), ozonation followed by filtration through granular activated carbon (GAC) is more effective in removing geosmin than either treatment is alone. A treatment scheme incorporating both of these options was incorporated into the overall design Since quantitative data on geosmin in Rangia clams is presently unavailable, the actual sizing of this treatment loop is not complete. Experimental studies on the separate components in addition to the combination will be performed to determine to determine the optimum geosmin reduction unit.

Water, flowing from the upflow sand filter to a 10-gallon constant head chamber, is injected with an ozone/air mixture. Given the two gpm flow rate under normal operational conditions, the contact time is five minutes. From the head chamber, the water gravity feeds through the activated carbon column and UV unit and returns to the holding tank. The activated carbon absorbs organic impurities, while the UV unit eradicates potentially harmful coliform bacteria

ALGAE CULTURE SYSTEM

The development of an automated algae chemostat has been discussed in depth by Rusch et al , (1988), Rusch and Malone (1989) and Rusch (1989). Details of an algae growth chamber (70 liters) are illustrated in Figure 4. All routine operations of the chemostat (harvesting, feeding, disinfecting, data recording) are controlled by a Zenith Z-184 Supersport laptop micro-computer Computer control and monitoring are discussed in a later section.

The algae chemostat to support feeding in the Rangia clam recirculating holding system was sized based on algae growth kinetics and Rangia clam consumption studies The kinetic studies in combination with a computer model predict that 5.08 grams of algae (Chlorella minutissima) can be harvested from each chamber daily while maintaining relatively constant culture concentrations of 158 to 190 mg/L. This harvesting regime implies a 38 percent volume removal per chamber per day for clam feeding

Results from the clam consumption studies indicate that Rangia clams consume 0.42 percent of their wet meat body weight per day. To ensure feeding to satiation, this consumption rate was increased to 1.00 percent. Based on an average wet meat body weight of 3 7801 grams, one bushel of Rangia clams consume 30.2 grams of Chlorella minutissima daily. Six chemostat chambers harvested twice daily would provide 30.5 grams of algae per day, enough to feed one bushel of Rangia clams. A 50-gallon settling basin (large enough to hold a daily harvest) is used to reduce the amount of water added to the recirculating system.

DISCUSSION

SYSTEM OPERATION

This section discusses one possible operational mode for the holding system design. During normal operation, water is pumped at two gpm from the sump through the upflow sand filter to the head chamber. Ozone is injected into the water before reaching the head chamber. The water is gravity-fed through the GAC column and UV unit and returned to the holding tank. The entire water volume is circulated through the biological and geosmin removal unit in one hour.

Twice daily, the upflow sand filter is backwashed. The first backwashing occurs just prior to feeding. The flow rate to the filter is increased to 13 gpm, expanding the bed to 50 percent and releasing and trapped solids. A bypass line is utilized during backwashing to maintain the two gpm flow to the constant head chamber. During feeding, the filter is maintained at 50 percent expansion for approximately six hours and the geosmin treatment unit is bypassed to ensure that the algae

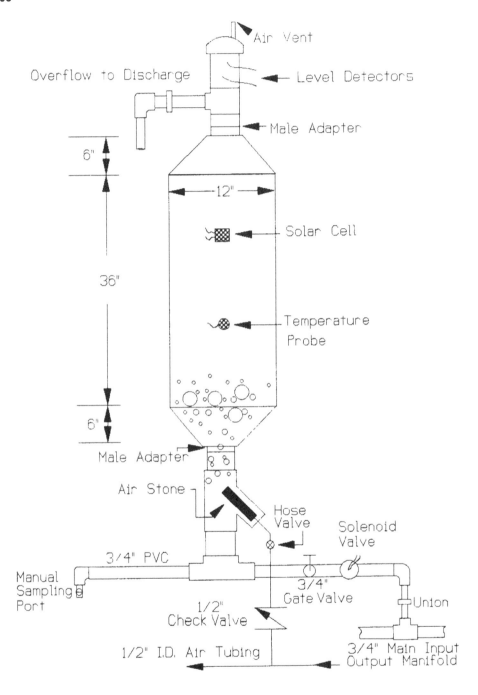

FIGURE 4. Details of a single algae growth chamber.

is neither trapped nor destroyed. Water and algae from the chemostat is pumped to the settling tank, the algae is allowed to settle, and the concentrate plus an additional 14 gallons is gravity fed to the holding tank. This additional water volume added daily to the system balances that lost during filter backwashing. After the feeding cycle is completed the filter is returned to normal operation, trapping the excess algae. The filter is then backwashed, releasing the algae and discharging it to a drain.

COMPUTER CONTROL AND MONITORING

The individual components of both the Rangia clam recirculating holding system and the algae chemostat are independent of each other but controlled by one Zenith Z-184 supersport laptop micro-computer interfaced to monitoring devices with a Remote Measurement Systems ADC-1 data and acquisition unit (Remote Measurement Systems, Inc., 1983). Precision measurement devices including temperature probes, solar cells, level detectors, a fluorometer, and a velocity meter are connected to the analog input channels of the ADC-1 board. The analog signals are converted to digital signals and relayed to the computer to make procedural decisions. Pumps and solenoid valves are controlled by BSR Type X-10 control units (Heathkit, Model No. BC-227). The computer sends and 'on/off' signal to the ADC-1 transmitter which then generates and sends a digital code superimposed over the AC wiring to the BSR units.

The control program, written in Turbo Pascal, contains a supervisor stack sequence and is flowcharted in Figure 5. The supervisor mode contains two processes. First, the supervisor watches the internal clock and the time associated with the command at the top of the stack. When the two times are equal, the supervisor calls the procedure associated with the command. The activated procedure executes instantaneous operations, adds delayed operations to the stack, and relinquishes control back to the stack supervisor. Second, the stack is loaded through a stack sorting procedure which prioritizes execution chronologically. With this organization, each control block can be programmed independently utilizing time-of-day execution and condition verification loops to avoid conflicts. Therefore, the same program can control both the chemostat and the holding system.

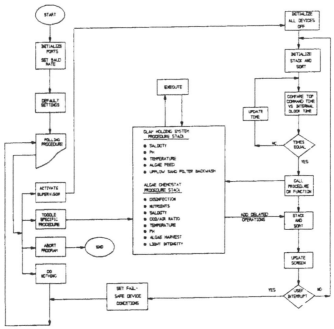

FIGURE 5. Flowchart of the control program for the aglae chemostat and Rangia clam recirculating purging/taste enhancement sytem.

PRESENT STAGE OF RESEARCH

The authors recognize possible limitations and problems associated with the present design of the algae chemostat and the Rangia clam recirculating purging/taste enhancement system, therefore suggesting the need for more research and development.

The chemostat has been in operation for seven months with only a few minor problems. First, while the chemostat is capable of producing 190 mg/L Chlorella minutissima, preventing the growth of Nitzschia sp. (diatom) on the chamber sidewalls has been a continual problem. This growth reduces the amount of light available to Chlorella. Second, due to the configuration of the bottom port of each chamber, a dead area is created allowing some algae settling. Additionally, disinfectant does not reach this area, allowing any contaminant present to remain in the system.

The bench scale recirculating purging/taste enhancement design, as presented in this paper, suggests bypassing the geosmin/bacterial removal unit during feeding The authors recognize this to be in violation of depuration guidelines and that further investigation of alternative operational modes is necessary Additionally, the chemostat to support feeding for the recirculating purging/taste enhancement system was very conservatively designed based on the available food consumption data This was done to compensate for the unknown locational and seasonal variations in the feeding pattern of the clams. Refinement of this component of the overall system is necessary.

CONCLUSIONS

The automated recirculating system was designed to purge one bushel of Rangia clams of bacteria and to remove the off-flavor agent, geosmin. The algae chemostat system designed to support feeding, was tested and is capable of maintaining Chlorella minutissima concentrations between 158-190 mg/L. Preliminary tests showed that an upflow sand filter is capable of supporting Rangia clams in a recirculating system. Additionally, the upflow sand filter has been successful in maintaining adequate water quality in recirculating crawfish and crab systems (Burden, 1988). The computer has been successful in controlling the chemostat and the addition of necessary procedures to control the recirculating system poses no problems.

RECOMMENDATIONS

Recommendations for future work include:

1) A bench scale prototype system must be constructed and tested to determine the maximum holding time required until geosmin is no longer detectable by either sensory or chemical methods and bacteria have been eradicated.
2) The holding capabilities of the recirculating system must be determined. This might include the assurance of adequate water quality, the ability of the system to reduce or remove geosmin, the ability to maintain sufficient feeding.
3) The ozone/activated carbon treatment scheme requires testing for determination of the best treatment unit/size.
4) Optimization studies on the algae chemostat must be performed in order to reduce the number of chambers required for feeding.
5) Specific operational criteria for the elimination of off-flavors in Rangia clams must be developed.

ACKNOWLEDGMENTS

This work was funded by the Louisiana Sea Grant College Program, an element of the National Sea Grant College Program, under the auspices of NOAA, U.S. Department of Commerce.

REFERENCES

Andrews, L.S. 1988. Relaying Louisiana brackish water clams, Rangia cuneata, to improve microbial quality. Master's Thesis, Louisiana State University, Baton Rouge, LA.

Burden, Daniel G. 1988. Development and design of a fluidized bed/upflow sand filter configuration for use in recirculating aquaculture systems. Ph.D. Dissertation, Louisiana State University, Baton Rouge, LA.

Furfari, S.A 1966. "Depuration Plant Design". National Shellfish Sanitation Program, U.S. Department of Health, Education, and Welfare, Washington, D.C

Hsieh, T.C.-Y., Tanchotikul, U and Matiella, J.E 1988. Identification of geosmin as the major muddy off-flavor of Louisiana brackish water clam (Rangia cuneata). Journal of Food Science 53(41):228-1229

Malone, R.F. and Burden, D.G. 1988. Design of recirculating soft crawfish shedding systems. Report for Louisiana Sea Grant College Program, Baton Rouge, LA.

National Shellfish Sanitation Program (NSSP). 1990a. Manual of Operations, Part I: Sanitation of Shellfish Growing Areas Public Health Service, U.S Food and Drug Administration, Washington, DC

National Shellfish Sanitation Program (NSSP). 1990b. Manual of Operations, Part II Sanitation of the Harvesting, Processing and Distribution of Shellfish, Public Health Service, U.S. Food and Drug Administration, Washington, DC.

Remote Measurement Systems, Inc. 1983. ADC-1 Owner's Manual. Seattle, WA.

Richard, Y. and Fiessinger, F. 1972. Empoi complementaire des traitements ozone et charbon actif. International Ozone Institute, Congress, May, Paris.

Rusch, K.A., Zachritz, W.H., Hsieh, T.C.-Y. and Malone, R.F. 1988. Use of automated holding systems for initial off-flavor purging of the Rangia clam, (Rangia cuneata). Proceedings of Ocean '88, Vol. I, Baltimore, MD.

Rusch, K A. and Malone, R.F. 1989. Development of an automated chemostat for continuous algal production. Invited Paper, Special Proceedings of 20th Annual World Aquaculture Society Meeting, Los Angeles, CA, February 12-16.

Rusch, K.A. 1989. Development of an algal chemostat to support feeding in a recirculating Rangia clam (Rangia cuneata) purging system. Master's Thesis, Louisiana State University, Baton Rouge, LA.

CONTAINER-RELAYING OF OYSTERS:
AN ALTERNATIVE TO DEPURATION AND RELAYING

John Supan

INTRODUCTION

Utilizing the oyster's natural cleansing ability is of great interest to many in the oyster industry. There are many reasons for such interest, including access to closed 'reefs' during shortages from approved waters and attaining obvious marketing benefits by providing a safer oyster for raw consumption.

In the U.S., shellfish harvesting and processing are regulated by each state's shellfish control agency, under the guidance of the National Shellfish Sanitation Program (NSSP). There are three acceptable methods of cleansing contaminated oysters. Relaying is the transfer of shellfish from restricted waters to approved waters for natural biological cleansing, using the ambient environment as a treatment system. Depuration, or controlled purification, is a process that uses a controlled aquatic environment to reduce the level of bacteria and viruses in live shellfish Container-relaying is the transfer of shellfish from restricted areas to approved or conditionally approved areas for natural biological cleansing in a container using the ambient environment as a treatment system (NSSP, 1990a,b). Which method is utilized generally is dependent upon a person's knowledge, skills, oyster and/or lease availability, and financial situation

PROCESS DESCRIPTIONS

RELAYING

Relaying is the most traditional method of oyster cleansing in the U S., since the technique has been utilized in oyster farming for over a century (Pausina, 1988) Dredging is a common commercial harvesting method (Figure 1), but tongs are utilized in many states with public oyster resources The oysters are harvested from restricted shellfish growing waters (waters receiving known detrimental impacts and has a fecal coliform MPN value of $\leq$ 88/100 ml) and relaid to approved or conditionally approved growing waters (waters receiving infrequent detrimental impacts and a fecal coliform MPN value of $\leq$ 14/100 ml) for 14 consecutive days when environmental conditions are suitable for purification unless shorter periods are demonstrated to be adequate (NSSP, 1990a,b). High losses can occur during relaying, since the oysters are essentially thrown overboard onto the waterbottom using shovels or high-pressure water pumps These losses may be due to, 1) physiological stress, 2) shell damage inflicted by relaying, 3) smothering and clogging by sediments, 4) predators, 5) incomplete second harvests and 6) theft.

DEPURATION

Depuration is a process that provides a clean sea water environment in which shellfish may actively cleanse themselves This usually means the installation of tanks with disinfected or otherwise treated sea water (Furfari, 1966). Oysters are harvested from restricted shellfish growing waters and placed in tanks for usually 48-72 hours Degree of cleansing is verified by laboratory analysis In the U.S., an acceptable depuration process is contingent upon the state shellfish control agency exercising very stringent supervision over all phases (NSSP 1990a,b), including planning, construction, harvesting, operation, maintenance, verification and recordkeeping

FIGURE 1   Louisiana oyster vessel in Mississippi Sound with above deck side boards, dredge tables and water "pumps" on bow, all common adaptations for relaying.

There is a great deal to consider prior to building and operating a depuration facility (Furfari, 1966; Canzonier, 1984). The main considerations include; 1) oyster physiology, 2) location, 3) water quality, 4) system design (i.e., flow-through or recirculating, ultraviolet light or ozone treatment, tank size and capacity, container type, handling and array, pumps, plumbing, etc.), and 5) laboratory analysis. Other considerations include oyster availability, personnel qualifications, training and marketing.

CONTAINER-RELAYING

Container-relaying involves the use of a "relaying device" (e.g., raft, rack, etc.) (Figures 2 and 3) to hold or suspend containers (Figures 4 and 5) of oysters on or off the bottom in approved waters until they are cleansed. After 14 consecutive days, harvesting is accomplished by simply lifting the containers from the water. A shortened time period is possible if intensive monitoring of critical environmental parameters indicates effective purification (NSSP, 1990a,b). Container-relaying alleviates some of the problems of onbottom-relaying while reducing bottom suitability as a limiting factor. It may also be more cost effective than depuration. If approved water quality is consistently available in the relaying area, the factors that determine successful cleansing are: (1) oyster physiology, (2) the design of the container, and (3) its method of use

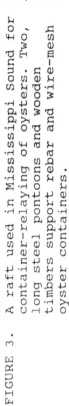

FIGURE 3. A raft used in Mississippi Sound for container-relaying of oysters. Two, long steel pontoons and wooden timbers support rebar and wire-mesh oyster containers.

FIGURE 2. On oyster rack design used in in Mississippi Sound. With welded angle-iron from supports for 48 oyster containers.

208

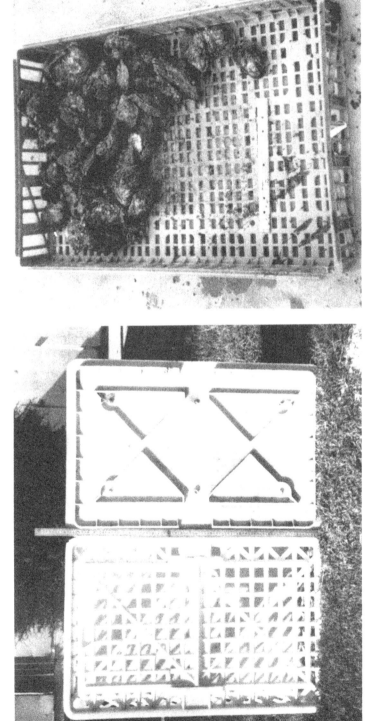

FIGURE 4. Modular containers (chicken coops) used in container-relaying activities in Mississippi Sound. Container has a formated top and sides. The top has a hinged lid. The container measures 3 x 2 x 1 ft^3.

FIGURE 5. A formainated container (chicken coop bottom) used in an oyster rack in Mississippi Sound

PROCESS EVALUATIONS

SITE REQUIREMENTS

Oyster cleansing may fluctuate, particularly in the relaying waters, if heavy rainfall and the resulting land runoff increases bacterial levels to allow recontamination (Cook, 1969). Such impact is less critical if a depuration plant is designed and proven to handle such variation. All three processes are site specific, with success relative to harvest area proximity and availability of consistently suitable water quality. Since depuration utilizes more complex technology, close proximity to support services (e g., power source, hardware, repair services, etc.) and transportation is also important. Prior to selecting a site, watershed impacts should be assessed, especially utilizing the sanitary and shoreline surveys conducted by state shellfish control agencies (NSSP, 1990a,b).

Availability of appropriate waterbottom can be limiting in many coastal areas Nearly 60% of transplanted oysters land upside down during relaying (Gunter and McGraw, 1973). Smothering and gill clogging are usually the result of relaying oysters onto unsuitable or inconsistent waterbottom (i.e., mud, sand and/or silt) (Supan, 1981). Bottom suitability has little effect on container relaid oysters if the containers are held off the bottom using appropriate devices. Hard reef is necessary, however, for onbottom placement of containers (Supan and Cake, 1982; DuPaul and Oesterling, 1987).

POST-TREATMENT HARVESTING

Most commercial oystermen are quite capable with a dredge or tongs, however, not all onbottom relaid oysters are recovered during the second harvest. One problem that local oystermen are facing is a shortage of experienced captains and deckhands for their vessels (Dugas et al , 1981) The most critical problem which relates to private ownership of leased waterbottoms is thievery (Leffler, 1986). Thievery is a dual problem contributing to loss of resource, and reduced incentive to relay oysters onto private grounds (Pausina, 1988). Containerization can eliminate such losses and improves handling, when relaying devices are used that hinder or eliminate theft. This method provides more complete second harvests with greatly reduced mortality (Supan and Cake, 1982) (Figure 3).

CONTAINERS

Containers used for depuration and container-relaying must allow free flow of water to the shellfish. Becker (1977) conducted an offbottom, containerized cleansing study using stainless steel baskets (35.5 cm/side). He found that indicator bacteria were effectively eliminated from single-layer oysters in 24 hours when ambient water quality met or exceeded the criteria for approved shellfish growing waters Bottom layers of oysters in half-full baskets purged satisfactorily in 96 hours, but at a slower rate that did single-layer oysters. Bottom layers of oysters in full baskets increased their bacterial content and failed to purge satisfactorily during the 96-hour test periods. The failure was attributed to the higher weight of the overlying oysters and the restriction of water flow through the stacked shellfish. Depuration operations in the U S. are required to load oysters into containers no deeper than 3" (76mm) (NSSP, 1990b).

A broad, flat container shape increases the surface/volume to allow commercial quantities to be held in one container with a reduction in the internal mass of shellfish (Supan and Cake, 1982) Plastic chicken coops (polyethylene structural foam) measuring 3 x 2 x 1 cu. ft. (91 x 61 x 30 cm^3) are very suitable for relaying and depuration (Figure 5). Containers measuring 4 x 4 x 1 (1.2 x 1 2 x 0 3m^3) are used to relay clams in Virginia with successful results (DuPaul and Oesterling, 1987) Many container shapes and materials have been used in depuration systems over the years based on identical criteria (Furfari, 1966).

The most attractive attribute in depuration is the relatively quick turnover rate of the containers of oysters (48-72 hours) as opposed to container-relaying (14 days or less) Reduction of the relaying period and product quality verification equal to depuration can be achieved with laboratory analyses. The cost of such analyses may be less for container-relaying since sampling intensity would be less (NSSP, 1990a,b) and an onsite laboratory would not be necessary in some states (LA State Code, 1989)

The container array will depend upon the relaying device or tank dimension Racks can be design to hold the containers apart (Figure 2) or to simply support stacked, modular containers Rafts (Figure 3) and other suspension techniques provide floating support, but may expose the shellfish to freshwater runoff from land, increasing the potential for recontamination. Longlines or floatlines of individual containers must be placed on hard bottom and may need to be elevated such as with the two 6 x 6in^2 (15.2 x 15.2 cm^2) wooden skids reported by DuPaul and Oesterling (1987). A floatline system used in Mississippi Sound allows for the harvesting of four containers (chicken coops) simultaneously using common handline attached to a steel frame that lifts the containers by their four-corner guylines (Burrage, 1989) A vessel in motion from a fixed anchorage can utilize dual longlines of individual containers (Figure 6) to bring the containers to the vessel for removal plus allowing deployment of other containers. Longline and floatline methods can greatly reduce the cost of hardware and specialized vessels (Figures 7 and 8), but increases the chance of theft. Direct contact with the bottom allows more intra-container sedimentation (Supan and Cake, 1982) and may expose the shellfish to viruses deposited on the sediment (Smith et al., 1978; Ellender et al., 1980). Greater surveillance is required to decrease theft and gear conflicts with trawlers.

Depuration and container-relaying can provide improved shelflife by utilizing the containers for more efficient post-treatment refrigeration. Better cooling methods have been demonstrated (Moody and Bankston, 1989, unpublished data), utilizing a fan-induced air circulation system within a walk-in cooler. Shellfish containers are stacked in a radiator-like fashion and a large fan draws the refrigerated air through the shellstock With properly sized refrigeration units, this method can quickly reduce shellstock temperature, controlling the bacterial quality and increasing shelflife and marketability.

OTHER COST CONSIDERATIONS

Depuration obviously costs more to achieve successful cleansing than relaying Tanks, pumps and plumbing, and influent/effluent treatment systems (i.e., ultraviolet light or ozone treatment and sedimentation and/or filtration) must be properly engineered, operated and maintained (Furfari, 1966; Canzonier, 1982 and 1984). Laboratory analysis is required and personnel qualifications and training are more critical. Plant managers may soon discover that the operational requirements are so onerous that they are either circumvented or the process is abandoned altogether (Canzonier, 1988).

Improper tank design and hydraulics will lead to inconsistent cleansing Water quality parameters will deteriorate in the tank from the influent end to the effluent end, particularly with regards to dissolved oxygen. The reduced water quality will stress the shellfish, resulting in reduced pumping rates and inconsistent cleansing. Tank hydraulics are important, therefore, not to simply achieve required minimum flow rate of 1 gal (3 78L/min/bu, but to provide good water quality (Furfari, 1966) equally throughout the tank of containerized shellfish.

FIGURE 6   A low cost longline device for onbottom, container-relaying.

FIGURE 7   A specialized vessel for rack handling and transport used in Mississippi Sound.

FIGURE 8   A spud barge, typical oilfield support vessel suitable for container-relaying.

Good oyster physiology is another important cost factor specific to site selection, and successful cleansing in general. Close proximity to the harvesting area(s) not only helps in obtaining oysters for plant operation, it will reduce the oyster cleansing response time. Time-temperature abuse (i.e., the exposure of shellfish to high temperatures after harvesting, particularly during the summer months) is not only a common cause of oyster stress and mortality during relaying and depuration, it is often the reason for increased microbial levels in shellfish (Cook and Ruple, 1989) Oyster physiology is also affected by spawning, predation, disease, high temperatures in shallow water, improper dredging and culling, exposure to urban/industrial pollutants, etc. Most of these effects can be eliminated with proper site location and planning.

Each state with depuration and relaying activities has individual regulations as part of their shellfish sanitation program. Louisiana regulations require a $5,000 cash performance bond and immediate supervision of harvesting activities by a commissioned municipal, parish (county), or state police officer or a bonded security guard, payment responsibility by the permittee. Regulations regarding the depuration building's ground, floors, walls, ceilings, lighting, heating and ventilation, water supply, pest control, plumbing and sewage disposal must comply with regulations (NSSP, 1990b).

CONCLUSIONS

Some state shellfish control agencies utilize public oyster reefs to conduct relaying programs (Etzold, 1975). Easley and Seabolt (1981) economically compared public and private relaying efforts, finding relaid oysters have two potential prices at final harvest; one if placed in public bottom, and a second, higher value if they were privately owned. Prices and incomes were lower when oysters were harvested from public bottoms than from leased bottoms Price differences of $0.05 and $0.21/lb of meat were noted. The higher oyster prices for lease production was attributed to better timing of harvests. They summarize the larger economic yields from the resource are possible, but the key is private ownership (leasing). If relaying only to public waters is continued and free entry prevails, there is no reason to expect any economic benefit for fishermen In the long run, we would expect little if any significant improvement in real per capita net income (Easley and Seabolt, 1981) Leasable waterbottom must be available to utilize any container-relaying techniques

Containerization will greatly reduce the losses associated with onbottom relaying, depending on the relaying device. Racks provide more security and improved handling but increase the need for specialized vessels (Figures 8 and 9).

Depuration and container-relaying will improve shellfish handling and quality but the market price must support the added costs Which cleansing method is more appropriate depends upon whether good water quality is adjacent to land, and largely on a person's abilities and financing. Container-relaying generally provides a simpler means of cleansing with less regulatory requirements, but its success depends upon sustainable water quality at the relaying site.

Economic success of any cleansing technique depends on a consistent supply of shellfish. One cannot simply assume that just because there are oysters in restricted areas they can be consistently supplied to the cleansing operation. Oyster fishermen may have unique personalities and work habits that must be diplomatically and monetarily dealt with to ensure reliable harvesting, particularly when shellfish are available for harvest from approved areas.

Finally, clean growing waters are the most important aspect to shellfish production and shellfish industry members should take an active role in supporting appropriate provisions for reducing water pollution to levels that will not compromise the sanitary quality of their product (Canzonier, 1988)

REFERENCES

Becker, R.E 1977 A basket relaying study on the coast of Alabama: Reduction of coliform bacteria as a function of time and basket loading. Wilt, D.S. (ed). Proc. 10th Nat. Shell. Sanitation Workshop; 1977 June 29-30; Hunt Valley, MD. U.S. Dept. Health, Educ., and Welfare Publ. No. (FDA) 78-2098:174-181.

Burrage, D. (ed) 1989. Mississippi oyster industry; past, present and future. Proc of Conference held in Biloxi, MS, Dec. 1-2, 1988. Miss-Ala Sea Grant Consort Publ No. MASCP-88-048 (in print).

Canzonier, W.J. 1982. Depuration of bivalve molluscs - What it can and cannot accomplish and some practical aspects of plant design and operation. Intl. Seminar Management of Shellfish Resources. Tralee, Ireland, 27 p.

Canzonier, W.J. 1984. Technical aspects of bivalve shellfish production and marketing. J. Shellfish Research 7(2): 261-266.

Cook, D.W. 1969. A study of coliform bacteria and Escherichia coli on polluted and unpolluted oyster bottoms of Mississippi and a study of depuration by relaying. U.S. Bur. Comm Fish Comp. Rep , Fed. Adi Proj (PL-88-309) No. 2-25-R: 63 p (Available: Gunter Library, GCRL, Ocean Springs, MS 39564).

Cook, D W and Ruple, A.D 1989 Indicator Bacteria and Vibrionacease multiplication in post-harvest shellstock oysters J of Food Protection, 52(5)· 343-349.

Dugas, R.J., Pausina, R.V. and Voisin, M. 1981. The Louisiana oyster industry. In: K.K. Chew (ed.) Proceedings of the North American Oyster Workshop. Louisiana State University, Division of Continuing Education. Baton Rouge, LA 101-111p.

DuPaul, W.D. and Oesterling, M.J. 1987. Evaluation of a containerized system for the relaying of polluted clams (Mercenaria mercenaria). Proc. Trop. and Subtrop. Fish. Technol. Soc of Am , 1987. p. 135-143.

Easley, J.E. and Seabolt, J.D. 1981. Shellfish relay: A preliminary review of potential gains from alternative property rights in southeastern North Carolina UNC Sea Grant Publ. No. UNC-SG-WP-81-1 14 p.

Ellender, R.D , Mapp,J.B., Middlebrooks, B.L , Cook, D.W. and Cake, E.W. 1980. Natural enteroviruses and fecal coliform contamination of Gulf coast oysters. J. Food Prot. 43(2):105-110.

Etzold, D.J. 1975 Estimated annual cost for the transplanting project. Univ. Southern Miss., 8 p. (mimeo)

Furfari, S.A. 1966 Depuration Plant Design, U.S. Public Health Serv. Publ. No. 999-FP-6 119 p.

Gunter, G and McGraw, K A 1973 Basic studies on oyster culture. How do single oysters land on the bottom when planted? Proc. Natl. Shellfish. Assoc 47: 31-33.

Leffler, M. 1986. Bringing up oysters. Maryland Sea Grant Publ. No. UM-SG-RS-87-08. 7 p.

Louisiana Sanitary Code (LSC). 1989. Chapter IX. Proposed Rule for Shellfish Depuration and/or Onshore Wet Storage. Dept. of Health and Hospitals. Seafood Sanitation Branch, New Orleans, LA.

Moody, M. and Bankston, D. 1989. Unpublished data. Louisiana State University, Baton Rouge, LA

National Shellfish Sanitation Program (NSSP). 1990a. Manual of Operations, Part I: Sanitation of Shellfish Growing Areas. Public Health Service, U.S. Food and Drug Administration, Washington, DC.

National Shellfish Sanitation Program (NSSP). 1990b. Manual of Operations, Part II: Sanitation of the Harvesting, Processing and Distribution of Shellfish, Public Health Service, U.S. Food and Drug Administration, Washington, DC.

Pausina, R. 1988. An oyster farmer's perspective to the past, the present and the future of the Louisiana oyster industry. J. of Shell. Res., 7(3): 531-534.

Smith, E.M, Gerga, C.P. and Melnick, J.L. 1978. Role of sediment in the persistence of enteroviruses in the estuarine environment. Appl. Environ. Microbiol. 35.685-698.

Supan, J.E. 1981. A comparison of onbottom and containerized relaying of contaminated oysters (Crassostra virginica [Gmelin]) in Mississippi Sound. Hattiesburg, MS: Univ. Southern Miss. 90 p. Thesis.

Supan, J.E. and Cake, E.W. 1982 Containerized-relaying of polluted oysters (Crassostrea virginica [Gmelin]) in Mississippi Sound using suspension, rack and onbottom-longline techniques J Shell Res 2(2)·141-151

ACKNOWLEDGEMENTS

I wish to thank Mr. Ron Becker, Assoc. Director, Ms. Elizabeth Coleman and Mr John Brown, Louisiana Sea Grant College Program, and Mr. John Wozniak, Louisiana Cooperative Extension Service for their assistance and support. I also thank, Dr. Ken Roberts, Mr. Becker and Ms. Coleman for their editorial review.

Monitoring the Accumulation and Depuration of Paralytic Shellfish Toxins in Molluscan Shellfish by High-Performance Liquid Chromatography and Immunological Methods

Allan D. Cembella and Gilles Lamoureux

INTRODUCTION

Paralytic shellfish poisoning (PSP) is a serious, and occasionally fatal, syndrome caused by the consumption of toxic shellfish by humans (Prakash et al., 1971). In most temperate waters, the organisms responsible for PSP contamination are usually marine dinoflagellates belonging to the genus Alexandrium (alternatively known as members of the Protogonyaulax catenella/tamarensis species complex, and formerly as species of Gonyaulax) and Gymnodinium catenatum; in the tropics, Pyrodinium bahamense is often implicated (Taylor, 1984; Shimizu, 1987). The serious public health risk posed by PSP has restricted the commercial and recreational exploitation of wild shellfish stocks, as well as the development of mariculture, in many areas of the world. Furthermore, many countries, particularly in North America, northern Europe, and southeast Asia, have been forced to implement comprehensive shellfish monitoring programs as a protection for domestic consumers and to ensure continued access to export markets.

The worldwide financial losses suffered by the molluscan shellfish industry due to PSP have been substantial. For example, in eastern Canada, an estimated one-quarter of all potentially harvestable wild molluscan stocks cannot be exploited at present, as PSP toxin levels consistently exceed regulatory limits (80 μg saxitoxin equivalents [STXeq]/100 g shellfish tissue). Commercial bivalve culture operations are similarly affected by PSP, particularly during the summer and early autumn, when high consumer demand often coincides with the peak period of toxic dinoflagellate blooms and consequent PSP toxicity.

This constraint has led to urgent requests from the shellfish industry for active depuration and/or passive detoxification methods which could be applied to harvested stock after substantial PSP contamination has occurred Unfortunately, no panacea has been developed to deal adequately with this problem, especially for shellfish with high PSP toxin loads. Ozonation has been proposed as a possible depuration technique for PSP toxins (Blogoslawski and Stewart, 1978; Blogoslawski et al., 1979), but the results have been equivocal, and further work on the efficacy of this process is obviously required. In experiments involving direct ozone treatment of PSP toxins extracted from Alexandrium spp., and present in natural algal bloom populations of these toxic dinoflagellates, ozonation did seem to be effective in reducing toxic activity, although the efficiency appeared to be pH dependent (Dawson et al., 1976; Blogoslawski, 1988). Treatment with ozonized seawater for two weeks was shown to rapidly enhance detoxification in PSP-contaminated surf clams Spisula solidissima (Blogoslawski and Stewart, 1978). In a study on the depuration of PSP toxins in the soft-shell clam Mya arenaria, ozone treatment of seawater was also found to greatly accelerate toxin loss (Blogoslawski et al., 1979). However, subsequent work (White et al., 1985) indicated that no detoxification of Mya from the same location was achieved under similar conditions at comparable toxin levels. The latter authors suggested that the discrepancy may be attributable to differences in the bivalves' prior history of toxin exposure. Detoxification was presumed to be effective when the toxins were recently acquired, but ineffective when the toxins had been retained for extended periods of time. PSP toxins are known to be subject to bioconversion over time, and can also be transferred to and bound within specific tissue compartments, thereby altering the kinetics of toxin release (Sullivan et al , 1983; Bricelj et al , 1990a,b)

The most widely used method of PSP toxicity monitoring is based upon the original mouse bioassay technique developed in the 1930s for the analysis of shellfish contaminated by the chain-forming dinoflagellate Alexandrium catenella in California (Sommer and Meyer, 1937). With minor modifications in subsequent years, the mouse bioassay, which involves replicate intraperitoneal injections of crude shellfish tissue extracted in 0.1 N HCl, has been accepted as a standard method by the American Association of Analytical Chemists (AOAC, 1984), and is used routinely in most regulatory laboratories throughout the world. Although simple in its execution, the mouse bioassay suffers the inherent drawbacks of many other bioassays: relatively low sensitivity, lack of toxin specific response, and low precision (±20%) (Sullivan and Wekell, 1984, 1986). These limitations are compounded by the need for the rearing and maintenance of large numbers of mice of the appropriate strain, sex, and age. Along with the requirement for trained personnel to perform reproducible assays, these factors have tended to restrict the use of the mouse bioassay to centralized regulatory facilities. This can result in bottlenecks and possible delays in the dissemination of toxicity reports due to overload of bioassay laboratories.

To compare alternative methods for the analysis of PSP toxins in molluscan shellfish, approximately 80 samples of the blue mussel Mytilus edulis were obtained from experimental myticulture sites in the Bay of Gaspé, Quebec, Canada during the summer of 1988 and tested for toxicity. The Gaspé region of eastern Canada is rapidly expanding as an area of mariculture development. A commercial ozonization plant has recently been established for the depuration of PSP toxins in cultured bivalves, requiring rapid and reliable quality control methods to determine toxin elimination rates. The mussel samples used to evaluate the alternative methods were from a study on the kinetics of in situ uptake and detoxification of PSP toxins in shellfish contaminated by natural Alexandrium blooms. Details of this study will be published separately in a forthcoming article.

The following review deals with two analytical alternatives to the conventional mouse bioassay, high-performance liquid chromatography (HPLC) and immunological techniques The discussion illustrates how these methods may be successfully incorporated into the monitoring of PSP toxins in commercial depuration facilities or used in conducting detoxification experiments

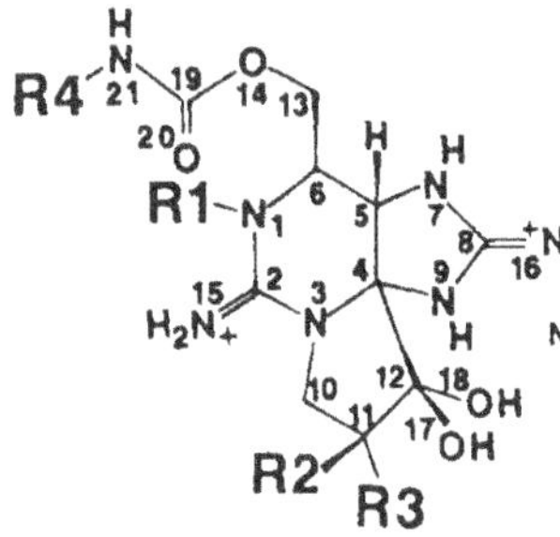

	R1	R2	R3	R4=H	R4=SO_3^-
SAXITOXIN GROUP	H	H	H	STX	B1
	H	H	OSO_3^-	GTX_2	C1
	H	OSO_3^-	H	GTX_3	C2
NEOSAXITOXIN GROUP	OH	H	H	NeoSTX	B2
	OH	H	OSO_3^-	GTX_1	C3
	OH	OSO_3^-	H	GTX_4	C4

FIGURE. 1 Structures of saxitoxin and various carbamate (R4 = H) and N-sulfocarbamoyl (R4 = SO_3^-) derivatives found among PSP-producing dinoflagellates. Certain decarbamoyl analogues may also be present in contaminated shellfish. Saxitoxin = STX; neosaxitoxin = neoSTX; gonyautoxins 1,2,3,4 = $GTX_{1,2,3,4}$

ANALYTICAL METHODOLOGIES

HIGH-PERFORMANCE LIQUID CHROMATOGRAPHY

Most analytical methods for the determination of PSP toxin levels in shellfish have been based upon the oxidation of these tetrapurine toxins (Fig. 1) to fluorescent pyrimido-purine derivatives under alkaline conditions (Bates and Rappoport, 1975; Shoptaugh et al., 1981). An early version of this approach involved a manual fluorescence assay (Bates and Rappoport, 1975) using total fluorescence as a measure of total PSP toxin content. The general applicability of this assay has been hampered by the fact that the various PSP toxin components differ markedly in fluorescence yield per unit specific toxicity. In later developments, fluorescence-based techniques evolved from low-pressure chromatography "toxin analyzers" (Buckley et al., 1978), which separated rather crude PSP-toxin fractions, to highly sophisticated HPLC systems (Oshima et al., 1984; Sullivan and Wekell, 1984; 1986, Sullivan et al , 1985; Nagashima et al., 1987), capable of resolving and quantifying most of the individual PSP toxin components.

The available HPLC methods for the analysis of PSP toxins differ considerably in minor details (type of analytical column, solvent flow rates, reaction temperature and pH, choice of oxidant, etc.), however, they are fundamentally similar in basic design and operation (Oshima et al., 1984; Sullivan and Wekell, 1984, 1986; Nagashima et al., 1987). All involve the separation of toxins on an analytical column, followed by post-column oxidation at an elevated temperature (Fig. 2). Detection of fluorescent derivatives is accomplished using a sensitive fluorescence detector equipped with independent adjustable monochromators for excitation and emission wavelengths. The application of such an HPLC system in a PSP monitoring program has been extensively reviewed by Sullivan and co-workers (Sullivan and Wekell, 1984, 1986; Sullivan et al , 1985)

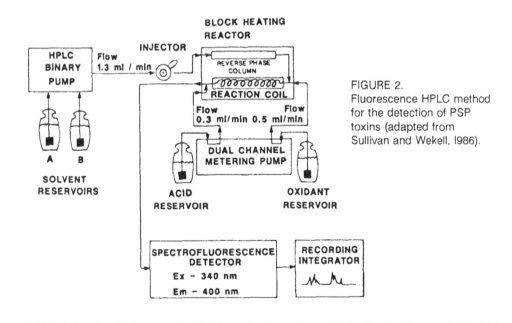

FIGURE 2.
Fluorescence HPLC method for the detection of PSP toxins (adapted from Sullivan and Wekell, 1986).

IMMUNOLOGICAL TECHNIQUIES

Although an antibody to STX, one of the major PSP toxins, was produced over two decades ago, activity was poor and relatively unstable (Johnson et al., 1964). Since that time, significant advances in the development of immunological assays for the detection of PSP toxins, particularly STX, have been reported These techniques have included both radioimmunoassays (RIA)(Carlson et al , 1984; Davio et al., 1985, Hurst et al., 1985), where the toxins are labelled with a radioactive isotope, and enzyme-linked immunosorbent (ELISA) assays (Davio et al., 1985; Chu and Fan, 1985), in which the presence of the toxins is usually indicated by a simple colour change reaction.

Previous attempts to determine the cross-reactivity of STX-antibodies have usually indicated much less affinity for the N-1 hydroxy derivatives, such as members of the neoSTX group (Fig. 1), than for STX (Carlson et al., 1984; Chu and Fan, 1985). Since both the causative dinoflagellates and PSP-contaminated shellfish usually contain a variety of N-1 hydroxy and sulfate derivatives (Sullivan and Wekell, 1986; Cembella et al., 1987; Bricelj et al., 1990a,b), it is imperative that cross-reactivity be considered in the development of immunological methods for PSP toxins.

Researchers at the Institut Armand-Frappier (IAF) have recently succeeded in developing a novel combination absorption-inhibition and ELISA method (Poulin, 1988; Cembella et al., 1990), which displays cross-reactivity towards some of the major PSP toxins. In its present format, the IAF STX-kit is based upon a polyclonal antibody produced from rabbit serum, following a series of injections of STX coupled to a synthetic polypeptide.

To test the specificity of this antibody, the cross-reactivity has been evaluated against a number of purified derivatives of STX, including neoSTX, GTX_2, GTX_3, and a mixture of N-21-sulfocarbamoyl compounds (C_{1-4}) (Fig. 1) (Cembella et al., 1990). The STX-antibody was shown to possess a high affinity for STX, while cross-reacting to various degrees with the other derivatives tested No cross-reaction was evident with the sulfocarbamoyl derivatives, nor against the non-tetrapurine biotoxins, domoic acid (amnesiac shellfish poison, ASP), okadaic acid (a diarrheic shellfish poison, DSP) and Staphylococcus enterotoxin B. The relatively broad antigen specificity of this polyclonal STX-antibody when incorporated into an ELISA kit indicates that it may be used as a rapid diagnostic assay for PSP toxins in contaminated shellfish. Although the test kit was designed principally for the detection of these toxins in bivalves, it has also proven to be effective for use with PSP-producing dinoflagellates.

MATERIALS AND METHODS

Mussel extracts for PSP analysis were prepared according to the AOAC protocol for mouse bioassay (AOAC, 1984) Aliquots were sent for mouse bioassays under "double blind" conditions to two independent laboratories (BIOASS1 and BIOASS2). Replicate samples were analyzed by reverse-phase ion-pair HPLC, followed by fluorescence detection of the post-column reaction products, according to the method of Sullivan and Wekell (1986) (Fig. 2). In summary, the method involved binary gradient separation of PSP toxins on a 4.1 X 150 mm polystyrene divinylbenzene analytical column (PRP-1, Hamilton Co., Reno, NV), with hexane- and heptane-sulfonate serving as ion-pair reagents. Mobile phase A was 1.5 mM ammonium phosphate buffer, with 1.5 mM of each ion-pair reagent (pH 6.7); mobile phase B was 6.25 mM ammonium phosphate buffer, containing 1.5 mM ion-pair reagents and 25% V/V acetonitrile (pH 7.0). The separated toxins were oxidized with 5.0 Mm alkaline periodate in 100 Mm sodium phosphate buffer (pH 7 8), followed by neutralization with 0.75 M nitric acid. Fluorescent derivatives were detected with a spectrofluorometer (Perkin-Elmer LS-4, Norwalk, CT) equipped with a 3 μL flow cell.

For comparison, replicate AOAC mussel extracts were analyzed using the "Saxitoxin-Test" immunological kit (Institut Armand-Frappier, Laval, Quebec). The protocol employed was adopted as provided in the instructional pamphlet included in the test kit. Briefly, the technique involved immobilized STX fixed to polystyrene dipsticks, which binds free STX-antibody from the toxic sample/antibody incubation mixture. In the second step, the dipsticks were incubated in Protein A-horseradish-peroxidase conjugate, then a colored reaction product was finally developed in cuvettes containing the enzyme substrate solution. The optical density of the colored product was determined spectrophotometrically at 450 nm, with the color intensity varying inversely with the STX concentration (i.e , zero optical density signifies complete STX absorption, >64 μg STXeq/100 g shellfish tissue). The toxin content of highly toxic samples was determined by serial dilution of the AOAC extracts in the reaction buffer.

RESULTS AND DISCUSSION

The PSP toxicity levels in the mussel samples varied from undetectable to 6,200 μg STXeq/100 g shellfish tissue, according to the mouse bioassay results The wide toxicity variation found in these samples spans the range that might typically be encountered in a depuration facility operating on the Atlantic coast of North America. In terms of detection limits (Table 1), there is an obvious advantage in replacing mouse bioassays by either the HPLC or immunological methods. It should be noted that the IAF ELISA kit was deliberately configured for optimal detection in the range of 5 to 64 μg STXeq/100 g shellfish tissue, with the possibility of extending the upper or lower ranges of detection to >80 or <1 μg STXeq/100 g, respectively. Thus, merely by adjusting the dilution factors, the immunological technique can achieve a sensitivity of at least 10,000 times greater than the HPLC method, and several million times that of the AOAC mouse bioassay, particularly when STX is the major toxin analogue present.

TABLE 1. A comparison of HPLC, IAF ELISA kit and AOAC mouse bioassay detection limits (μg STXeq/100g shellfish tissue) for molluscan shellfish samples prepared according to the AOAC (1984) protocol for PSP toxin extraction.

Toxin	HPLC[1]	IAF ELISA[2]	Mouse Bioassay[3]
B_1	8 4		566
B_2	26.8		566
C_1	0.1	ND	325
C_2	0.1	ND	325
GTX_1	12.3		44
GTX_2	0.5	1.4	84
GTX_3	0.3	1.2	53
GTX_4	4.9		44
neoSTX	2.7	1.6	42
STX	2.3	1.0	42

[1]detection limit defined as twice the baseline noise deviation, based upon peak height; 20 μL injection volume.
[2]detection limit as determined from cross-reaction studies on purified toxins using 250 pg toxin per assay; binding affinity estimated after 5 minute incubation. ND = toxin not detectable; missing values are due to a lack of purified reference standards for these derivatives.
[3]based upon a 1 mL intraperitoneal injection of shellfish extract.
Toxin specific factors for the calculation of toxicity (μg STXeq/100 g) from concentrations (μmol/L) were adopted as compiled by Sullivan and Wekell (l986) A conversion factor of 1 mouse unit = 0.215 μg STXeq was used to interpret the bioassay results

The Spearman correlation coefficient matrix (Table 2) reveals a highly significant correlation between values obtained by these alternative methods. Although the ELISA immunological test is optimized for the detection of STX, and cross-reacts to a lesser extent with certain other PSP toxins (Cembella et al., 1990), the correlation between the results of the immunoassay and the HPLC values, as calculated from total toxin composition, is not significantly different from when the concentration of STX alone is considered. This indicates that in samples from this series of uptake and detoxification experiments under natural conditions, the STX concentration generally tracks the total PSP toxicity.

The application of HPLC techniques to the study of PSP uptake and depuration in molluscan shellfish has several advantages over the conventional mouse bioassay. First, since the toxin spectrum, rather than the integrated toxicity, is elucidated, in depth analysis of the kinetics of uptake, assimilation, and elimination of particular PSP toxin components is possible In this way, the comparative aspects of metabolic versus treatment-assisted toxin conversion and elimination can be investigated. If desired, total toxicity (in μg STXeq or mouse units [MU]) of each sample may be calculated from known toxin specific conversion factors (in μg STXeq/μmol). Second, unlike the mouse bioassay, where non-PSP toxic artifacts can result in inaccurate determinations of total PSP levels, the identification and quantification of PSP toxins by HPLC is usually sufficiently conclusive. The analytical precision of the HPLC method ($\pm$10%; Sullivan and Wekell, 1984) is superior to the bioassay, yet quantitative results, when converted to toxicity units, generally agree well with values obtained by bioassay (Sullivan et al., 1985).

The sensitivity of the HPLC method depends upon toxin-specific fluorescence yield and hardware characteristics (detector sensitivity, pump stability, etc). Nevertheless, detection limits for all PSP toxins are substantially lower than by mouse bioassay (Table 1) Although the samples must be analyzed sequentially, the HPLC system may be automated for virtually continuous operation for extended periods. As only small volumes (<20 μL per injection) are required for PSP analysis, the toxin levels in individual specimens, and even within separate tissue compartments, may be determined

TABLE 2. Correlation matrix of values of the Spearman correlation coefficient ($p < 0.001$) comparing PSP toxicity in cultivated mussels as determined by alternative analytical methods. HPLC(TOT): HPLC, total toxicity; HPLC(STX): HPLC, saxitoxin, ELISA immunoassay; BIOASS2, mouse bioassay, laboratory 1; BIOASS2; mouse bioassay, laboratory 2; n = number of samples.

	HPLC(STX)	HPLC(TOT)	BIOASS1	BIOASS2
HPLC(TOT)	0 95 n=82			
BIOASS1	0 92 n=76	0.93 n=75		
BIOASS2	0.89 n=81	0.91 n=80	0.98 n=81	
ELISA	0.86 n=81	0.88 n=80	0.91 n=79	0 90 n=84

Results of recent studies on PSP toxin uptake, assimilation and detoxification in Mytilus edulis and the hard-clam Mercenaria mercenaria under controlled conditions in small-scale microcosms (Bricelj et al., 1990a,b), have shown the effectiveness of the fluorescence-HPLC method as applied to the study of toxin transfer kinetics. By extending this method to larger depuration operations, it should be possible to evaluate the efficiency of various treatments, and also to model depuration rates of individual toxin components under a variety of regimes.

In principle, immunological techniques are attractive alternatives to the mouse bioassay and HPLC methods for monitoring the uptake and depuration of PSP toxins in shellfish. Such techniques can generally yield assay results that are highly specific for the metabolites of interest, with excellent sensitivity and a rapid analysis time. When configured as an immunological test-kit, they require little detailed training to apply and can often be performed outside of a conventional well-equipped laboratory. Diagnostic test kits are expensive to produce initially, however, when manufactured in quantity, the cost per analysis becomes very competitive with other methods Immunological techniques are thus ideally suited to the needs of aquaculturists, field technicians, and government regulatory laboratories. Such methods could fulfil a critical role in the monitoring of PSP toxins in industrial-scale depuration plants and in smaller experimental facilities.

For a commercial shellfish depuration facility, a rapid turnaround time for analytical results is of crucial importance in maintaining the financial viability of such operations. The ELISA immunological technique may be particularly useful in these circumstances, as a screening technique for PSP toxins. Since such tests can be applied quickly, and qualitative positive results can be observed visually with the naked eye, or more precisely quantified by means of an inexpensive spectrophotometer, the time-course of toxin depuration can be followed without delay or disruption of operations. The IAF immunological test kit is ideally suited for monitoring PSP levels in depuration experiments on molluscan shellfish, where knowledge of the specific toxin spectrum is not critical The test can be applied to crude extracts in <15 minutes and requires a minimum of hardware for toxin detection

Shellfish found to be strongly toxic by the immunological technique may be immediately rejected as unfit for human consumption, or subjected to additional depuration treatment. Weakly toxic samples (<80 μg STXeq/100 g shellfish tissue) or those which yielded negative results by immunoassay could be tested by an alternative broad-spectrum method, such as the bioasssay, to minimize the risk of intoxication by non-PSP toxin components.

Immunoassays should also prove to be valuable to mariculturists and regulatory personnel to identify the in situ presence of PSP toxins in shellfish, finfish and natural assemblages of toxic phytoplankton. By monitoring the phytoplankton for signs of a positive antigenic response, the immunological test can serve as an early warning before significant toxin accumulation in shellfish can occur, thereby reducing the cost of subsequent depuration Taxonomic characterization of potentially toxic species by microscopy is often difficult, due to the lack of equipment or training in commercial mariculture operations and regulatory facilities. Immunological methods may be of particular significance where toxic and non-toxic strains of certain phytoplankton species, which cannot easily be distinguished upon morphological criteria, are known to co-occur.

In summary, a critical weakness of previous studies on depuration of PSP toxins in bivalves has been the dependence on the relatively insensitive and artifact-prone bioassay. Many of the observed discrepancies in the results of ozone depuration experiments may be attributable to the previous lack of reliable alternative detection methods. Both the fluorescence-based HPLC method and immunological techniques can be applied with high precision to yield unambiguous identification of the presence and quantity of PSP toxins. The effectiveness of depuration and "relaying" techniques in reducing PSP toxins to acceptable levels, and the influence of the prior history of the bivalves exposure to Alexandrium on this process remains to be rigorously tested under controlled laboratory conditions For studies on the partitioning of PSP toxins in various molluscan tissues and the relative depuration rates of each compartment, the quantity of each tissue type that may be sacrificed for analysis may be severely limited. Methods such as HPLC and immunoassays which can be applied to small tissue volumes are especially useful in this context

ACKNOWLEDGEMENTS

The authors thank C. Poulin, D. Jones, and Y. Parent (Institut Armand-Frappier) for the production and testing of the STX-antibody. The "STX-Test" immunological kit is manufactured and distributed by Nouveau Concept Technologique, Inc , Terrebonne, Quebec, Canada R. Larocque (Maurice Lamontagne Institute) and M. Desbiens (Ministère d'Agriculture, des Pêcheries et de l'Alimentation du Québec [MAPAQ]) collaborated actively on the field intoxication study and provided the mussel samples for analysis. Mouse bioassays were performed through the courtesy of E. Todd (Microbial Hazards Division, Dept. of Health and Welfare, Canada) and M. Bilodeau (MAPAQ).

REFERENCES

Association of Official Analytical Chemists. 1984. Procedure 18.086-18.092 Official Methods of Analysis, 14th edition, W. Horowitz, ed., Washington.

Bates, H A. and Rapoport, H. 1975. A chemical assay for saxitoxin, the paralytic shellfish poison. J. agric. Food Chem 23: 237-239

Blogoslawsi, W.J. 1988 Ozone depuration of bivalves containing PSP: pitfalls and possibilities J Shellfish Res 7 702-705

Blogoslawsi, W.J. and Stewart, M.E. 1978. Paralytic shellfish poison in Spisula solidissima: anatomical location and ozone detoxification Mar. Biol. 45: 261-264.

Blogoslawski, W.E., Stewart, M.E., Hurst, J W. Jr , and Kern, F.G III. 1979. Ozone detoxification of paralytic shellfish poison in the softshell clam (Mya arenaria). Toxicon 17: 650-654.

Bricelj, V.M., Lee, J.H , Cembella, A.D. and Anderson, D.M. 1990a. Uptake kinetics of paralytic shellfish toxins from the dinoflagellate Alexandrium fundyense in the mussel Mytilus edulis. Mar. Ecol. Prog. Ser. 63: 177-188.

Bricelj, V.M., Lee, J.H., Cembella, A D , and Anderson, D.M 1990b. Uptake of Alexandrium fundyense by Mytilus edulis and Mercenaria mercenaria under controlled conditions. In. "Toxic Marine Phytoplankton", E Granéli, B. Sundström, L Edler and D.M. Anderson, eds , Elsevier, New York, pp 269- 274.

Buckley, L.J., Oshima, Y. and Shimizu, Y. 1978. Construction of a paralytic shellfish toxin analyzer and its application. Anal. Biochem 85· 157-164.

Carlson, R.E., Lever, M.L., Lee, B.W. and Guire, P.E. 1984. Development of immunoassays for paralytic shellfish poisoning, a radioimmunoassay for saxitoxin. In: Seafood Toxins, ACS Symposium Series 262, E P Ragelis, ed., Amer. Chem. Soc., Washington, pp. 181-192.

Cembella, A D., Sullivan, J.J., Boyer, G L., Taylor, F.J.R., and Andersen, R,J,. 1987 Variation in paralytic shellfish toxin composition within the Protogonyaulax tamarensis/catenella species complex Biochem. System. Ecol. 15: 171-186.

Cembella, A D , Parent, Y., Jones, D. and Lamoureux, G. 1990. Specificity and cross-reactivity of an absorption-inhibition enzyme-linked immunoassay for the detection of paralytic shellfish toxins. In· "Toxic Marine Phytoplankton" E. Granéli, B. Sundström, L. Edler and D.M. Anderson, eds., Elsevier, New York, pp. 339-344.

Chu, F.S. and Fan, T.S.L l985. Indirect enzyme-linked immunosorbent assay for saxitoxin in shellfish. J. Assoc. Off. Anal. Chem. 68: 13-16.

Davio, S.R., Hewetson, J.F. and Beheler, J.E. l985. Progress toward the development of monoclonal antibodies to saxitoxin; antigen preparation and antibody detection. In: Toxic Dinoflagellates, D.M. Anderson, A.W. White and D G Baden, eds., Elsevier-North Holland, New York, pp. 343-348.

Dawson, M A , Thurberg, F.P., Blogoslawski, W.J., Sasner, J.J., Jr., and Ikawa, M l976 Inactivation of paralytic shellfish poison by ozone treatment. In: Proc. Food-Drugs from the Sea Conf , H.H. Webber and G.D Ruggieri, eds., Marine Technology Society, Washington, pp. 152-157.

Hurst, J.W., Selvin, R., Sullivan, J.J., Yentsch, C M., and Guillard, R.R.L. l985. Intercomparison of various assay methods for the detection of shellfish toxins. In: Toxic Dinoflagellates, D.M. Anderson, A W White and D.G. Baden, eds., Elsevier-North Holland, New York, pp. 427-432.

Johnson, H M , Frey, P.A., Angelotti, R , Campbell, J.E , and Lewis, K.H l964 Haptenic properties of paralytic shellfish poison conjugated to proteins by formaldehyde treatment (29599). Proc Soc Exp Biol. Med 117· 425-430

Nagashima, Y , Maruyama, J., Noguchi, T., and Hashimoto, K. l987. Analysis of paralytic shellfish poison and tetrodotoxin by ion-pairing high performance liquid chromatography. Bull Japan. Soc. Sci Fish 53. 819-823.

Oshima, Y., Machida, M., Sasaki, K., Tamaoki, Y., and Yasumoto, T. l984. Liquid chromatographic-fluorometric analysis of paralytic shellfish toxins. Agric. Biol. Chem 48: 1707-1711

Poulin, C. l988. Production d'anticorps contre la saxitoxine et elaboration d'un essai immunologique pour diagnostiquer la toxicité des mollusques M Sc thesis, Inst Armand-Frappier, Laval, Quebec. 174 pp.

Prakash, A , Medcof, J.C., and Tennant, A D. 1971. Paralytic shellfish poisoning in eastern Canada Fish Res. Bd Can , Bull 177 87 pp

Shimizu, Y. 1987. Dinoflagellate toxins. In: "The Biology of Dinoflagellates", F.J.R. Taylor, ed., Blackwell Scientific Press, Oxford, pp. 282-315.

Shoptaugh, N.H , Carter, P.W., Foxall, T.L., Sasner, J.J., Jr. and Ikawa, M. 1981. Use of fluorometry for the determination of Gonyaulax tamarensis var excavata toxins in New England shellfish J. agric. Food Chem. 29 198-200.

Sommer, H. and Meyer, K.F l937. Paralytic shellfish poisoning. Arch. Pathol 24· 560-598.

Sullivan, J.J., Iwaoka, W.T., and Liston, J l983 Enzymatic transformation of PSP toxins in the littleneck clam (Protothaca staminea) Biochem. Biophys Comm. 114: 465-472.

Sullivan, J.J. and Wekell, M.M. 1984. Determination of paralytic shellfish poisoning toxins by high pressure liquid chromatography. In: Seafood Toxins, E.P. Ragelis, ed., ACS Symposium Series 262, Amer. Chem. Soc., Washington, pp. 197-205.

Sullivan, J.J. and Wekell, M.M. l986. The application of high performance liquid chromatography in a paralytic shellfish poisoning monitoring program. In: Seafood Quality Determination, Developments in Food Science, Vol 15, D.E. Kramer and J. Liston, eds , Elsevier, New York, pp. 357-371.

Sullivan, J.J , Jonas-Davies, J. and Kentala, L L. l985. The determination of PSP toxins by HPLC and autoanalyzer. In: Toxic Dinoflagellates, D.M. Anderson, A.W. White and D.G. Baden, eds., Elsevier-North Holland, New York, pp. 275-280.

Taylor, F.J.R. l984. Toxic dinoflagellates: taxonomic and biogeographic aspects with emphasis on Protogonyaulax. In: Seafood Toxins, ACS Symposium Series 262, E.P. Ragelis, ed , Amer Chem. Soc., Washington, pp. 77-97.

White, A.W., Martin, J L , Legresley, M and Blogoslawski, W.J. l985. Inability of ozonation to detoxify paralytic shellfish toxins in soft-shell clams In: Toxic Dinoflagellates, D.M. Anderson, A.W White and D G. Baden, eds., Elsevier-North Holland, New York, pp. 473-478

THE DEPURATION OF PACIFIC OYSTERS
(Crassostrea gigas)

G.C. Fletcher, P.D. Scott and B.E. Hay

INTRODUCTION

In 1978, a major outbreak of viral gastroenteritis in New South Wales, Australia resulted in the stringent requirement that all oysters sold in that state be depurated. In that Australia has always been New Zealand's largest export market for farmed oysters, this gave the initial impetus for a series of studies on the effectiveness of depuration of Pacific oysters (Crassostrea gigas) grown under New Zealand conditions.

METHODS

LABORATORY SCALE BACTERIAL DEPURATION

A closed system of five 40L insulated glass tanks (Fig. 1) was used for laboratory depuration trials (Buisson et al., 1981). The water was sterilized by ultraviolet (UV) light, passed through a 25 micron filter (Johns Manville P5 ES-1) at temperatures between 5°C and 25°C. The tanks were artificially aerated. Oysters were intentionally contaminated by holding them for at least a week in local areas known to have high fecal coliform levels. The surfaces were washed lightly before being placed in the tanks. In the tanks the oysters were held on trays 10 cm above the tank floor to allow feces to fall away to the tank floor.

Salinity and oxygen levels were regularly recorded in each trial using YSI meters. Salinity was kept within 20% of that of the growing waters and ranged from 24 to 33.3 mg/g. Dissolved oxygen levels ranged from 6.5 to 11.8 mg/kg and always exceeded 50% saturation. The experimental temperatures were never more than 3°C different from those of the growing water. The oyster density of the experimental tanks was kept constant with replacement oysters from a reserve tank. The efficiency of the UV lamp was regularly evaluated by passing water inoculated with 10^3 to 10^8 cells of Escherichia coli/100 mL through the lamp in ten successive passes. This always showed a decline in viable bacteria of at least 3 orders of magnitude after the first three passes. The efficiency of the lamp was such that fecal coliforms in the water were always recorded at less than 2/100 mL during the trials.

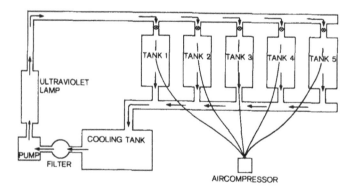

FIGURE 1. Laboratory scale depuration plant (Buisson et al., 1981)

One hundred and thirty one trials were preformed over the course of 15 months. Trials were carried out to determine the effects of: different flow rates through the tanks, different loading levels in the tanks (0.5, 1, 1.2, or 3 oysters/liter), different water temperatures (5.3 to 25.0°C), and oysters with microbial loadings ranging from 20 to 210,000 most probable number (MPN) per 100g. Each trial lasted 3 days. Fecal coliforms were determined in samples of 5 oysters every 24 h, using the modified A1 procedure (Hunt and Springer, 1978).

BACTERIAL MONITORING OF A COMMERCIAL PLANT

The first commercial depuration plant in New Zealand (Tide Farm, Mahurangi), was monitored for the first year of its operation. The plant was carrying out depuration on a 36 hour cycle. Samples of 12 oysters were randomly taken at the beginning and end of each run. A more intensive monthly program was also carried out where samples were taken from different locations in the tanks every 12 hours. Twelve oysters were pooled and tested for fecal coliforms. Eleven runs were tested for Vibrio parahaemolyticus (Fletcher, 1985) using enrichment in salt polymyxin broth, selection of typical colonies from Thiosulfate Citrate Bile Salts Sucrose Agar (TCBS) and confirmation in the medium of Kaper et al (1980).

VIRAL EXPERIMENTS

Cricket paralysis virus (CrPV), an insect picornavirus, was chosen as an easily assayed model virus to study the uptake and depuration of viruses from oysters. It has many physicochemical and biological properties in common with the vertebrate enteroviruses. CrPV was replicated in Drosophila melanogaster cells. Virus RNA was labelled with ^{32}P-orthophosphate (Scotti, 1976), or with ^{3}H-uridine (Scotti et al., 1983) and virus coat protein with ^{125}I (Hay & Scotti, 1986 using the methods of Fraker & Speck, 1978).

VIRUS UPTAKE AND DEPURATION

Ten live oysters were placed in one 10 L tank of clarified seawater and 10 oysters killed by freezing were placed in another. One hour later, ^{32}P-labelled CrPV was added to each tank along with bovine serum albumin (25 μg/ml) to reduce nonspecific binding of the virus to container walls. The amount of ^{32}P remaining in the seawater was monitored using a liquid scintillation spectrophotometer.

For uptake experiments, infectious CrPV was added to the laboratory scale depuration tanks, each of which contained up to 100 live oysters. The number of infective virus was assayed (Scotti, 1977) using the sample preparation method described in Scotti et al. (1983). Viral depuration studies, using this assay method, were also carried out in a commercial sized tank.

VIRAL LOCATION

In a separate study, autoradiography was carried out to determine the location of radioactively-labelled virus RNA and protein in the oyster tissues before depuration and after treatment for 12, 24, 36, and 64 h (Hay & Scotti, 1986).

RESULTS

LABORATORY SCALE BACTERIAL DEPURATION

Depuration was most rapid during the first 24 h with the rate of purification leveling off markedly after this time (Fig. 2). Minimum levels of fecal coliforms were usually reached within the first 24 h. Attempts to enhance depuration during the second and third days by feeding were unsuccessful. The amount of depuration was related to the initial level of contamination and the water temperature (Figs. 2 & 3). More fecal coliforms could be depurated at higher temperatures.

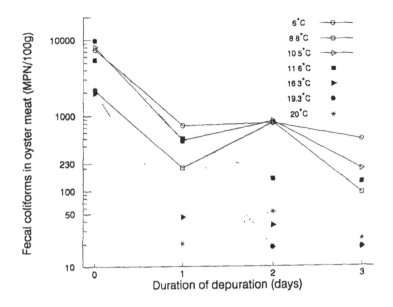

FIGURE 2. Depuration of oysters with fecal coliform levels
between 1000 and 10,000 MPN/100 g.

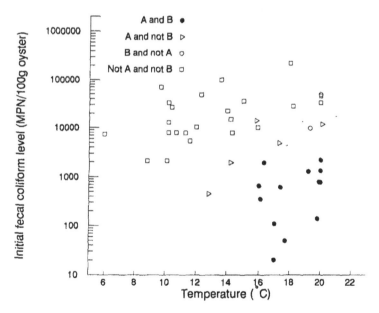

FIGURE 3 The effect of initial fecal coliform levels and
temperature on the success of depuration. A = fecal
coliform levels below 230 MPN/100g in 24 h; B = fecal
coliforms below 50 MPN/100g in 48 h. Each point
represents 1 to 4 tanks.

Maximum levels of fecal coliforms that could confidently be depurated at different temperatures can be derived from Figure 3.

BACTERIAL MONITORING OF A COMMERCIAL PLANT

Although 92 commercial runs were monitored, only 20 of these had initial fecal coliform levels exceeding the limit of 230/100g. When the plant was properly run, all depurated samples had fecal coliform levels less than 50 MPN/100g. The results of the laboratory scale trials were confirmed.

Vibrio parahaemolyticus was detected in oysters entering the plant on 9 of the 11 occasions tested (Table 1). Although depuration generally resulted in a reduction of the levels of V. parahaemolyticus, the reductions were not large and, in run 77, an increase from 230 to 462 MPN/100g was observed. This happened when water temperatures were particularly high and V. parahaemolyticus would be multiplying rapidly.

Viral uptake
The amount of radioactive CrPV particles remaining in the seawater indicated that approximately 50% of the added virus was taken up by the live oysters compared with approximately 20% by the killed oysters. Uptake reached a maximum within 12 h of virus being added to the tanks. Using infectious virus which was assayed in tissue culture, confirmed the rapid uptake of virus. The amount of CrPV in the oysters in pooled samples after 16 h of uptake was similar to that after 6 h. Individually examined, similarly treated oysters took up variable numbers of infectious particles (10^4-10^6/oyster). However, the weight of the individual oysters was not related to the amount of virus taken up.

Table 1. Effect of commercial depuration on Vibrio parahaemolyticus
and fecal coliform levels (MPN/100 g). (Fletcher, 1985)

Analysis	Run Number										
	57	58	61	66	67	68	77	79	80	87	89
Temp °C	16	18	20	--	--	19	24	18	--	16	15
Vibrio p. MPN/100g											
0 h	>40	20	>40	110	9.1×10^6	390	230	<100	160	91	<30
36 h	13	10	20	49	314	<30	462	<100	60	36	<30
Fecal coliform MPN/100g											
0 h	130	3500	20	<18	20	110	130	45	20	330	330
36 h	44	776	<18	18	<18	<18	28	<18	<18	<18	<18

Vibrio p. = Vibrio parahaemolyticus

Viral depuration
Although the UV-sterilizing system inactivated CrPV in the water to below detectable levels within a few hours, the amount of infectious virus in pooled oysters was reduced only 20-fold after 5 days depuration in the laboratory scale tanks (Fig. 4). Assaying individual oysters from the laboratory scale tanks showed a wide range of depuration levels (Fig. 5). These ranged from oysters showing little if any depuration after 10 days to those showing approximately 100-fold depuration.

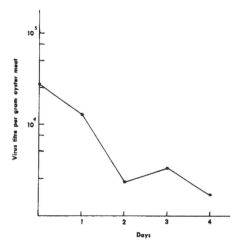

FIGURE 4   Titre of CrPV during depuration - pooled
samples (Scotti et al. 1983).

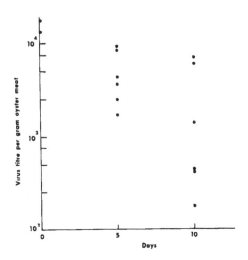

FIGURE 5.   Titre of CrPV during depuration - indivdual
samples (Scotti et al. 1983).

In a trial using a commercial sized tank, viral depuration was negligible over 9 days although fecal coliforms were depurated very effectively (Fig. 6). This demonstrates the potential problems, both in relying on depuration to purify shellfish with respect to viruses and in using fecal coliforms as indicators of virus levels in depurated shellfish.

Viral location after uptake

Autoradiographs consistently showed radioactive label concentrated in the same areas of tissue of each oyster High concentrations were present in the mucus throughout the digestive tract (Fig. 7) This indicates that the virus enters the oyster by bonding to the mucus. Lower concentrations were also detected within the epithelial cells of the digestive diverticula tubules (Fig. 8) and in the epithelium of the midgut (Fig. 9). In some instances, label was concentrated along the basement membrane of the epithelium (Fig. 10). Thus a small proportion of the virus was released from the mucus in the stomach. The coarsely granular distribution of label in the epithelial cells of the digestive diverticula tubules suggests that the virus was ingested by endocytosis. There was also label scattered throughout the connective tissue surrounding the digestive tract, particularly around blood sinuses and arteries (Fig. 11). This indicates that the virus was not discharged into the lumen of the tubules with the waste products of intracellular digestion but passed out through the basement membrane of the epithelial cells into the connective tissue and haemocoel. The distribution of the label within the tissues was the same regardless of whether virus genome or coat was labelled. This implies that the virus was not uncoated when it was absorbed intracellularly.

Viral location after depuration

After 64 h depuration, label was still evident in the epithelium of the digestive diverticula tubules, along the basement membrane of the ascending midgut and in the connective tissues (Fig. 10). By this time, the amount of label in the mucus had visibly decreased and was not detected in the ducts leading to the digestive tract (Fig. 12).

These results indicated that the virus entered the oyster and bound to the mucus. A large proportion of this passed from the stomach through the intestine, to be excreted with the feces. A smaller amount of virus was released from the mucus in the stomach and some of this was ingested by endocytosis. It is hypothesized that these were the viral particles that were not removed by depuration.

CONCLUSIONS

Parameters for the successful depuration of fecal coliforms from Pacific oysters have been determined. Although the depuration of fecal coliforms is assumed to correlate with the depuration of other Enterobacteria, the same cannot be said of marine vibrios, specifically V. parahaemolyticus.

Virus particles are rapidly accumulated by Pacific oysters although the rate and degree of accumulation varies from oyster to oyster. Although viruses are initially attached to the mucus in the digestive tract, some are absorbed intracellularly into epithelial cells of the midgut. These viruses appear to pass out through the basement membrane of the epithelial cells into the connective tissue and haemocoel. Depuration of virus from Pacific oysters is unreliable, apparently due to the intracellular absorption of intact virus particles.

Depuration is therefore not recommended for oysters that have been harvested from areas likely to have been subjected to viruses which pose a threat to human health. Depuration may be a useful technique for reducing bacterial pathogens to safe levels in water contaminated solely from animal sources. It is not a suitable means of eliminating hazards due to viruses in waters contaminated by human sewage.

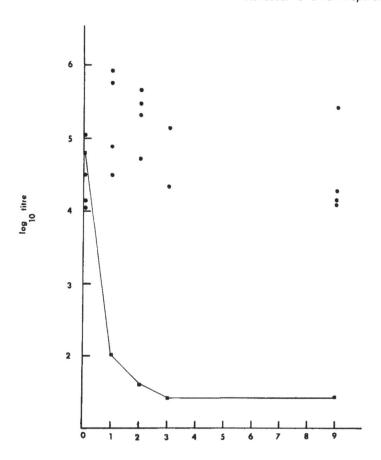

FIGURE 6. Depuration rates of E. coli and CrPV - individual samples
per day; ■=bacterial/mL, •=CrPV infectious units/mL

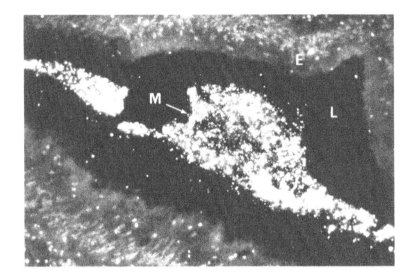

FIGURE 7. Cross-section of the acending mid-gut of C. gigas showing virus in mucus in lumen after 2 hours uptake E = Epithelium; L = Lumen; M = Mucus

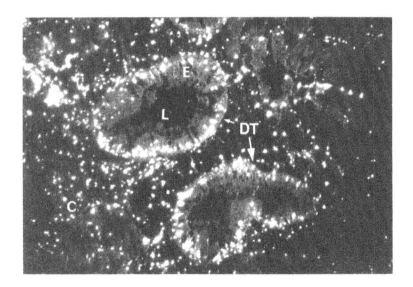

FIGURE 8. Virus in epithelium of digestive diverticula tubules and in surrounding connective tissue after 12 hours uptake. C = Connective tissue; DT = Digestive diverticula tubule; E = Epithelium, L = Lumen

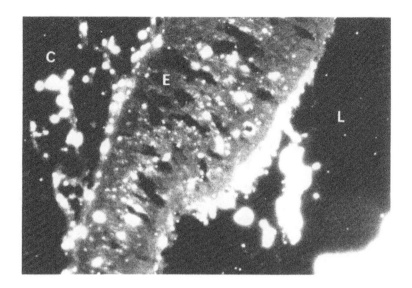

FIGURE 9. Section through wall of mid-gut showing the presence of
virus in the epithelium C = Connective tissue; E = Epithelium,
L = Lumen.

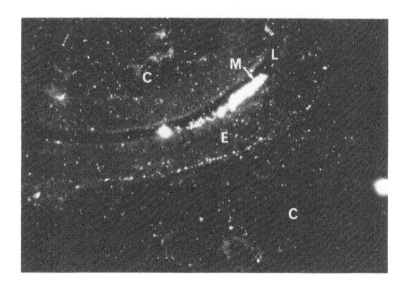

FIGURE 10 Virus in muscus within the ascending mid-gut, along the
basement membrane of the epithelium and in the surrounding
connective tissue after 12 hours depuration. C = Connective
tissue; E = Epithelial cells; L = Luman; M = Mucus

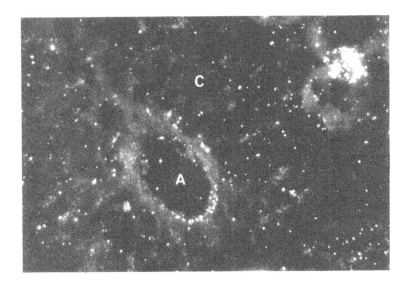

FIGURE 11 Virus concentrated in the cells lining an artery and in the surrounding connective tissue after 12 hours of uptake, A = Artery, C = Connective tissue.

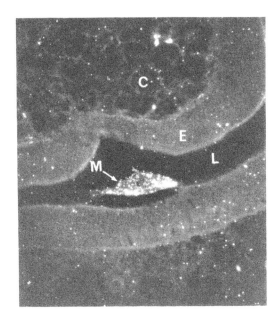

FIGURE 12. Virus in mucus of ascending mid-gut after 64 hours depuration. C = Connective tissue; E = Epithelium of mid-gut; L = Lumen of mid-gut; M = Mucus.

REFERENCES

Buisson, D H., Fletcher, G. C., and Begg, C. W. 1981:Bacterial depuration of the Pacific oyster (Crassostrea gigas) in New Zealand. N.Z.J.Sci. 24:253.

Fletcher, G. C. 1985: The potential food poisoning hazard of Vibrio parahaemolyticus in New Zealand Pacific Oysters. N.Z.J.Mar.Freshwater.Res. 19:495.

Fraker, P. J., and Speck, J. C. Jr. 1978: Protein and cell membrane iodinations with sparingly soluble chloramide, 1,3,4,6-tetrachloro-3α,6α-diphenylglycouril. Biochemical and biophysical research communications 80:849.

Hay, B., and Scotti, P. 1986: Evidence for intracellular absorption of virus by the Pacific oyster, Crassostrea gigas. N.Z J.Mar.Freshwater Res. 20:655.

Hunt, D. A., and Springer, J. 1978: Comparison of two rapid test procedures with the standard E C test for the recovery of faecal coliform bacteria from shellfish growing waters.　JAOAC. 16(6):1317.

Kaper, J. B., Remmers, E. F., Colwell R. R. 1980: A medium for presumptive identification of V. parahaemolyticus. J Food Protect. 43(12):936-938.

Scotti, P. D. 1976: Cricket paralysis virus replicates in cultured Drosophila cells. Intervirology 6.333-342.

Scotti, P. D 1977: End point dilution and plaque assay methods for titration of cricket paralysis virus in cultured Drosophila cells. Journal of General Virology 35:393

Scotti, P. D., Fletcher, G. C., Buisson, D. H., and Frederickson, S. 1983: Virus depuration of the Pacific oyster (Crassostrea gigas) in New Zealand. N.Z.J.Sci. 26:9.

DETOXIFICATION OF BIVALVE MOLLUSCS NATURALLY CONTAMINATED WITH DOMOIC ACID

David J. Scarratt, M.W. Gilgan, R. Pocklington, and J.D. Castell

INTRODUCTION

In November 1987, serious but unusual illnesses were reported by persons eating cultivated mussels, Mytilus edulis, from Prince Edward Island, Canada., (P.E.I.) and controls were imposed on the industry while the cause was determined. When it was determined that these illnesses were caused by mussels contaminated by a hitherto unknown neurotoxin, it was agreed that samples of toxic mussels would be taken at regular intervals from one of the leases known to be affected, and transferred to the quarantine unit at the Halifax Laboratory. This would permit the toxin content to be monitored over time in controlled conditions, and an estimate made of the rate of their depuration It would also allow for monitoring the subsequent development and mortality of the mussels in field and experimental conditions During the course of the emergency some 153 persons suffered symptoms ranging from mild nausea and vomiting to severe illness resulting in persistent memory loss. Three fatalities were attributed to this cause (Perl, 1989), which is now referred to as Amnestic Shellfish Poisoning (ASP).

By mid-December 1987 the contaminant had been identified as domoic acid, (Bird et al. 1988, Bates et al. 1988) and analytical methodology using HPLC was developed for its detection and quantification (Quilliam et al. 1988).

In January of 1988, the source of the toxin was shown unequivocally to be the pennate diatom, Nitzschia pungens (RAO et al., 1988; Bates et al. 1989). Institution of a coast-wide monitoring program for domoic acid in shellfish lead to progressive re-opening of the shellfisheries by April 1988. Subsequently, domoic acid concentrations which exceeded the newly established acceptance level for human consumption of 20ug/g soft tissue were detected in clams Mya arenaria and mussels Mytilus edulis, and resulted in closure of clam flats in Passamaquoddy Bay, N.B. in the late summer of 1988 (Burridge, 1989). In the Cardigan Estuary, P.E.I. in Nov-Dec. 1988, domoic acid content of mussels exceeded the newly established control level of 20 μg/g. Both incidents allowed for some of the earlier depuration experiments to be repeated and extended.

Besides concern for mussels, there was some apprehension that other commercial species might become toxic either through direct ingestion of the source organism, or indirectly by consuming toxic mussels. Lobsters support an important fishery in the area and include mussels in their diet (Scarratt, 1980). Hence it was decided to attempt to feed mussels to lobsters, and monitor the accumulation, if any, and study depuration as a means to remove domoic acid.

METHOD

All depuration experiments were conducted at the Halifax Laboratory quarantine unit. Filtered water from Halifax Harbour is available at more of less ambient temperature, 15 C and 20 C. Mussels (or clams) from the contaminated areas were loosely strewn in fibre glass tanks liberally supplied with running water which was subsequently collected into a series of holding tanks and subjected to a minimum 30 minutes contact with chlorine at 15 ppm. This minimized the chance of introducing N pungens into Halifax Harbour. Controls from local, uncontaminated sources were placed in adjacent tanks. In the earliest runs there was considerable pressure on the analytical capabilities, and sampling intervals were long and irregular. There was, subsequently, clear evidence that initial depuration rates were very high, and all later runs were done with initial sampling intervals of 24 hours.

Lobsters (approx. weight 500 g) were held at 13 C and fed either contaminated or toxin free mussel flesh. At first feeding lobsters ate toxic mussels readily but at the second and subsequent feedings they avoided the digestive gland and only ate it if starved. Lobsters can clearly identify and avoid domoic acid contaminated mussels, or the toxic parts thereof. Lobsters were sacrificed at regular intervals, and tail muscle and digestive gland analyzed for domoic acid. Later, in a more carefully designed experiment, 20 g lobsters were fed a standard crustacean diet formulation (Castell, et al. 1989) with the normal crab protein constituents replaced with equivalent amounts of freeze dried toxic or non-toxic mussels.

For each analysis, ten mussels were dissected and the meats pooled and blended with a polytron. Initially, extractions were done according to the approved analytical method for PSP testing by mouse biassay (Lawrence et al., 1989), which requires careful pH control, filtering, and bringing to standard volume. This is unnecessary for domoic acid and later extractions were done simply by boiling an aliquot of the blended meats with distilled water, followed by centrifugation and washing, and the extracts made up to a standard volume. Samples were prepared for HPLC by treatment with an equal volume of methanol followed by centrifugation and filtering at 0.2μm. Extracts of lobster hepatopancreas and tail meat were treated similarly. Aliquots were diluted with water and analyzed by HPLC with a ultraviolet (UV) detector at 242 nm. Chromatographic procedures were equivalent to those described by Bird et al , (1988). Clean-up and analysis of boiling water extracts were by the methods of Quilliam et al. (1988). At the time of the observations, no reliable domoic acid standard was available, therefore levels shown are not necessarily absolute. All results are given as μg/g wet weight so that there may be direct comparison with the standards used for regulating the sale of toxic shellfish.

RESULTS AND DISCUSSION

WINTER 1987-88

Early values for domoic acid content are reported in the 500-600 μg/g range during early-mid November 1987 (Gilgan, unpubl. data, 1989). Our earliest pooled sample of mussels from Cardigan River had a value of 160 μg/g while individual mussels frozen at that time and analyzed in early February ranged from 20-290 μg/g, with a mean of 145 μg/g. Twelve mussels taken in Cardigan on January 11 were assayed individually. Domoic acid content ranged from 0-70 μg/g with a mean of 30. Individual mussels either take up or lose domoic acid at different rates.

Depuration curves (Fig. 1) clearly show a more rapid depuration of mussels held in the Quarantine Unit than in the field. Domoic acid content of native mussels from Cardigan was 30 μg/g in early March 1988, as opposed to zero in lab-depurated specimens. This may be due to the continuing prescence of Nitzschia in the plankton throughout that time (Bates et al. 1988), and perhaps a lower elimination rate in the colder water.

LATE SUMMER 1988

Domoic acid content of clams and mussels held at 13 C dropped from 43 to 15 ppm in 24 hours (Fig. 2), but traces still remained as much as 6 days after harvesting. Depuration at 7 C was somewhat slower. Mussels appear to depurate somewhat more rapidly than do clams but may also have higher initial contamination levels.

FALL-WINTER 1988-89

Two runs were completed (Figs. 3, 4). Domoic acid content dropped from 130 to 20 ppm in 4-6 days at 15 c, and in 7-9 days at 7 C. Traces remained after 14 days depuration at the cooler temperature. The question arises whether the presence of food other than Nitzschia would accelerate the depuration process.

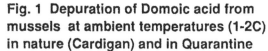

Fig. 1 Depuration of Domoic acid from mussels at ambient temperatures (1-2C) in nature (Cardigan) and in Quarantine

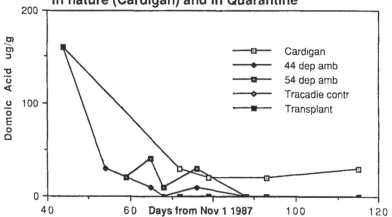

Fig. 2 Depuration of Domoic acid from starved Mussels and Clams. Sept 88

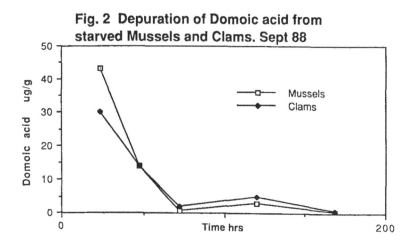

Fig.3 Depuration of Domoic acid from starved mussels, Experiment 1 Dec 1988.

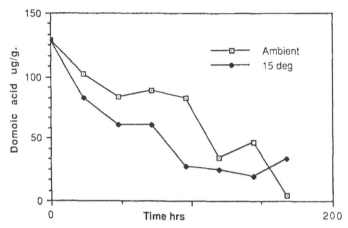

Fig. 4 Depuration of Domoic acid from Mussels. Experiment 2. Dec.1988

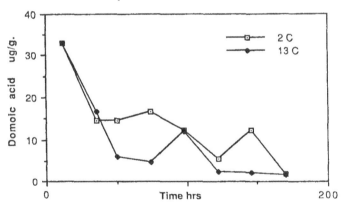

LOBSTERS

After 8 weeks feeding on a exclusive diet of domoic acid-contaminated whole mussels, the concentration of domoic acid reached 70 μg/g in the digestive gland, but was not detectable in the tail meat. Direct observation showed that after an initial meal of contaminated mussels, adult lobster preferentially ate the gills, mantle, gonad and musculature of the mussels and either avoided the digestive gland or consumed it last. In the experiment with juvenile lobsters of known history fed the formulated diet there was a marked reluctance to eat the toxic formulation, and the condition of lobsters fed this formulation deteriorated markedly in contrast to controls: growth, molt frequency and vigour were all lower; and mortality was higher (Castell, unpubl. data, 1989).

UPTAKE OF DOMOIC ACID BY MUSSELS

Attempts to have mussels concentration domoic acid from solutions in filtered sea water or from a C. gracilis culture spiked with domoic acid at a concentration sufficient to yield 500 μg/g in mussel flesh were not successful. The mussels were in all respects healthy, ready to spawn, and had fed heavily, but no domoic acid was detected in them. Mussels fed cultures of N. pungens have been shown to take up domoic acid (Martin, 1989; Wohlgeshaffen, 1989; and Scarratt, unpubl data, 1989), but Nitzschia is proving difficult to culture and, in culture at least, only seems to produce domoic acid after log-phase of growth is complete (de Freitas, 1989; Bates, 1989).

Now that a monitoring program has been established which regularly examines shellfish from both nearshore and offshore areas, it has been detected in other species at a number of locations far from the initial crisis area of eastern Prince Edward Island, and including sea scallops, Placopecten magellanicus from Georges bank. This is of no commercial concern provided the tradition of shucking scallops at sea and marketing meats only is retained. Domoic acid is largely restricted to the digestive gland, with only minor contamination of the roe, while the meats are never contaminated (Gilgen, unpubl data, 1989). In clams the total body burden of domoic acid has at times exceeded the 20 μg/g control limit, sufficient to require closure of the commercial fishery. As for cultured shellfish, most mussel culture operations will likely have the option of waiting until a Nitzschia bloom has passed and the shellfish depurated naturally. Canadian growers of Bay Scallops Argopecten irradians may not be so fortunate as the shellfish are marketed whole, as a cash crop marketed in late fall. Bay Scallops will not overwinter in Canadian East Coast waters, therefore there is danger of losing a crop if depuration techniques are not developed.

CONCLUSIONS

Domoic acid is initially eliminated extremely rapidly, probably by reason of its solubility in water, therefore commercial depuration should be feasible. While Canadian regulations only permit the direct sale of molluscan shellfish from areas approved under the National Shellfish Sanitation Program, it seems likely that depuration of domoic acid could be accomplished using relatively simple procedures, although the protocols are not yet established. For most species the best procedure will simply be to wait for the bloom to pass Mussels and clams appear to be none the worse for exposure to domoic acid; however, for seasonal species such as cultivated bay scallops which are marketed in November-December, depuration may be the only means of salvaging a crop.

Lobsters appear to have a natural fastidiousness which will protect them against contamination by domoic acid in nature, and the requirement for intervention in the fishery in the event of outbreaks of domoic acid contamination in molluscs seems unlikely.

REFERENCES

Bates, S.S., Bird, C.J., Boyd, R.K., deFreitas, A.S W., Falk, M., Foxall, R.A., Hanic, L.A., Jamieson, W.D., McCulloch, A.W., Odense, P., Quilliam, M.A , Sim, P.G., Thibault, P., Walter, J A., and Wright, J.L.C. 1988. Investigations into the source of domoic acid responsible for the outbreak of amnesic shellfish poisoning (ASP) in eastern Prince Edward Island. Atlantic Research Laboratory Tech. Rept. 57, NRC 29086.

Bates, S.S., Bird, C.J., Boyd, A.S.W., de Freitas, M., Foxall, R.A., Gilgan, M.W., Hanic, L.A., Johnson, G.E., McCulloch, A.W., Odense, P., Pocklington, R., Quilliam, M.A., Sim, P.G., Smith, J.C., Subba Rao, D.V., Todd, J.A., Walter, J., and Wright, J.L.C. 1989. Pennate diatom Nitzschia pungens as the primary source of domoic acid, a toxin in shellfish from eastern Prince Edward Island, Canada. Can. J. Fish. and Aquat. Sci. 46, 12202-1215.

Bird, C.J., Boyd, R.K., Brewer, D., Craft, C., de Freitas, A.S.W., Dyer, E.W., Embree, D.J., Falk, M., Flack, M.G, Foxall, R.A., Gillis, C., Greenwell, M., Hardstaff, W.R., Jamieson, W.S., Laycock, M.V., Leblanc, P., Lewis, N.I., McCulloch, A.W., McCully, G.K., McInerney-Northcott, M., McLnnes, A.G., McLachlan, J.L , Odense, P., O'Neill, D., Pathak, V.P., Quilliam, M.A., Ragan, M.A., Seto, P F., Sim, P.G., Tappan, D., Thibault, P., Walter, J.A., Wright, J.L.C., Backman, A.M., Taylor, A.R., Dewar, D., Gilgan, M., and Richard, D.J.A. 1988. Identification of domoic acid as the toxic agent responsible for the P.E.I. contaminated mussel incident. Atlantic Research Laboratory Technical Report, #56, NRCC #29083

Burridge, L. 1989. Domoic acid in Mussels Mytilus edulis from Passamaquoddy Bay. In: Workshop on Harmful Marine Algae, Moncton N.B. Sept. 27-28, 1989. eds. SS. Bates and J. Worms, Can. Tech. Rep. Fish. and Aquat. Sci. 1712, pp. 58.

Castell, J.D., Kean, J.C., D'Abramo, L.R., and Conklin, D.E. 1989. Standard reference diet for crustacean nutrition research. J. World Aquac. Soc. 20(3)100-106.

Castell, J.D. 1989. Unpublished data. Dept Fisheries and Oceans, Halifax, Canada.

de Freitas, A.S.W. 1989. Domoic acid: some biological and ecological considerations on the production of this neurotoxic secondary metabolite by the diatom Nitzschia pungens. In: Workshop on Harmful Marine Algae, Moncton N.B. Sept. 27-28, 1989. Unpublished.

Gilgan, M.W. 1989. Unpublished data. Dept. Fisheries and Oceans, Halifax, Canada.

Lawrence, J.F., Charbonneau, C.F., Menard, C., Quilliam, M.A., and Sim, P.G. 1989. Liquid chromatographic determination of domoic acid in shellfish products using the AAOC paralytic shellfish poison ectraction procedure. J. Chromatogaphy 462, 329-256.

Martin, J. 1989. Personal communication Dept. Fisheries and Oceans, St. Andrews, New Brunswick.

Perl, T. 1989. Human health aspects of marine phycotoxins. In: Canadian Workshop on Harmful Marine Algae, Moncton N.B. Sept. 27-28, 1989. Can. Tech. Rep. Fish. and Aquat. Sci. 1712, pp. 58.

Quilliam, M.A., P.G. Sim. A.W. McCulloch and A.G. McInnes. 1988. Determination of Domoic acid in shellfish tissue by high performance liquid chromatography. Atlantic Research Laboratory Tech. Rept. 55. NRC 29015, 17p.

Rao, Subba, D.V. Quilliam, M.A. and Pocklington, R. 1988. Domoic acid - a neurotoxic amino acid produced by the marine diatom <u>Nitzschia pungens</u> in culture. Can. J. Fish. and Aquat. Sci. 45, 2076-2079.

Scarratt, D.J. 1980. The food of lobsters. In: Proceedings of the Workshop on the Relationship Between Sea Urchin Grazing and Commercial Plant and Animal Harvesting. Can. Tech. Rept. of Fish. Aquat. Sci. No. 984, pp 66-91.

Scarratt, D.J. 1989. Unpublished data. Dept. of Fisheries and Oceans, Halifax, Canada.

Wohlgeshaffen, G. 1989. Personal communications. Dalhousie University, Halifax, Canada.

POTENTIAL OF IRRADIATION TECHNOLOGY FOR IMPROVED SHELLFISH SANITATION

John C. Mallett, Leon E. Beghian, and Theodore Metcalf

INTRODUCTION

Food processing by ionizing irradiation has been shown to be safe, effective and efficient for the purposes of food preservation, reduction of post-harvest storage losses, prolongation of shelf-life, pest disinfestation, inactivation of food-borne parasites, and destruction of microbial pathogens. The wholesomeness of various foods exposed to doses of 10 kiloGray (kGy) of ionizing irradiation has been endorsed by the U N. Food and Agriculture Organization (WHO/FAO, 1988), the American Medical Society, U S Department of Health and Human Services (FDA), U.S Department of Agriculture, and Codex Alimentarius Commission (Crawford, 1990) The joint FAO/International Atomic Energy Agency Division of Isotope and Radiation Applications (1988) lists 32 nations which accept over 40 classes of irradiated foods as safe for general trade and consumption. The technology is presently applied to fruits, vegetables, dried commodities such as raisins, spices and seasonings, whole and mechanically recovered meats and poultry, fish, froglegs, and frozen shrimp.

The processing techniques use energy in the form of X-rays, gamma rays, or high energy electron beams. The ionizing radiation dose absorbed by the food mass determines the effectiveness of the treatment and is itself determined by specific conditions of radiation intensity, distance from the source, duration of exposure and resistance to penetration offered by the target The radiation dose or quantity of radiant energy absorbed by the food mass is defined in units of the "Gray"(Gy) which is equal to one joule absorbed per kilogram of food mass Dose ranges recommended for foods by the U N Codex Alimentarius Commission extend up to a maximum of 10,000 Gy (10 kGy). In absolute terms, the energy applied in the process is very low; the maximum recommended dose exposure of 10 kGy is the thermal equivalent to only 2.4 calories With such dose exposures it physically is not possible to induce any radioactivity in the foods processed, thus the food does not become radioactive in any way.

The biological effectiveness of the process results from energizing of orbital electrons in the food target material. The energy transfer results in formation of free radicals (particularly the water derived peroxides and hydroxy radicals) which interact with essential macromolecules The radiation chemistry of water soluble food components has been reviewed by Simic (1983), and Nawar (1983) has reviewed the radiation chemistry of nonaqueous food components

Irradiator facilities employ one of several sources of ionizing radiation which result in essentially identical effects of preservation and disinfection. So-called "fixed" sources consist of concentrations of man-made, gamma emitting radionuclides such as cobalt-60 or cesium-137 These isotopes, sealed in aluminum cassettes or "plaques", continuously emit gamma rays as a result of nuclear decay. The plaques are permanently isolated from the food products by physical barriers, such as aluminum partitions, which allow gamma transmission, thus it is not possible for foodstuffs to come into contact with the source. Gamma rays have good penetration capacity, but these sources have disadvantages in that they are limited in availability and continually are undergoing loss of emission through decay (Co -60 has a half-life of 5.3 years and Cs -137 has a half-life of 30.2 years) Fixed sources uninterruptedly emit intense radiation and they require dense, protective shielding during down-time and storage as well as operation. Additional security measures must be taken to protect the plaques from theft, misuse, mishandling, or hostile activities.

Machine sources include large, industrial scale electron accelerators and X-ray generators The X-ray emission mode has not proven to be cost effective as it involves an inherently inefficient transfer of the electron beam energy to emitted x-ray energy.

An alternative operational mode of the machine source is the liner electron accelerator or LINEAC in which a high energy electrons beam is used to irradiate the product. The electrons do not have the penetration capacity of gamma rays, however, for 10 MeV electrons, the penetration distance does not effectively exceed 3.6 cm of materials having a density equivalent to that of water. LINEAC irradiation can be very effective for thin layer streams of grains, small diameter vegetable materials or for conveyor belt streams of prepared foods of limited thickness, (e.g., items as individually prepared food trays for hospital or airline service), but it is not necessarily appropriate for bulk irradiation of foods in larger packets (Beghian, 1987).

Irradiation cost estimates vary with such factors as radiation source (fixed or machine), mix of products being irradiated, special handling or packaging requirements, product volume, and transport of products to and from the irradiator site. Cost estimates range between $.02 and $.40 per kilogram (WHO/FAO, 1988), with LINEAC irradiation offering definite economic advantages over fixed isotope sources at higher output volumes

Additional economies may be achieved with mobile machine sources that can be moved from one site of seasonal food production to another, eliminating periods of idle down time, redundant packaging and repackaging or additional transport to a distant facility. Shielding can be permanently installed at processing centers and the beam accelerator can be transferred from site-to-site via a specially constructed tractor trailer or railroad car, rolled into a shielded position, connected to the power source and activated by simple switch.

HISTORY OF SHELLFISH IRRADIATION

Irradiation was explored almost three decades ago to achieve extension of storage life and preservation of shucked clam meats and products. Nickerson (1963) exposed soft-shelled clams and clam meats to doses equivalent to 8 kGy and detected no significant differences between irradiated and control samples after 40 days storage at 6°C. Slavin et al. (1963), Ronsivalli and Slavin (1965), and Ronsivalli et al. (1965) reported clam meats exposed to 4.5 kGy to be of equal quality to untreated controls. Chowder prepared from clams irradiated at this dose and stored at 0.6 - 1.7°C for 15 days, was well accepted by expert and non-professional taste panels. Clam meats stored well at 4°C up to 28 days after exposure. Nickerson (1963) irradiated soft-shelled clam meats at doses up to 8.0 kGy and reported no difference in taste after deep fat-frying of samples stored for 40 days at 6 1 - 7 2 C. Connors and Steinberg (1964) submitted clam meats irradiated at doses of 2 5 3.5, 4 5 and 5.5 kGy to professional taste panels and reported no significant alterations in taste Gardner and Watts (1957) however had reported a "grassy" off odor after 5 hours of low intensity irradiation

CURRENT STUDIES

In our current study food irradiation technology was applied to live shellfish to evaluate the process in four general areas of concern.

1.) To determine the effects of low dose irradiation exposure on post-irradiation survival of three shellfish species

2) To quantify the effects of gamma irradiation on the appearance, aroma, taste, and texture of the product

3) To estimate gamma doses necessary to achieve log cycle reduction (D-10) of selected bacterial contaminants in the soft-shelled clam

4.)   To determine the D-10 inactivation response of selected viruses to a range of radiation exposures.

Irradiation survival studies of Mya arenaria, Mercenaria mercenaria, an Crassostrea virginica employed the University of Lowell/D.O.E. 0 8 megaCurie Co source. Shellfish were packaged in two-liter polyethylene containers and irradiated over a dose range from 0.2 to 7 kGy. The small containers, rotated 180° mid-way during the irradiation, minimized sample density and provided relatively uniform dose distributions throughout the batches. Minimum/maximum dose ratios, determined by film dosimeters placed throughout the samples showed a less-than 15% loss from maximum dose across the sample in M. arenaria and a 20% loss in C. virginica.

Post-irradiation viability was tested by use of the siphon withdrawal reflex in clams, the valve closure reflex in quahogs, and adductor muscle resistance in oyster. Each individual was challenged at 24 hour intervals until response failure. All cohorts were followed until the last survivor failed to respond. Survival curves were plotted for irradiated batches and corresponding controls Direct comparisons of mortality at day 6 post-irradiation were made by Mann-Whitney U-test for ordered events or matched pair -tests (Sokol and Rolf, 1981) The six day survival period corresponds to a reasonable expected shelf-life of products in retail markets

Soft-shell clam (M. arenaria) samples showed no adverse effects at doses below 1.5 kGy Over the 0.6 o 1.0 kGy range irradiated clams showed a slight, but significant, prolongation of shelf-life Low dose exposures may be somewhat effective in reducing spoilage bacteria. Doses in excess of 1.5 kGy produced a significant acceleration in mortality.

Survival of the quahog (M. mercenaria) was tested in two trials cohorts of 30 specimens each ,exposed to doses of 0.5, 1.0, 3.0 5.0 and 7.0 kGy revealed an acute exposure range between the 1.0 and 3.0 kGy doses Cohorts of 50 quahogs each at doses of 0 5, 1 0, 2.0 and 3 0 kGy showed the most critical dose increment to be between 1.0 and 2.0 kGy. No increases in mortality attributable to irradiation were observed at six days post-exposure between the control group and either 0.5 kGy or 1 0 kGy exposure groups.   Both the 0 5 and the 1.0 kGy exposure groups achieved median survival times of 15 days post exposure when held at 4°C in air, whereas the non-irradiated control group achieved median survival of 17 days. The 2.0 kGy and 3 0 kGy groups experienced accelerated mortality schedules. Matched pair t-test comparisons of daily mortality between exposure vs. control groups showed significant differences (p=0.05) between controls and those exposed to 2.0 and 3.0 kGy. It appears that doses in excess of 1.0 kGy significantly jeopardize viability.

Samples of 50 oysters (C. virginica) each were exposed to doses of 1.0, 1.5, 2.0, 2 5, 3.0, 5 0, and 7.0 kGy in each of two trials. Viability was tested daily by challenging adductor muscle tension. Cohort survival was recorded until the death of the last individual and survival curves were plotted. Oysters appear to be exceptionally resistant to acute radiation effects.   No significant differences in six-day survival times were observed at doses upto 2.5 kGy. Median post-irradiation survival was in excess of 25 days for doses of 2.5 kGy and lower Batches exposed to 3.0, 5 0 and 7.0 kGy doses displayed earlier onset of radiation effects with median response failure around the eleventh day post exposure for the three groups, suggesting acute lethal dose at 3.0 kGy.

ORGANOLEPTIC PROPERTIES OF LIVE IRRADIATED SHELLFISH

Irradiated clams, quahogs and oysters were tested for relative sensory appeal and palatability (Table 1). Live samples were irradiated at exposures of 1 0, 2.0 and 3 0 kGy in the Lowell reactor and with nonirradiated controls were transported in plastic bags, on ice, to the NMFS - Northeast Fisheries Center where they were prepared and served to professional taste panels of six to ten persons. Each panelist evaluated three servings at each of the radiation dose levels and compared these treated portions to a known non-irradiated reference control sample. Soft-shelled clams (Mya) were prepared and presented to a professional organoleptic evaluation panel as

Table 1. Organoleptic Evaluations of Irradiated Shellfish, Mean (+SD).

FRIED CLAMS

	Control	0.5 kGy	1.0 kGy	2.0 kGy	3.0 kGy
Appearance	8.0 (0.94)	8.0 (0.47)	7.8 (0.42)	7.7 (0.67)	7.5 (0.97)
Odor	8.5 (0.53)	8.0 (0.67)	8.0 (0.67)	7.9 (0.74)	7.9 (0.74)
Taste	8.0 (0.94)	7.3 (1.06)	7.6 (1.17)	7.4 (1.26)	7.1 (1.37)
Texture	8.2 (0.92)	6.8 (1.23)	7.4 (0.97)	7.4 (1.07)	6.3 (1.83)

STEAMED CLAMS

	Control	0.5 kGy	1.0 kGy	2.0 kGy	3.0 kGy
Appearance	8.8 (0.41)	8.7 (0.82)	8.8 (0.41)	8.5 (0.55)	8.5 (0.55)
Odor	9.0 (0.00)	8.5 (0.84)	8.8 (0.41)	8.5 (0.55)	8.3 (0.52)
Taste	9.0 (0.00)	8.3 (0.52)	8.2 (0.98)	8.2 (1.33)	7.3 (1.86)
Texture	8.3 (0.41)	8.5 (0.55)	8.3 (1.21)	7.3 (2.25)	7.0 (2.45)

BAKED OYSTERS

	Control	0.5 kGy	1.0 kGy	2.0 kGy	3.0 kGy
Appearance	8.5 (0.71)	8.3 (0.95)	8.2 (0.63)	8.4 (0.70)	8.2 (0.63)
Odor	8.3 (0.82)	8.0 (1.05)	8.2 (0.79)	7.9 (1.10)	7.7 (0.94)
Taste	8.4 (0.70)	7.9 (1.29)	8.3 (0.67)	7.8 (1.14)	7.8 (0.92)
Texture	8.2 (0.79)	8.18 (1.10)	8.2 (0.79)	8.0 (0.94)	8.0 (0.82)

RAW OYSTERS

	Control	0.5 kGy	1.0 kGy	2.0 kGy	3.0 kGy
Appearance	8.0 (0.70)	8.2 (0.83)	7.6 (0.98)	7.9 (1.17)	7.9 (1.12)
Odor	8.3 (0.87)	7.4 (1.86)	7.8 (1.12)	7.7 (1.05)	7.9 (1.12)
Taste	8.0 (1.04)	7.1 (0.99)	7.3 (0.64)	7.0 (1.77)	6.6 (1.30)
Texture	8.2 (0.83)	7.6 (0.98)	7.6 (1.04)	7.5 (1.12)	7.3 (1.45)

BAKED CHERRYSTONE CLAMS

	Control	0.5 kGy	1.0 kGy	2.0 kGy	3.0 kGy
Appearance	8.8 (0.44)	8.4 (0.73)	8.6 (0.53)	8.6 (0.73	8.3 (0.87)
Odor	8.6 (0.73)	8.4 (0.73)	8.3 (0.71)	8.3 (0.71)	7.8 (0.97)
Taste	8.3 (0.71)	8.1 (1.05)	7.8 (1.39)	8.3 (0.87)	7.3 (1.32)
Texture	8.2 (0.67)	7.6 (1.13)	7.9 (0.78)	8.1 (0.78)	8.0 (0.87)

RAW CHERRYSTONES

	Control	0.5 kGy	1.0 kGy	2.0 kGy	3.0 kGy
Appearance	8.3 (0.82)	7.5 (1.22)	8.0 (0.89)	7.8 (1.17)	7.3 (0.82)
Odor	8.7 (0.82)	8.0 (0.89)	8.2 (0.75)	7.7 (1.51)	7.8 (0.75)
Taste	8.5 (0.84)	7.5 (0.84)	7.7 (0.82)	7.0 (1.55)	7.0 (1.26)
Texture	8.8 (0.41)	7.8 (1.17)	8.8 (0.89)	7.5 (1.22)	7.8 (1.17)

steamed clams and fried clams; quahogs were prepare as raw cherrystone clam on the half-shell and as baked quahogs; oysters were evaluated raw (oysters on the half-shell), and as baked oysters.

Products were judged sequentially for appearance, odor, flavor,and texture on a nine point hedonic scale. Evaluations ranged from "Inedible" at a score of one point through "very poor", "poor" "Slightly Poor' "Borderline" "Fair", "Good" "Very Good" and "Excellent" at nine points.

In fried, softshell clam trials the controls were ranked as being of very good quality in appearance, aroma, taste and texture (scores of 8 0, 8.5, 8.0 and 8.2 respectively). The irradiated product displayed a slight decrease in quality at the textural properties in the 0.5 to 3.0 kGy exposures which were given respective score of 6.8 and 6.3 in the fair range. Mean scores of 7.4 were recorded for both the 1.0 and 2.O kGy samples. All other qualitative assessments of irradiated fried clam averaged in the "good" to "very good" range.

Irradiated steamers were ranked as "good" to "very good" across the dose range with the lowest scores being noted in the textural category where clams exposed to the 3 0 kGy dose had scores in the "good" range, averaging 7.0

Quahogs, evaluated as baked and raw preparations, scored well with the samples exposed to the most intense doses (3 0 kGy receiving scores of 8.3 for appearance, 7 8 for odor, 7.3 for taste and 8.0 for texture. The raw cherrystones were well received with the highest dose treatment being awarded average scores of 7 3, 7.8, 7.0 and 7.8 for appearance, aroma taste and texture, respectively.

Oysters also proved to be quite resistant to radiation induced sensory changes Baked oysters were graded with averages in the high sevens and eights for all categories and all doses. Raw oysters on-the-half-shell were awarded lower scores. The least appealing taste scores fell to an average of 6.6 for those receiving the 3 0 kGy exposures. It should he noted that scores in this range none-the-less are regarded as being of fair and acceptable quality

BACTERIAL INACTIVATION IN MYA

Bacterial inactivation by gama treatment was investigated in the soft-shell clam, Mya arenaria (Mallett et al., 1985; Jonsson, 1986). Freshly shucked clams were homogenized and individual aliquots were inoculated with cultures of E. coli, Salmonella typhimrium, Staphylococcus aureus or Stretococcus faecalis at densities in excess of 10^6 cells per gram of meat and irradiated over a dose range of O to 3.60 kGy in the NMFS Products Development Irradiator at the Northeast Fisheries Center Laboratory. Serial dilutions were made and bacterial populations were determined by standard methods Data were plotted a log bacterial survivors vs. exposure dose. The negative inverse of the regression slope was calculated as the specific D_{10} log decrement value for each bacterial species (Figure 1).

E. coli was reduced by a factor of 10 for each 0 37 kGy dose received, Salmonella typhimurium populations were decremented by a log cycle with a dose of 0.51 kGy and Staphylococcus aureus was calculated to be reduced by an order of magnitude with exposure to 0 42 kGy. Streptococcus faecalis was decremented by a factor of ten in response to a dose equivalent to 0.7 kGy. These results indicate effective bacterial inactivation is possible at relatively low doses.

VIRAL INACTIVATION

Poliovirus 1, simian rotavirus SA-11, and hepatitis A virus were injected into live oysters and quahog and irradiated at doses up to 10 kGy to determine D_{10} inactivation doses SA-11 was

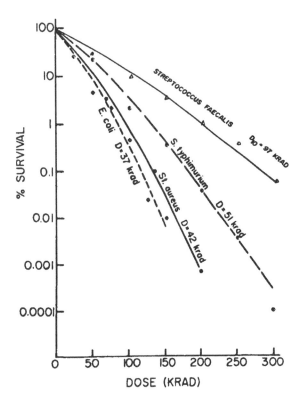

FIGURE 1.   Inactivation of bacterial species in <u>Mya</u> <u>arenaria</u>
homogenate by gamma irradiation (KRAD equals
.01 kGy). Data from Mallett et al. (1985) and
Jonsson (1986).

cultured in fetal rhesus monkey kidney (FRhK) cell culture, and poliovirus 1 was derived from
buffalo green monkey (BGM) cell culture. HAV was harvested from a persistently infected African
green monkey kidney cell line developed by Simmonds et al. (1985). Cultures were maintained at
Baylor College of Medicine and shipped frozen to the University of Lowell where they were thawed
and inoculated into quahogs and oysters for irradiation. Following treatment, shellfish were
shucked, individually frozen, and transferred to Baylor for viral recovery and enumeration.

SA-11, quantified plaque forming units (PFUs) in FRhK (MA-104) cell culture, was
determined to have a D-value of 2.4 kGy. Polio virus was assayed as PFUs in BGM cell culture as
described by Rao and Melnick, 1986). Poliovirus 1 D calculations averaged 3.1 kGy in oysters and
quahogs. Hepatitis A virus was assayed by several techniques; immunofluorescent focus (IFF) was
used was a direct assay and radioimmunoassay (RIA), nucleic acid probe (RNA probe)
hybridization assay, and a newly developed cytopathic effect (CPE) assay were used as quantal
methods.

The log inactivation dose for Hepatitis A in quahogs, as determine by immunofluorescent focus IFF techniques was 1.66 kGy. A similar IFF assay of HAV inactivation in oysters determined the D_{10} value to be 1.48 kGy. HAV virus deactivation doses in oysters by insitu RNA hybridization resulted in higher a D_{10} estimate of 3.16 kGy. The cytopathic effect (CPE) analysis utilizing a rapidly growing strain of HAV provided a D_{10} estimate of 2.06. An average estimate of HAV inactivation derived from analysis of pooled data from all assays resulted in an overall estimate of 2.02 kGy for the virus (Figure 2).

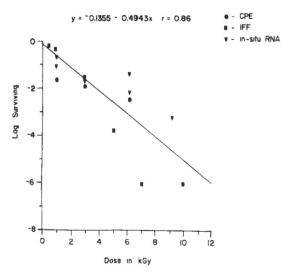

FIGURE 2.   Inactivation hepatitis A virus in live <u>Crassostrea virginica</u> by gamma irradiation. D_{10} = 2 02

DISCUSSION

The capacities of intertidal bivalve mollusks to sustain intense radiation exposure are exceptional. It is speculated that their unique anatomical and physiological adaptations to life in the intertidal zone predispose these bivalves to sustain relatively intense radiation exposures Adaptations which collectively may contribute to radioresistance are:

1.   Facultative anaerobic metabolic capacities

2.   Minimalization of body lipid stores

3   Anatomically simplified nervous system

4.   Metabolic production of radioprotective agents

5.   Low metabolic rate and low cell division rate

Tissues are less sensitive to the effects of irradiation in the absence of oxygen As bivalve tissues become deprived of oxygen during lowtide periods of stagnation or air exposure, alternate glycolytic pathways enable stored energy reserves to be utilized anaerobically. When they enter into the state of anaerobiosis, the subsequent lack of free molecular oxygen presumably would eliminate any "oxygen enhancement effect" of radiation damage of tissues.

A physiological correlative to anaerobic metabolism is seen in the reduction of body lipid stores. Energy reserves in facultative anaerobes are stored preferentially as shellfish glycogens Lipids, the preferred energy storage molecules of obligate aerobes, require aerobic oxidation to reconstitute a suitable carbohydrate substrate to enter into the normal glycolytic pathway. Conversion of lipids to carbohydrate is not possible under conditions of anaerobiosis; therefore, lipids are sparse in the intertidal bivalves. Oysters have total lipid contents of only 2.4% and clams maintain only 1.2% of tissue weight as lipid (Cataldo et al., 1989). In the absence of significant quantities of lipids the problems of radiation-induced lipid peroxidation are lessened.

In anaerobic glycolysis, marine mollusks employ anoxic pathways which feature a reversal of certain portions of the tricarboxylic acid cycle. Phosphoenolpyruvate (PEP) from the glycolytic sequence is converted to oxaloacetate (OXA) and reduced through malate and fumarate to succinate which acts as a metabolic hydrogen acceptor (Hoar, 1975) Succinate is accumulated in anaerobic tissues as a stored metabolic end product. Succinate has been reported in other organisms to serve as a radioprotective agent, acting as an inhibitor of membrane lipid peroxidation (Ronai et al. 1987) Intertidal bivalves may thus synthesize quantities of radioprotective agents as normal by-products of facultative anaerobic metabolism.

Preservation of organoleptic qualities in radiation processed shellfish may also be related to their facultative anaerobic status. The absence of significant quantities of lipids combined with the possible prophylactic effects of elevated succinate may preclude the formation and accumulation of lipid peroxidation products that degrade sensory appeal in products high in fats

Neurological simplicity may confer some degree of radio-insensitivity upon these species The relatively simple nervous system of the bivalves is of probable advantage in sustaining high radiation insult. The primitive, non-centralized nervous system consisting of relatively few ganglia and distribution trucks appears to be less sensitive to ionizing radiation and less likely to incur damage.

Low metabolic rate and low cell division rates may provide additional capacity to sustain these high radiation doses. Radiation induced damage to nucleic acids and mechanisms of cell division is not as likely to immediately incapacitate the less active molluscan cells. Infective microbial agents however, are obliged to undergo accelerated cell division schedules and would be influenced by dose effects upon initiation of cell division or viral particle replication. Thus a radiation dose delivered to shellfish may result in rapid decimation of its microbial populations without causing immediate damage to the host.

ROLE OF RADIATION IN SHELLFISH SANITATION

Irradiation presents distinct advantages for enhancing microbial quality in live shellfish when applied as an adjunct to existing sanitation programs. Bacterial clearance, after 48 hours in recirculating depuration systems approximates reductions obtained by irradiation at 1.0 kGy. Hartlant and Timoney (1979) indicated 20 hrs of flow-through depuration were necessary for a log cycle reduction of $\underline{E}$. coli and $\underline{S}$. typhimurium in hard shelled clams and oysters held at 6°C in sea water. Piel et al. (1974) showed after UV depuration, fecal coliforms from soft-shelled clams were reduced by 8.0×10^{-2} after 48 hrs of treatment. The present study indicated that irradiation, even at low doses, compares well with the bacterial reductions obtained after 48 hrs of treatment in recirculating UV depuration.

Ionizing treatment can provide highly effective safeguards and backups to traditional depuration treatments. Flow through depuration ultimately depends upon the ability of the mollusks to irrigate large volumes of clean water through the mantle cavity and gill regions Rate of bacterial clearance in this process is dependent upon the pumping rate of physically intact and physiologically healthy specimens. These prerequisites for effective depuration are not always met in practice. Stock with broken shells, physiological stress, or harvest induce mortality can not depurate effectively. Flow-through depuration also assumes that all microbial species respond similarly; assumptions that all pathogens have clearance rates equivalent to those of the indicator organisms are open to criticism. The vibrio species and the viral pathogens may not depurate as efficiently as the coliforms

Under ideal conditions soft-shelled clams depurate bacterial populations by about two logs in a 48 hour period. Ionizing radiation doses in the range of 2 kGy can achieve bacterial reductions in the range of 10^{-4} for Staphylococcus and 10^{-2} for Streptococcus pathogens. The serious hazards posed by Vibrio cholera, may be effectively eliminated by ionizing treatment as the vibrios are among the most radiation sensitive for bacteria. Some D_{10} values as low as 0.1 kGy have been reported for V. parahaemolyticus, and complete elimination of V. cholera was reported from oyster homogenate which was inoculated with 10^7 cells per gram and irradiated at 1.0 kGy (Bandekar et al. 1987). Aeromonas hydrophile, which has been implicated in oyster-borne illness (Abeyta et al., 1986) also appears to be highly susceptible to ionizing radiation; the D_{10} values of 0.14 to 0 22 kGy were reported by Palumbo et al., (1986). Total elimination of these species may be possible with low dose treatments (Table 2).

Table 2. Radiation inactivation values of selected microbial contaminants of seafood.

MICROBIAL SPECIES	MEDIUM	D10	Author
Aeromonas hydrophila	Chilled Fish	0.16 kGy	Palumbo et al., 1985
Campylobacter jejuni	Raw Beef	0.15 kGy	Tarkowski et al. 1984
Clostridium perfringens	Reduced buffer	0.37 kGy	Gombas and Gomez, 1985
Escherichia coli	Minced Clam Meat	0.37 kGy	Mallett et al., 1985
Salmonella typhimurium	Minced Clam Meat	0.51 kGy	Mallett et al., 1985
Shigella dysenteriae	Frozen Shrimp	0.22 kgy	Mossel, 1985
Shigella flexneri	Frozen Shrimp	0.41 kGy	Mossel, 1985
Staphlococcus aureus	Minced Clam Meet	0.42 kGy	Mallett et al. 1985
Streptococcus faecalis	Minced Clam Meat	0.97 kGy	Jonsson, 1986
Vibrio parahaemolyticus	Frozen Shrimp	0.10 kGy	Bandekar et al., 1987
Yersinia enterocolitica	Phosphate Buffer	0.10 kGy	El-Zawahry and Grecz, 1981
Hepatitis A Virus	Live Oyster	2.02 kGy	This Study
Poliovirus I	Live Quahog	3.30 kGy	This Study

Viral carriage in shellfish is a serious, ill-defined, health risk. Metcalf (1982) and others have stated the dangers from HAV and Norwalk virus. Viruses are only presumed to be adequately depurated by traditional techniques, but mechanisms of concentration, translocation to internal tissues, and clearance of viruses are not well understood. The suspected capacity of shellfish to retain viruses for several weeks has special significance when considering the effectiveness of traditional depuration treatment. It appears however that HAV populations in shellfish can be reduced approximately by a factor of 10 following treatment with 2.0 kGy. Whether proportional reductions in infectivity would be achieved by irradiation treatments remains to be considered.

CONCLUSIONS

Raw and undercooked shellfish, even when harvested from areas meeting regulatory standards for microbial water quality, have been implicated as a serious public health threat. Monitoring and depuration programs have not been totally effective in ridding harvests of enterpathic microbes of anthropogenic and natural sources. Irradiation processing, when used as an adjunct to conventional depuration modes, holds high potenital for significantly enhancing the sanitary status of shellfish But it must be clearly and unequivocable stated that the irradiation process must not be used in an attempt to market stocks of inferior microbial quality. Potential abuse or mis-application of the process would result if, by chance or design, shellfish which did not meet established sanitary standards were allowed to be irradiation processed. Irradiation processing should be viewed as an adjunct process to provide increased consumer safety in supplies which were previously capable of meeting sanitary standards (such as those of the National Shellfish Sanitation Program or the European Economic Community proposal on bivalve molluscan safety).

ACKNOWLEDGEMENTS

We thank the U S. Dept. of Energy for support through Grant No. DE FG04-87AL45784 and the U S. Department of Commerce, National Marine Fisheries Service, Northeast Fisheries Center for technical support

REFERENCES

Abeyta, C., Kayser, C.A., Wekell, M M., Sullivan, J. and Stelma, G.N. 1986. Recovery of *Aeromonas hydrophila* from oysters implicated in an outbreak of foodborne illness. J. Food Protection. 49(3):643-646

Bandekar, J.R., Chandler, R. and Kerkan, D.P. 1987. Radiation control of *Vibrio parahaemolyticus* in shrimp. Jour. Food Safety

Beghian, L.E. 1987. The Electron Beam - An Alternative Approach. In. Food Safety Towards the 21st Century Wordsworth & Co., London. pp. 69-72.

Cataldo, C.B., Nyenhuia, J.R. and Whitney, E.N. 1989. Nutrition & Diet Theraphy, Principles and Practice. West Pub Co , St Paul, MN 703 pp.

Crawford, L.M. 1990. Personal communication. U.S. Department of Agriculture, Washington, D.C.

El-Zawahry, Y.A. and Grcz, N. 1981. Inactivation of *Yersinia enterocolitica* by radiation and freezing. Appl. Envir. Microbiolo. 42(3):464-468.

FAO/IAEA, 1987. Food Irradiation Newsletter, No. 11, No. 2 September, 1987.

FAO/IAEA, 1988. Food Irradiation Newsletter, No. 12, No. 1 April, 1988

Gardiner, E A and Watts, B.M. 19875 Effect of Ionizing Radiations on Southern Oysters Food Technology. June 1957.329-331

Gombas, D.E. and Gomez, R.F. 1978. Senitization of *Clostridum perfringens* spores to heat by gamma irradiation. Appl. Environ Microbiol. 36:403.

Hartland, B.J. and Timoney, J.F. 1979. In-vivo clearance of enteric bacteria fromt he hemolymph of the hard clam and American oyster Appl. Environ. Microbiol. 37(3):517-520.

Hoar, W.S 1975. General and Comparative Physiology. Prentice-Hall, Inc. Englewood Cliffs, NJ 848 pp.

Jonsson, I M. 1986. Bacterial depuration of the soft-shelled clam, Mya arenaria, by means of gamma irradiation. M.S. Thesis, Univ. of Lowell Library, Lowell, MA 62 pp.

Mallet, J.C., Kaylor, J.D. and Licciardello, J.J. 1985. Depuration of live and shucked softshelled clams by gamma irradiation. Proceedings of the IEAE/FAO Conf. on Food Irrad. pp. 241-242.

Metcalf, T.G 1982 Viruses in shellfish growing waters. Environ. Internat. 7:21-7

Mossel, D.A.A. 1985. Irradiation: An effective mode of processing food for safety. Food Irradiation Processing, Proceedings of IAEA/FAO Symposium in Washington, D.C., Internal. Atomic Energy Agency, Vienna p. 251-279.

Nawar, W.W. 1983. Radiolysis of Nonaqueous Components of Food. In Josephson, E.S. and M. S. Peterson (eds.) Preservation of Food by Ionizing Radiation. C.R.C. Press, Boca Raton, FL. pp. 75-124.

Nickerson, J.T.R. 1963. The storage life extension of refrigerated marine products by low-dose radiation treatment. Exploration of Future Food Processing Techniques MIT Press, Cambridge, MA

Palumbo, S A., Jenkins, R K , Buchanan, R L. and Thayer, D.W. 1986. Determination of irradiation D-values of Aeromonas hydrophila Journal Food Protection 49(3):189-191.

Piel, E., Ceurvels, A R , Hovanesian, J. Der and Pow, J. 1974. Analysis of depuration for soft-shelled clams at Newburyport, MA and a program for bacterial standards. Chesapeake Sci. 15(1):49-52.

Rao, V.C. and Melnick, J.L. 1986 Environmental Virology. American Society of Microbiology. Washington, E.C. 88 pp.

Ronai, E., Tretter, L , Szabados, G. and Horvath, I. 1987. The inhibitory effect of succinate on radiation-enhanced mitochondrial lipid perioxidation Int. J Radiat. Biol. 51(4):611-617

Ronsivalli, L.J. and Slavin, J.W. 1965. Pasteurization of Fishery Products with gamma rays from a Cobalt-60 sources. Comm Fisheries Rev. 27(10):1-8.

Ronsivalli, L.J., Steinberg, M.A., and Seagram, H.L. 1965. Radiation Preservation of Fish of the Northwest Atlantic and the Great Lakes. Radiation Preservation of Foods, Publ. 1273, National Res. Council, National Acad. Sci., Washington, D.C.

Simic, M.M. 1983. Radiation Chemistry of Water Soluable Food Components. In Josephson, E S. and M.S. Peterson (eds). Preservation of Food by Ionizing Radiation C.R C. Press, Boca Raton, FL pp. 1-73.

Simmonds, R.S., Szucs, G., Metcalf, T G , and Melnick, J.L. 1985. Persistantly infected cultures as a source of hepatitis A. virus Appl Environ. Microbiol 49(4) 749-755.

Sokol, R.R. and Rolf, F J. 1981. Biometry (2nd Ed.). W.H. Freeman and Co., San Franciso, CA 859 pp.

Tarkowski, J.A., Stoffer, S.C.C, Beumer, R.R., and Kampelmacher, E.H. 1984 Low dose irradiation of raw meat. I. Bactriological and sensory quality effects in artifically contaminated samples. Internat. Journal Food Microbiol. 1(1):13-23.

WHO/FAO 1988. Food Irradiation. A Technique for Preserving and Improving the Safety of Food World Health Organization, Geneva. 84 pp.

International Settings

A PRELIMINARY REPORT OF INTERNATIONAL
TRADE IN MAJOR MOLLUSCAN SHELLFISH

Myles Raizin

This paper constitutes a preliminary investigation into the international markets for fresh and frozen clams, mussels, and oysters with a special section devoted to United States imports of fresh and frozen clams and oysters. The first section provides the reader with information on the production and movements of shellfish using data obtained from the Food and Agriculture Organization of the United Nation's Yearbook of Fishery Statistics, 1977 to 1987 The following section serves to examine the quantity and countries of origin for fresh and frozen clams and oysters coming into United States ports. The data for this section were obtained from the database for U.S. imports and exports which is managed by the headquarters of the National Marine Fisheries Service, Washington, D.C and is complied from data furnished by the Department of Commerce, Bureau of Census Unfortunately, the Bureau of Census is only now beginning to record import and export data for mussels. Therefore, the species under discussion in this section are limited to clams and oysters. The final section of the paper discusses potential impacts on public health arising from trade in these major molluscan shellfish products. The text excludes values for imports because this report is not economic in nature, but may serve as a basis for further research.

GLOBAL PRODUCTION AND TRADING PATTERNS

FRESH AND FROZEN CLAMS

PRODUCTION

In 1987 China was the major producer of clams in the world with a production level of 889,951 metric tons(mt). This amount represents increases of 33 percent above the 1986 level of 669,139 mt and 88 percent above the 1985 level of 473,013 mt. The United States ranked second in clam production at 400,698 mt in 1987 with major species being the ocean quahog (Arctica islandica) and surf clam (Spisula solidissima). Other countries of notable production in 1987 were Japan at 191,729 mt, the Republic of Korea at 188,714 mt, and Canada at 23,537 mt

EXPORTS

Given that China was the world's largest producer of clams in 1987, it is not surprising to discover that this country was also the world's leading exporter of fresh and frozen clams. In 1987, China exported 13,483 mt of fresh and frozen product (Figure 1). However, this level of export represented a decline from previous periods in relation to production. The 1987 export level was 1.5 percent of production as compared to 2.1 percent in 1986 and 2.3 percent in 1985. Although market constraints may have been a factor, this relative decline was most likely a function of China's inability to rapidly increase infrastructural requirements for export such as port and cold storage facilities.

The Democratic People's Republic of Korea ranked second among exporters of fresh and frozen clams at 7,266 mt in 1987. Although production levels are not available for this country, it may be assumed that at least a portion of this export was in the form of re-exports of class produced in China.

Canada ranked third among the countries exporting fresh and frozen clams and remained the leading non-Asian exporter with a total of 6,570 mt of product in 1987. Canada's export is characterized by slow and stable growth void of political or institutional impediments. The United States, which is Canada's major trading partner in aggregate terms, was also the major purchaser of Canadian clams accounting for approximately 75 percent of Canadian exports.

IMPORTS

Japan and Thailand were certainly the largest importers in the world of fresh and frozen clams at 1987 levels of 20,946 mt and 20,391 mt, respectively. Given that Japan produced 191,729 mt of product in 1987, this level of import reinforces the well-known fact that the Japanese have a voracious appetite for seafood. We have not been able to verify Thailand's import of 70,466 mt in 1979 (Figure 2) and should regard this as an outlier of the data set. It is clear, however, that the major global clam markets are located in Asia.

FRESH AND FROZEN MUSSELS

PRODUCTION

Spain was the world's leading producer of mussels with a production level of 206,706 mt in 1987 down from 246,955 mt in 1986 China ranked second with a 1987 production of 151,768 mt. The Netherlands produced 98,367 mt in 1987 followed closely by Italy at 85,400 mt, Denmark at 78,009 mt, and France at 56,455 mt. Italy produces the Mediterranean mussel (Mytilius galloprovincialis) while other European countries concentrate on the Blue mussel (Mytilius edulis)

EXPORTS

The Netherlands was the leading exporter of fresh and frozen mussels with a 1987 export of 45,668 mt (Figure 3). In fact, this country has been the largest exporter of these products over the eleven-year period, 1977 to 1987. Spain ranked second in exports at 27,787 mt in 1987 and has held this position over the period with the exception of the years 1984 thru 1986 when Denmark challenged Spain in the global market. However, Denmark's export level fell to 8,106 metric tons in 1987 from 29,406 mt in 1986.

IMPORTS

France was the leading importer of fresh and frozen mussels with an import level of 38,381 mt in 1987. With the exception of 1984, France has held this position since 1077 (Figure 4) Belgium ranked second among mussel importers at 24,889 mt in 1987 followed closely by the Netherlands at 22,975 mt. The rather large import of mussels by the Dutch in 1984 was coincidental with a drop in production from 119,643 mt in 1983 to 60,149 mt in 1984 The data reflect that much of this shortfall in supply may have been eliminated by the importation of Danish mussels. However, the global market for fresh and frozen mussels is clearly centered in Europe.

FRESH AND FROZEN OYSTERS

PRODUCTION

The Republic of Korea was the world's leading producer of oysters with a production level of 303,223 mt in 1987 followed by Japan at 258,776 mt. The Pacific cupped oyster (Crassostrea gigas) is the major species produced by these countries. The United States was the third major producer of oysters at 217,632 mt in 1986 with major species being the Pacific cupped oyster and

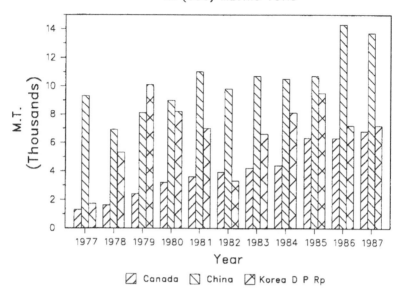

FIGURE 1. LEADING CLAM EXPORTERS
IN (000) METRIC TONS

Canada · China · Korea D P Rp

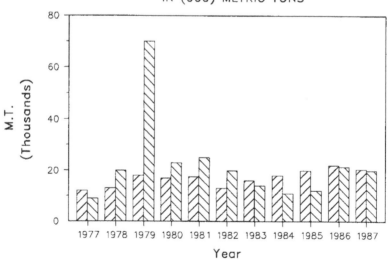

FIGURE 2. LEADING CLAM IMPORTERS
IN (000) METRIC TONS

Japan · Thailand

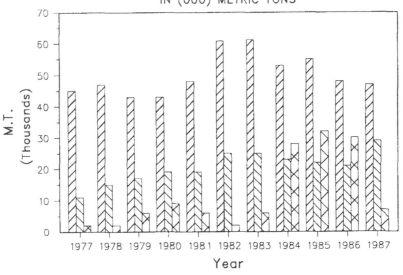

FIGURE 3. LEADING MUSSEL EXPORTERS
IN (000) METRIC TONS

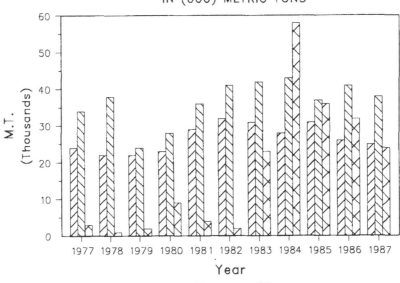

FIGURE 4. LEADING MUSSEL IMPORTERS
IN (000) METRIC TONS

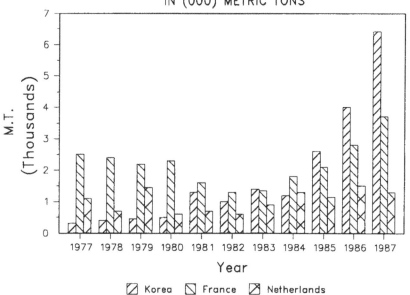

FIGURE 5. LEADING OYSTER EXPORTERS
IN (000) METRIC TONS

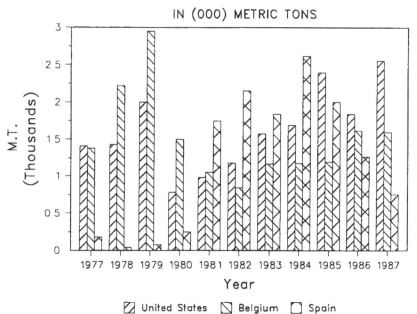

FIGURE 6. LEADING OYSTER IMPORTERS
IN (000) METRIC TONS

the American cupped oyster (Crassotrea viginica). France was also a major producer of oysters with a production level of 123,162 mt in 1987.

EXPORTS

The world's largest producer of oysters is also the leading exporter of fresh and frozen oysters. The Republic of Korea exported 6,288 mt of product in 1987 (Figure 5) which represented a 54 percent increase over a 1986 export of 4,075 mt and a 149 percent increase over the 1985 level of 2,518 mt. The rapid expansion of Korea's exports may indicate that excess demand exists for these products. France, which had been the leading exporter until 1985, ranked second in exports in 1987 with a level of 3,863 mt. The Netherlands was the third leading exporter in 1987 with an export of 1,308 mt followed closely by Italy at 1,262 mt and the UK at 1,244 mt.

IMPORTS

The major importer of fresh and frozen oysters in 1987 was the United States at 2,538 mt in 1987 (Figure 6). This position has been held by the United States since 1985. Belgium was the leading importer from 1978 to 1980 and Spain lead the world in imports from 1981 to 1984. In 1987, Italy ranked second in imports at 2,308 mt followed by Belgium at 1,590 mt The global market for fresh and frozen oysters remains the smallest among the molluscan shellfish markets but appears to be the most competitive with several counts participating.

TABLE 1. U.S. FDA SHELLFISH MOU'S (Memorandum of Understandings).

Effective Date	Country	Type	Remarks
1948	Canada	International Agreement	
1962	Japan	International Agreement	
1972, 1987	Rep of Korea	International Agreement	New MOU in 1987 requires shellfish to be frozen
1978	Iceland	MOU	
1980	New Zealand	MOU	
1981, 1987	Mexico	MOU	New MOU in 1987 for admin. purposes
1982	England	MOU	Now in abeyance due to 1984 depuration failure Norwalk virus informally ceased all export of shellfish
1986	Austrailia	MOU	Requires shellfish to be frozen
1989	Chile	MOU	

Table 2. U.S. Imports of Fresh and Frozen Clams (metric tons) by MOU Status Per Country; MOU = Memorandum of Understanding.

	Countries with MOU									Countries w/o MOU				Total (% of total)	
	Canada	Mexico	Ice-land	United Kingdom	Korea	Japan	Austra-lia	New Zealand	Chile	Nether-land	Thai-land	China	Taiwan	Countries with MOU	Countries w/o MOU
1977	1273				61	8				1	1	3		1342 (99)	13 (1)
1978	1924				19	9					1	2		1952 (100)	5 (0)
1979	2017				23	7					1	1		2047 (100)	5 (0)
1980	1961			45	16	10		1				11		2033 (98)	34 (2)
1981	2980			681	29	16		50	2			9		3757 (100)	15 (0)
1982	3088	17		825	42	11		175			86	26	1	4158 (97)	122 (3)
1983	3732	9		630	40	12		112	1		23	4	4	4534 (98)	20 (2)
1984	4055		1	1	31	16		162	5	16	25			4267 (99)	62 (1)
1985	4079			2	72	10	1	54	17	61	23	8	20	4217 (96)	158 (4)
1986	4222	3			84	6		87	5	15	111	1	10	4401 (95)	221 (5)
1987	4879	9			88	6		198	1	280	86	20	21	5179 (92)	433 (8)
1988	4827	5			123	6		226	1	188	45	5	4	5187 (95)	260 (5)

Table 3. U.S. Imports of Fresh and Frozen Oysters (metric tons) by MOU Status Per Country; MOU = Memorandum of Understanding.

	Countries with MOU						Countries w/o MOU				Total (% of total)	
	Canada	Mexico	Korea	Japan	New Zealand	France	Thailand	China	Hong Kong	Taiwan	Countries with MOU	Countries w/o MOU
1977	16		834	560				5	7		1411 (98)	14 (1)
1978	23		1101	288			1	7	25		1413 (98)	34 (2)
1979	10		1657	326				9	6	2	1994 (99)	18 (1)
1980	8		484	206				5	7		698 (98)	16 (2)
1981	58		766	85	6		2	5	23	11	915 (95)	45 (5)
1982	63		1028	63	5			5	5		1159 (99)	11 (1)
1983	89		1222	164	16		14	40	13		1492 (95)	85 (5)
1984	382		1071	167	11	1	5	15	23	1	1631 (96)	67 (4)
1985	487		1568	255	4	20	1	10	63		2314 (96)	100 (4)
1986	505		1129	160	7			3	20	9	1801 (98)	39 (2)
1987	662	18	1699	101	30	5		5	14	1	2510 (99)	28 (1)
1988	781	76	1541	89	22		1	4	31		2510 (99)	38 (1)

U.S. IMPORTS OF FRESH AND FROZEN CLAMS AND OYSTERS

FRESH AND FROZEN CLAMS

In 1988 the United States imported 5,447 mt of fresh and frozen clams which represents a decline from the 1987 level of 5,612 mt. While the 1987 import level appears to be sizable, it represents only 1.4 percent of U.S. production of clams for that year. Canadian exports of fresh and frozen clams accounted for 88.6 percent of all imported product in 1988 and 93 percent of imports from countries which have agreements on quality and safety standards (MOU'S) (TABLE 1). Table 2 shows the clam import levels of countries by MOU status, both on an individual and aggregate basis. Of note are the substantial increases in imports beginning in 1985 from countries who have not signed agreements with the United States. These increases range from 4 percent of total imports in 1985 to 8 percent of total imports in 1987 with the Netherlands and Thailand accounting for the vast majority of non-MOU imports.

FRESH AND FROZEN OYSTERS

The analysis for oysters is drawn from the data appearing in Table 3 In 1988 the United States imported 2,548 mt of product which represents a small increase of 10 mt above the 1987 level. The 1987 level of import represents only 1.2 percent of U.S production for that year The importation of oysters from countries without agreements on quality or safety has decreased in relative terms from highs of 5 percent in 1981 and 1983 to only 1 percent in 1988. The Republic of Korea has been the leading exporter of these products to the U.S. over the period, 1977 to 1987.

DISCUSSION AND IMPLICATIONS

While the importation of fresh and frozen clams and oysters to the U.S. has increased over the period 1977 to 1988, quantities measured in terms relative to domestic production remain quite low. Therefore, it may be construed that any adverse action in regards to imports will have little effect on domestic markets for these products. If we assume that products coming into the U.S. from countries with MOU's have met the FDA standards regarding product safety, the risk of contaminated product in U.S. markets is diminished especially for oysters where the import from countries without MOU's was only one percent of total imports in 1988. However, the health risk associated with the import of class remains much higher when considering that five percent of total clam imports come from countries which have not signed agreements.

The large increases in the production of clams in China in the past few years should be of concern to clam producing and consuming countries. If China increases its export by only a small percentage relative to its very large production, this may cause economic stress to global clam producers and may dramatically increase health risk to consumers if proper safety measures are not put into effect.

CANADA DEPURATION

Dave Doncaster

INTRODUCTION

Depuration of shellfish is a new industry in Canada. In the Atlantic Provinces development of depuration technology started in the late 1970's with actual commercial production commencing in 1983 using container relaying. This technique was abandoned after three years as it was adversely affected by higher than anticipated costs and PSP blooms. In 1984, land based depuration was initiated using only the soft shell clam, Mya arenaria. Presently there are two registered production facilities in Atlantic Canada both of which are located in New Brunswick. Total production is approximately 5 to 10 thousand bushels annually with clams being obtained from Newfoundland, Prince Edward Island, Nova Scotia, and New Brunswick. A number of trial projects have been initiated in British Columbia but none presently exist as a commercial venture. Therefore, this discussion will be limited to activities in Atlantic Canada.

REGULATORY STRUCTURE

The Federal Department of Environment, Conservation, and Protection Service, is responsible for classifying all harvest areas in accordance with the standards established by the National Shellfish Sanitation Program (1990a,b). Depuration facilities obtain shellfish from areas designated as restricted. The classification of these areas is updated annually. The depuration facility must be licensed both with the appropriate Provincial Department of Fisheries and with the Federal Department of Fisheries & Oceans (DFO). The Provincial and Federal Regulations are very similar and in most cases Federal Inspectors also serve as Provincial Inspectors.

Harvesting of shellfish from contaminated areas is controlled by the Federal Sanitary Control of Shellfish Fishery Regulations and the New Brunswick Fishery Regulations (also a Federal Statue). These regulations provide for the closing of areas for direct harvesting and the issuance of special permits to harvest in these closed areas under specific conditions. Conditions attached to the permit require the company to abide by a specific protocol which is outlined in a Memorandum of Agreement or Understanding (MOU) signed by the company and the Department of Fisheries & Oceans. In addition to the MOU requirements, the processing plant must meet the requirements of the Federal Fish Inspection Act and Regulations. These regulations describe construction requirements for processing facilities, operating practices, record keeping, and packaging requirements.

PROCEDURES FOR ESTABLISHMENT OF A DEPURATION FACILITY

Interest in establishing depuration facilities is on the increase with active proposals in Newfoundland, Prince Edward Island, and Nova Scotia. In approving new facilities a number of criteria are considered, including volume of resource available, regulatory resources available for monitoring harvesting and production practices, and impact on traditional fisheries. Application for establishing a depuration facility can be made by any interested party. The following outlines the steps leading up to a fully operational facility:

(1)   The proponent contacts the Federal Department of Fisheries & Oceans and Provincial Officials for initial discussion and is provided with an information package.

(2)   A detailed proposal is submitted which outlines construction and operation of processing facility, area to be harvested, and controls to ensure that harvesting, transport, operational, and storage requirements are met. Likewise, the proposal outlines source of water, water quality parameters, testing procedures to be used and the planned schedule of construction.

(3)   The proposal is evaluated by Provincial and Federal agencies including Inspection, Enforcement, and Scientific groups.

(4)   When all concerns are resolved the proposal is accepted.

(5)   The building is constructed and inspections conducted to determine whether design and operation conform to Fish Inspection Regulations and NSSP requirements   A Memorandum of Agreement is signed and a harvesting permit issued.

(6)   The performance of the facility is evaluated during a commissioning period in which extensive samples are analyzed to ensure uniform hydraulic flow and depuration rates. The final product of a plant must achieve a geometric mean of 50 fecal coliform/100g and not more than 10% of the samples exceed 130 fecal coliform/100g for Mya arenaria

(7)   If the on site inspections and test results are acceptable, permission to export is granted.

OPERATING PROTOCOL

Companies depurating clams must abide by the provisions of a Memorandum of Understanding between them and DFO. If the requirements of the MOU are not followed their permit to harvest clams in restricted areas can be removed

Each week the company must submit a weekly harvesting plan to the DFO, Conservation & Protection Office in the Area   This plan will indicate what areas or portion of areas are to be harvested, when harvesting will occur and which diggers will be employed. All harvesting must occur during daylight hours. harvesting can be mechanical or by hoe or hack, with hand digging the most prevalent. All diggers must be registered with DFO as commercial fishermen. During harvesting all diggers must be directly supervised by a "digger representative" who is responsible for clearly identifying the geographical limits for harvesting, sealing and labelling the containers, and record keeping.

During transport to the depuration facility, the temperature of shellstock must be controlled to provide an optimum response to depuration. Shellfish are culled and washed prior to transportation

Extensive records must be maintained at each facility. These records document harvesting information, date and time of processing, bacteriological quality of depuration water (before and after sterilization), and other water parameters such as salinity, turbidity, temperature, and oxygen content. Records documenting the bacteriological quality of each lot of shellfish for zero hour and final hour results are also kept.

Each depuration company is required to analyze "0" hour and final hour samples for fecal coliform. The company can utilize an outside laboratory or their own with all current processors presently opting to establish their own laboratory. These laboratories are audited by DFO and participate in split sample programs

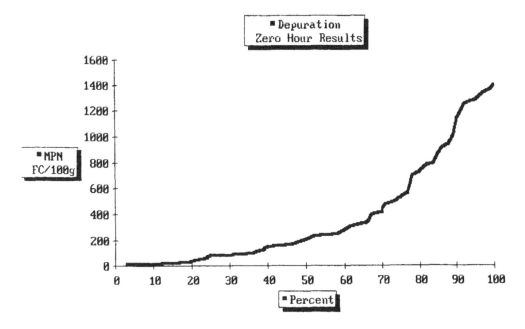

FIGURE 1   General zero hour fecal coliform/100 gram counts for <u>Mya</u> <u>arenaria</u>.

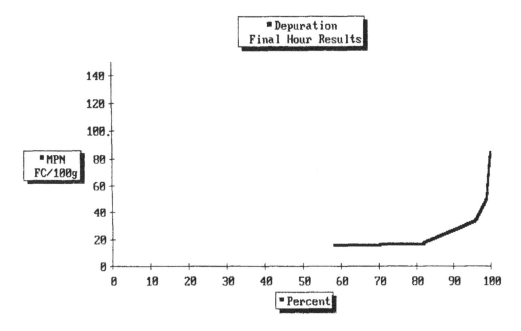

FIGURE 2.   General final (48 hr) fecal coliform/100 gram counts resulting for depuration of <u>Mya</u> <u>arenaria</u>.

A process is considered satisfactory if fecal coliform analyses of samples of cleansed clams result in a geometric mean of 50/100 grams of less and not more than 10% of the samples exceed a fecal coliform MPN of 130/100 grams. For each lot final hour samples are considered acceptable if no single sample exceeds of 170 fecal coliform MPN/100 grams or the arithmetic mean of duplicate samples does not exceed 125 fecal coliform MPN/100 grams

PERFORMANCE OF DEPURATION FACILITIES

Depuration facilities presently active have preformed very well. Zero hour counts (fecal coliform/100 gram) generally follow the pattern of in figure 1 with a 50 percentile of 200 and a 90 percentile of 1200 Little variation between harvesting areas has been noted. The general results of forty eight hour counts are shown in figure 2. These results demonstrate a consistent ability to satisfactorily depurate Mya arenaria shellstock.

Future success of depuration facilities and their ability to market a product without grit, undersized or broken clams, and to receive a premium price has generated increased interest in depuration. A number of new facilities are presently in the planning stage. With increased pressure for access to this resource, care must be taken not to over-harvest   As well this new fishery must be integrated with the traditional clam fishery with special emphasis on access to the resource.

REFERENCES

National Shellfish Sanitation Program (NSSP). 1990a. Manual of Operations, Part I: Sanitation of Shellfish Growing Areas. Public Health Service, U.S. Food and Drug Administration, Washington, DC.

National Shellfish Sanitation Program (NSSP). 1990b Manual of Operations, Part II· Sanitation of the Harvesting, Processing and Distribution of Shellfish, Public Health Service, U.S Food and Drug Administration, Washington, DC.

AN OVERVIEW OF THE BIVALVE MOLLUSCAN SHELLFISH INDUSTRY AND DEPURATION PRACTICES IN THE UNITED KINGDOM

Paul A. West

Although the United Kingdom (UK) is not as renowned as some of its European partners for the habits of seafood consumption, a small but thriving shellfish industry continues to exist at many locations around the coastline.

The development of the bivalve molluscan shellfish industry in the UK truly began many centuries ago. The 'native' oyster has been prized as a delicacy since Roman times. Last century, the abundancy of oysters made them the main food of the poor and destitute; an observation made famous by Charles Dickens in his book 'Pickwick Papers'. Cockle fishing in the outer estuary of the River Thomas, followed by cooking of cockle meat in the sheds at Leigh-on-Sea, have been undertaken on a commercial basis for several decades. Despite this long-standing history of trading, one of the greatest barriers to investment for enhanced production and sales is considered by many to be lack of widespread consumer confidence resulting from outbreaks of shellfish-associated gastroenteritis (Pain, 1986) This paper provides details of the UK bivalve molluscan shellfish industry and depuration practices (particularly in England and Wales), a technical account of commercial depuration, and assesses these public health concerns. It concludes with a personal insight to the future. Whilst the technical and scientific statements are valid, my views, opinions and observations on the UK industry are entirely personal. Prior to my recent move into the UK water supply industry, scientific duties in the aquatic environmental protection division of a government department in the Ministry of Agriculture. Fisheries and Food (MAFF) included liaison between public health regulatory agencies in England and Wales, and the shellfish industry, over control and prevention of food poisoning risk from shellfish, particularly through proper design and operation of depuration systems This overview draws on these experiences but does not represent any opinions and policies of MAFF or North West Water Ltd

SIZE AND GEOGRAPHICAL DISTRIBUTION

The UK (England, Wales, Scotland and Northern Ireland) has an extensive coastline. In some areas, the properties of the tides, sediments structure, and coastline configuration are simply not amenable for natural recruitment and growth of bivalve molluscan shellfish in commercially-exploitable quantities. In other more favorable areas, conditions may lead to extensive growth of these shellfish, such as mussels, which can form deep banks on the seabed. Increasingly there is scope for intensive mariculture exploitation using hatchery-treated juvenile bivalve molluscan shellfish particularly as the threat posed by anti-fouling paints containing tributyltin (TBT) now appears to be receding (Waldock et al , 1987, Edwards, 1988).

Commercially important bivalve molluscan shellfish found in the UK are oysters (The European flat or 'native' oyster, (O. edulis), and the heavier-shelled Pacific oyster, (C. gigas), mussels (M. edulis), cockles (C. edule), hard-shell clams (M mercenaria) and two species of scallops (P. maximus and C opercularis)

ENGLAND AND WALES

The North East coast of England is too rocky to support commercially-significant inshore bivalve molluscan shellfisheries although scallops are caught further out in the North Sea An area on the East coast known as 'The Wash' supports a major cockle and mussel fishery. Oyster beds are found along the Essex coastline along with cockles in the outer Thamas Estuary. The Solent on the South coast near the Isle of Wight is a major 'native' oyster fishery with Southampton Water harboring the largest natural hard clam fishery in the UK. Poole Harbour in Dorset is a significant clam and 'native' oyster fishery. Several estuaries and bays in Cornwall and Devon support Pacific oyster mariculture ventures and mussel fisheries. In the North West of England, cockle and mussel beds are located in the Ribble Estuary and Morecambe Bay. The Welsh coastline is abundant with cockles and mussels especially in the Menai Straits, Milford Haven and the Burry Inlet

Depuration is mandatory for bivalve molluscan shellfish taken from virtually all these coastal areas of England and Wales. The production value in 1988 for bivalve molluscan shellfish in England and Wales was approximately £1.1 million (Edwards, personal communication).

SCOTLAND

Due to the number of remote and sheltered inlets on this coastline, as well as access to clean seawater which does not receive any significant pipeline discharges of sewage or industrial effluent, the bivalve molluscan shellfish industry has grown rapidly in recent years with large commercial developments heavily investing in mussel, Pacific oyster, and scallop mariculture. Locally, there is some commercial harvesting of natural shellfish beds by individual fisherman. Due to the access to relatively clean seawater, shellfish rarely require depuration of relaying prior to sale for human consumption. The production value in 1988 for bivalve molluscan shellfish in Scotland was approximately £570,000 (Edwards, personal communication)

COMMERCIAL PRACTICES

The structure of the UK bivalve molluscan shellfish industry is fragmented. In Scotland and Northern Ireland, a few large commercial companies dominate the scene whereas in England and Wales, the industry is characterized by a large number of small (5 persons or less) independent family-based businesses each trading in isolation of each other, but often fishing the same local waters and competing for the same local market. Local rivalry is usually keen. Efforts for form local co-operatives with mutual benefit to all fishermen and traders are not generally successful

This fragmented structure has a significant bearing on the marketing strategies for the UK bivalve molluscan shellfish industry. Large commercial companies, particularly those specializing in mariculture, tend to concentrate finance and expertise on the production side of trade. The live, mature shellfish product is then passed to other businesses for 'added value' processing, storage and sale of the material. There are increasingly well-defined financial and operational boundaries between the production and marketing areas within the industry. In contrast, the small independent family-based business often undertake, without outside assistance, all the duties involved in shellfish harvesting, processing by depuration and transport to market. Overhead costs for vessel ownership, tenancy on buildings used for depuration, and transportation are often a significant financial burden Commercial marketing practices in the UK bivalve molluscan shellfish industry have also changed dramatically in the last decade. Historically, producers sent depurated material overland to major fish markets in cities like Manchester, Birmingham and the world-famous Billingsgate Market in London. These markets would sell live shellfish direct to the general public, and although these trade outlets continue to thrive, increasingly, public-health conscious customers are turning to multinational food chains for their bivalve molluscan shellfish purchases. These major food companies are able to provide an 'added value' product (e g. meat with the shell removed) in a frozen or chilled convenience style. Significantly, the small, independent family-based businesses cannot easily gain entry to the major food chain market. Only the large commercial mariculture developments in the UK bivalve molluscan industry appear at present to be able to satisfy the strict demands of multinational food chains for price, quality, quantity and continuation of supply This trend is likely to be accentuated in the next few years.

PUBLIC HEALTH CONCERNS

The unique ability of filter-feeding bivalve molluscs to accumulate human enteric pathogens from sewage-contaminated growing waters has long been recognized as the main characteristic which identifies these types of shellfish as vectors of communicable disease (Pain, 1986, Ballentine, 1987).

Public health concerns in the United Kingdom, particularly England and Wales, over shellfish-associated gastroenteritis have historically mirrored those of public health regulatory agencies in the United States. In essence, a historical perspective reveals that typhoid fever was the most significant illness in the UK spread by bivalve molluscan shellfish up to around the 1920's. After this, the widespread introduction and use of depuration technology has reduced to virtually zero the incidence of bacterial gastroenteritis and typhoid fever had an unknown non-bacterial aetiology, but were presumed to be viral from clinical symptoms In nearly 20 outbreaks, however, a small round structured virus or parvovirus-like particle was identified in the faeces of victims (Sockett, personal communication).

In general terms, the number of bivalve molluscan shellfish-associated outbreaks of illness in the UK over the period 1981-1988 exceeded that for the total in the previous 40 years. This rising trend of illness and public health concern over bivalve molluscan shellfish is due to a multiple of factors. Firstly, seafood consumption is now popular again in the UK Fish and shellfish are perceived by the public as contributing to a more healthy diet than meat and dairy products Research sponsored by the Government and the shellfish industry has also led to increased cultivation of certain bivalves, particularly Pacific oysters, by application of mariculture techniques. This recent expansion of the shellfish industry and the enhanced efforts to market a product which is traditionally eaten raw may be at least in part responsible for the predominance and number of illnesses associated with oysters. Finally, the increased recognition that bivalve molluscan shellfish are a significant vector for transmission of hepatitis A, a statutorily notifiable infectious disease in England and Wales, has led generally to associated with bivalve molluscan shellfish consumption in the UK. However, public health concerns now arise over the increasing frequency of bivalve molluscan shellfish-associated viral gastroenteritis (Sockett et al., 1985, Richards, 1987)

ILLNESS ASSOCIATED WITH BIVALVE MOLLUSCAN SHELLFISH

The national co-ordination of investigation and control of communicable disease in England and Wales in undertaken by the Communicable Disease Surveillance Centre (CDSC) whose headquarters are based at the Public Health Laboratory Service establishments in Colindale, London (Galbraith, 1982). A similar service exists in Scotland. The CDSC has for several years investigated the aetiology of illnesses due to shellfish consumption. Detailed reviews of shellfish-associated illness in England and Wales from 1941 were reported by Turnbull and Gilbert (1982) and from the CDSC by Sockett et al. (1985).

Nearly 80 outbreaks of foodborne illness associated with bivalve molluscan shellfish were substantively recorded in England and Wales in the period 1981-1988 Mussels alone were recorded as the vehicle of infection in only three outbreaks. Eleven outbreaks were due to cockles alone, over sixty outbreaks were due to oysters alone and, of 15 attributed to "mixed shellfish" food, many included cockles. Most outbreaks more rigorous epidemiological investigation of such outbreaks in recent years

PREVENTION OF ILLNESS

Regulatory control procedures for protection of shellfish consumers are enacted through the Public Health (Shellfish) Regulations of 1934 and 1948. These regulations give local authorities (port health authorities and borough or district authorities) in England and Wales statutory powers to control harvesting of shellfish for sale for human consumption from coastal areas with their justification   The regulations and their enforcement procedures have been described in detail by West et al , 1985

The written schedule available for public display describing the area, and the shellfish species subject to such regulations, is colloquially referred to as "the closing order".   Local authorities may make a "closing order" if there is direct evidence of communicable disease attributable to the consumption of shellfish taken from a harvesting area.   More likely, "closing areas" are implemented if, in the opinion of the Medical Officer for Environmental Health within the local authority, the consumption of shellfish from an area is likely to constitute a public health risk   Most commonly, "closing orders" are made when there is intermittent sewage contamination of a coastal area which results in variable sanitary quality of shellfish and so making it impractical for local health authorities to control harvesting times.

"Closing orders" are rarely short term.   Once in force, they permanently control the activities of the shellfish industry in an area.   The "closing order" may, on rare occasions, totally prohibit the removal of any bivalve molluscan shellfish for treatment and sale if sewage or chemical contamination of the coastal area is severe.   In most instances, however, the "closing order" permits harvesting of bivalve molluscs provided that such shellfish are treated and rendered fit before sale for human consumption.

There are two treatment options to remove sewage contamination. firstly, heat treatment to eradicate enteric pathogens by cooking; secondly, continuation of the natural filter-feeding activities after transfer to naturally cleaner seawater (i e , relaying) or after transfer to clean sea water and subject to depuration conditions (i e , self-purification in clean seawater systems).   Both relaying and depuration rely on filter-feeding actions to purge the sewage contamination from the gut and flesh of the live bivalve mollusc (Richards, 1988)

HEAT TREATMENT

Cooking procedures must conform to national hygiene regulations enforced by the local authorities in whose jurisdiction a processing plant is located (West et al., 1985).   Local authority environmental health officers spend time where necessary ensuring that bivalve molluscan shellfish, principally cockles, are cooked sufficiently to eradicate Hepatitis A virus.   The temperature and cooking time required to ensure adequate cooking compatible with tolerable loss of meat quality has been recently calculated for use by the UK shellfish industry (Millard et al , 1987)

RELAYING IN CLEAN SEAWATER

Relaying bivalve molluscan shellfish in non-polluted natural seawater for self-purification is rarely used by the UK shellfish industry despite extensive use in other countries, notably the United States.   Increasing, it is difficult to locate suitable inshore coastal waters which do not also receive waste discharges from significant domestic, agricultural or industrial sources.   More important, the movement of these shellfish to alternative sites is often limited by MAFF under the Molluscan Shellfish (Control of Deposit) Order 1974 to prevent the import and transfer of shellfish parasites and pathogens into otherwise disease-free areas.

DEPURATION IN CLEAN SEAWATER SYSTEMS

When a "closing order" is announced, shellfish merchants needing to make provision for depuration facilities generally seek technical advice from MAFF and industry-sponsored organizations such as the Seafish Industry Authority (SFIA).   In cooperation with local authority environmental health officers, technical advice on construction and operation of depuration tanks is provided to ensure that the essential requirements for successful depuration (i e , water temperature, salinity, oxygenation and system loading level) are implemented

A depuration establishment only receives technical approval from MAFF and the SFIA after it is shown to be capable of producing bacteriologically satisfactory shellfish capable of withstanding the stress of subsequent handling, transport and marketing before sale Technical approval of a depuration establishment is then conveyed to the Department of Health in London who issue a formal written seal of approval. When this has been given, plant operation may commence and routine public health sampling and inspection becomes the responsibility of the local authority (West et al., 1985).

DEPURATION SYSTEM DESIGN

HISTORICAL PERSPECTIVES

Commercial depuration of bivalve molluscan shellfish has been practiced in the UK for several decades Indeed, the particular association between typhoid fever and eating mussels, first described in 1985 by the Medical Officer for Health for the Borough of Brighten on the South Coast of England, directly led to the development of the first depuration plant at Conwy in North Wales around 1914. The depuration tanks at Conwy were pioneered by R.W. Dodgson whose classic report on investigations into the use of depuration for controlling shellfish sanitation was the foundation for commercial depuration tanks at Conwy are the only system in the UK to use chlorination for treatment of seawater. Seawater is sterilized with hypochlorite in large storage tanks, treated with sodium thiosulfate to remove residual chlorine, which inhibits shellfish filter-feeding activity, and then pumped into concrete tanks containing a single layer of mussels up to 15 cm deep. Depuration proceeds for 24 hour in static water before the water is drained off, the mussels hosed with water and the tanks replenished with freshly-treated seawater for a further 24 hours This method of depuration has not been more widely exploited in the UK due to the high construction costs of large concrete tanks and the cost and technical difficulties of chlorination and dechlorination for large volumes (approximately 90,000 gallons) of seawater

Since seawater for successful depuration must be treated to reduce the numbers of any sewage micro-organisms, other practical disinfecting procedures have been investigated. Ozone has generally been considered too expensive for depuration use and, without adequate control, could produce residuals which are toxic for shellfish. The use of germicidal ultraviolet light at 254 nm wavelength to cleanse seawater for shellfish depuration was introduced to the UK about 25 years ago by MAFF scientist P.C Wood. Irradiation with germicidal ultraviolet light enables seawater to be dosed continuously, to cope with variations in organic content and turbidity, and without leaving harmful residuals (Wood, 1961)

In the last two decades, depuration system designs in the UK have been refined in light of technology advances, increased understanding of the uptake, retention and elimination of sewage micro-organisms by bivalve molluscan shellfish, and in response to increasing land costs for sitting suitable facilities close to sources of seawater. Wood and Ayres (1977) described the use of artificial seawater mixtures for depuration at inland sites; Ayres (1978) introduced the recirculating constant volume design of depuration systems and West (1986) reported the operating criteria for such closed systems

RECENT DESIGN CRITERIA

The most common depuration systems used in England and Wales are designed to hold a constant volume of recirculating seawater which must be kept as microbiologically clean as possible and in a well-aerated condition. After contact with shellfish, seawater is repeatly recirculated past a source of ultraviolet light to maintain these sanitary conditions in the system, and then returned to the tank for use by shellfish. A circulation rate of one cycle of tank water per hour, and incorporation of cascades, provides adequate aeration and exchange of water in the depuration system. Most systems employ a shallow tank but, where space is limited, stacked tray systems over a slump tank are used (Ayres, 1978). Both designs operate to the same principle the provision of uniformly flowing clean, well-aerated, seawater over shellfish All depuration systems must be shore-based for ease of operation and for inspection by health officials

Shellfish in tank depuration systems are placed in trays or baskets raised at least 5cm off the base or floor. These containers are constructed of meshed walls and bottom to allow water to flow freely over the shellfish. This configuration allows detritus and faecal strands, eliminated from shellfish during depuration, to fall away from other shellfish and to be transported in the recirculation flow for irradiation by ultraviolet light or accumulate on the floor or base of the depuration system.

If shellfish containers are raised off the depuration system base or floor, rails or supports must run parallel to the axis of water flow to assist uniform water flow For ease of handling, individual trays should hold not more than approximately 200 hard clams, 300 oysters or 25 kg mussels unless mechanical lifting equipment is used. Ultraviolet light units are available from a variety of commercial sources. One 30 watt, totally enclosed lamp unit is sufficient to continually treat up to 2200 liters of seawater in a recirculating water system The dose of ultraviolet light to water should not be less than 10 mWcm^{-2}s^{-1}.

For mussel depuration, there should be a minimum of 8cm of water above a single 8cm layer of shellfish at the shallowest part when a tank design is used. The shellfish-to-water ratio for oyster depuration should not exceed 3 animals per liter of water. For hard clams, this ratio should not be more than 2 animals per liter.

It is important that depuration systems are built specifically for the intended purpose Attempts to convert swimming pools or crustacean holding tanks have been unsuccessful as there are often too deep for depuration purposes and encourage excessive stacking of baskets of shellfish in the belief that greater quantities can be treated. Tank length should not exceed three times the width to ensure as even a distribution of oxygenated seawater as possible. A tank base should slope at least 1:100 to assist draining and flushing between depuration cycles. The drain point should have a large diameter to ensure rapid emptying of the tank. Additional facilities for water temperature control and increased aeration of water can be added where appropriate.

Materials used for depuration system construction should generally have a smooth surface Often this is best achieved by coating concrete walls and floors with smooth, impervious and non-toxic epoxy paint. These surfaces limit build-up of the growth of marine life and slime, and are also durable and easy to clean. Accessories such as pipework and water pumps should be non-corrosive in seawater and not leach toxic products Metal fittings of bronze, brass, copper and iron should be excluded from use Plastic pipes of ABS (acrylo-nitrile butadiene styrene) or PVC materials are inexpensive and easy to assemble At strategic points, pipes should be fitted with removable caps or plugs to allow periodic cleaning of the inside of pipes with brushes or reamers Wood or plastic coated wire should be avoided, especially for shellfish containers and supports, since past experiences have shown that these are difficult to clean and accumulate slime easily

PLANT STRUCTURE AND LOCATION

Probably the most critical factor in choice of depuration plant site is an adequate supply of clean seawater Ideally, for ease of plant operation, depuration plants should be able to draw seawater of high salinity and low turbidity at all tidal states. Plants can be located in coastal areas of low salinity, in areas of unacceptable water contamination, or even away from the coast, if artificial seawater, prepared by dissolving the appropriate salts in mains taps water, is used (Wood and Ayres, 1977).

Sites chosen for location of depuration plants have easy access by vehicle, a piped and pressurized supply of potable water, a mains supply of electricity, and adequate waste disposal facilities. There is no standard design of depuration plant used in England and Wales, or indeed in the whole of the UK. After consideration of the availability of seawater, economic factors such as cost of acquiring the site, and the amount of shellfish to be handled, usually govern the location of an establishment. There are, however, statutory health regulations which relate to provision of toilet and washing facilities, cleanability of premises, waste disposal and control of pests. Site selection is generally made after appropriate consultation with local authority environmental health officers, MAFF and SFIA.

In each depuration plant, adequate space must be designated solely for handling and storage of untreated and purified shellfish as well as the depuration process itself. In order to avoid mixing and misidentification of batches of untreated and depurated shellfish, and to minimize cross-contamination, some physical separation of work areas is desirable. The plant should be laid out to ensure that material passes from the untreated area to the depuration system, then on to a clean area for the finished product

COMMERCIAL DEPURATION PRACTICES

SHELLFISH HARVESTING AND TRANSPORTATION TO DEPURATION SYSTEM

Harvesting, sorting, and handling techniques must not be too vigorous since shellfish with cracked shells or chipped flares often do not survive subsequent transport to shore-based depuration facilities Shellfish should be stored after harvesting in containers raised from the ground by duckboards in order to minimize contamination, and should be covered to protect from frost or direct sunlight. The temptation to "freshen" shellfish, held on a dockside or onboard boat prior to transporting onwards, by brief immersion in harbor water must be avoided.

In England and Wales, shellfish are usually transported to depuration plants in bags, at ambient temperature, on the day of harvesting, on open-topped vehicles. Shellfish can be damaged and unduly stressed if bags are too large or if loaded too densely causing animals at the bottom of a stack to be crushed. Large bulk containers with loose shellfish are not recommended since animals will be subjected to excessive abrasion during transport. Experience in the UK has shown that the survival time of shellfish, and their ability to regain filtration activity in depuration systems, is reduced if they are stored or transported for lengthy periods below 5°C or above 20°C. Shellfish must not be stored in direct contact with ice, or cold surfaces, or exposed to sudden large temperature variations during transportation, Shellfish should not be out of water for more than two days between harvesting and immersion in a depuration system and, wherever possible, they should be held in storage facilities with a controlled temperature of around 8-10°C

WASHING AND CULLING

This is an essential procedure before immersion of shellfish in depuration tanks Shellfish must be washed to remove mud and loose animals from the shells. Sorting of shellfish follows washing so that dead and damaged animals can be easily identified and removed. This is particularly important for clams and oysters but, due to sheer volume, can be impractical for large quantities of mussels.

Clean seawater or preferably, freshwater of drinking water quality must be used for washing animals before depuration. In smaller plants, animals are washed by hand but in larger plants, where mechanical devices are used, these processes should not stress animals by buffeting in rotary sieves or on rollers and conveyor belts.

PREPARATION OF THE DEPURATION SYSTEM

Before a depuration system is filled with shellfish and water, all detritus from the previous treatment cycle must be flushed out. Tank walls and base should then be scrubbed and flushed thoroughly with potable water. To prevent build-up of slime in the system, the tank and its pipework should be periodically cleaned and rinsed with a commercial sanitizing solution, and then flushed extensively with fresh water to remove traces of residual detergent and disinfectant.

LOADING THE DEPURATION SYSTEM

Few depuration establishments in England and Wales have mechanized loading facilities. Containers of shellfish are loaded manually be plant operators. In order to maintain adequate water circulation, and avoid areas of poor flow, containers holding oysters and hard clams at the approved loading rate must not be stacked more than three deep and with at least 3 cm space in between each container in s stack. Shellfish must not be loaded in baskets in deep layers. Accordingly, oysters must be laid as a single overlapping layer in containers to a density not exceeding approximately $500m^{-2}$. Hard shell clams vary in size but should not be loaded to a depth greater than 8 cm, which is about 70 kg m^{-2}.

Historically, containers of mussels must be laid in a single layer and not stacked. Mussels ar lighter-shelled and can be loaded to a depth of 8cm in containers without adversely affecting animals at the bottom of the pile This depth corresponds to a density of mussels of approximately 34 kg m^{-2}. More recent research by SFIA has shown encouragingly that some designs of "high-density" mussel depuration plant can permit stacking of layers of mussels in containers.

FILLING THE DEPURATION SYSTEM WITH SEAWATER

Seawater should be pumped into the system around high tide to obtain water with a salinity of about 25°/00. In order to minimize the time for acclimatization of animals in the tank, seawater for depuration purposes should in the range of ±20% of the salinity of seawater from which animals were harvested. In no circumstances should seawater be used in the salinity is below the minimum value which can be tolerated and in the UK these values are 19°/00 for mussels, 20°/00 for hard clams, 20.5°/00 for Pacific oysters and 25°/00 for 'native' oysters (Wood and Ayres, 1977) Excessively turbid water should not be used as it not only reduces the efficiency of ultraviolet light irradiation but can also cause clogging of shellfish gills and so adversely affect filtration rates. Such water should be either settled in holding tanks or passed through coarse sand and gravel filters to reduce exceptionally high turbidity levels before use.

During filling of the depuration system, water should be passed, at the same rate of flow that would be used once water is recirculating in the system, through the ultraviolet light units before cascading into the tank. It is sometimes necessary to build holding tanks for this purpose if it is not possible to achieve this flow rate during collection of seawater at high tide. Shellfish which have lost fluid from their shells during transportation replace it within a short time of re-exposure to seawater in the tank, so treated seawater is required at the outset.

STERILIZATION OF SEAWATER

Seawater is continuously dosed with ultraviolet light using compact, commercially-available, enclosed units. These are constructed of ultraviolet light-resistant, plastic cylinders with lamps surrounded by quartz sleeves, and are preferable to flow-through ('Kelly-Purdy') open units in which water flows under a series of lamps suspended above the water surface. Enclosed units are more efficient at killing micro-organisms, require less maintenance and are safer to use than the open design of units An accurate log of lamp operation must be kept and lamps replaced after 2500 hours use.

DEPURATION SYSTEM OPERATION

Once a depuration system, holding containers of shellfish, has been filled with treated seawater, the supply is closed off and the water is recirculated through the system for at least 36 hours. No shellfish must be added to, or removed from, the system at any stage. Water pumps and ultraviolet light water sterilizers must remain in operation throughout the depuration cycle Most tanks are operated for 44-48 hours over two days of business.

In the UK, to ensure that oysters and mussels undergo active filtration, water temperature should be maintained above 5°C throughout the cycle. The minimum water temperature for the depuration of hard clams is around 10-12°C for long periods since dissolved oxygen levels in seawater fall significantly above this temperature and spawning of shellfish may be induced.

Persistent low levels of dissolved oxygen severely inhibit filtration activity so levels in the water should always exceed 50% of the saturation value at the appropriate temperature and salinity. In recirculating systems, effective aeration results from the fall of water from the delivery pipe into the main body of water, but this can be enhanced by the use of venturi and nozzles attached to the delivery pipe. However, the cascade should not create undue turbulence in the water since this will adversely affect shellfish filtration activity. Trays of shellfish should not be place directly underneath cascades The use of a pumped air supply and diffuses in the depuration system is not recommended as this will disturb bottom debris, possibly recycling contamination back to shellfish.

REMOVAL OF SHELLFISH AFTER DEPURATION

At the end of a depuration cycle, the system should be drained within 15 minutes. In order to prevent recontamination, shellfish should not be handled until this is complete Shellfish must be washed or hosed thoroughly with potable water to remove faecal strands which may still adhere to shells. Hosing can take place in the system or after removal. Culling of dead shellfish often precedes packing for transport to market. There must be no delay in packing and no contamination should be introduced at this stage through use of dirty containers. Shellfish should be packed tightly in bags or boxes to help keep shells closed during transportation to markets.

QUALITY ASSURANCE

Regular monitoring and record keeping of depuration system use by plant operators and local authority environmental health officers is but one of several vital processes ensuring that the final product is safe for consumption. There is no visible means to demonstrate successful depuration other than by reduction of the numbers of bacterial indicators of sewage contamination Bacteriological examination of the final product contributes much to protection of public health but must not be used as the sole assessment of successful depuration. The application of the Hazard Analysis Critical Control Point (HACCP) concept to depuration systems has been pioneered for ensuring proper operation at all stages of the depuration cycle (West, 1986)

Whilst there are no statutory bacteriological requirements for bivalve molluscan shellfish sanitary conditions in the UK, a guideline of less than 230 <u>Escherichia</u> <u>coli</u> $100g^{-1}$ using a most-probable-number isolation procedure with modified-minerals-glutamate broth as primary enrichment, is increasingly accepted (West and Coleman, 1986). Routine bacteriological sampling of tank water and shellfish is recommended so that, as data accumulate, deviations in plant performance from expected values can be recognized at an early stage allowing rapid determination of factors responsible for a poor quality end product.

POSSIBLE FUTURE TRENDS

Since many estuaries and inshore coastal waters are subjected to sewage contamination from a variety of sources, it is likely that depuration systems will continue to be used extensively over the long term.

When properly undertaken, depuration greatly reduces the risk to public health from enteric diseases associated with raw bivalve molluscan shellfish consumption However depuration of shellfish may not always overcome the effects of sewage contamination of estaurine waters. Evidence from epidemilogical investigations and laboratory scale experimental operations demonstrates that depuration may not be effective for bivalve molluscan shellfish harvested from areas subject to chronic and severe contamination. Whilst numbers of E. coli in these shellfish can be reduced to acceptable levels, evidence is increasing that enteric viruses may persist much longer and cannot be removed by a short treatment period in depuration systems (Gill et al., 1983; Power and Collins, 1989). This is not a failure of depuration system design but is indicative that successful depuration depends on the initial level of sewage contaminatio (Richards, 1988). Accordingly, shellfish taken from areas receiving gross sewage discharges in close proximity should be moved (i e., relayed) to cleaner water for a period of several weeks to reduce the level of contamination before a final stage of depuration in shore-based systems. The degree of sewage contamination in growing waters above which such a relay operation is necessary has not yet been established in the UK, but is under review.

Such a requirement may be imposed via European Community legislation An expert working group convened by DG(VI) of the European Commission advised in 1988 that suitable control of shellfish sanitation can be made within the European Community by adaption of a legislative structure and framework based on regulatory experiences in the United States. The expert working group concluded that the degrees of sanitary treatment required for bivalve molluscan shellfish after harvesting should be graded on the basis of the extent of faecal contamination of shellfish flesh not the seawater in which the shellfish grow. This gradation of treatment needs, coupled with an as yet undetermined upper limit of faecal contamination beyond which depuration solely in shore-based systems can not be used, is in clear response to the increasing worldwide recognition that enteric viruses can persist in heavily-contaminated shellfish flesh even after routine depuration.

The impact of any European Community legislation on shellfish sanitation and the prosperity of the UK industry is difficult to gauge, particularly as the upper limit of faecal contamination which could prevent depuration by itself is not yet known. Whatever the outcome, it is likely that the UK shellfish industry will have to continue restructuring and rationalizing as discussed previously. The financial burden of meeting higher standards of food hygiene and customer-driven market demands is likely to force the smaller independent, family-based businesses may chose to locate at inland sites and use high-density depuration systems which operate with artificial seawater mixtures. These developments are being actively researched and funded through industry-based groups (e.g. SFIA) and probably represents the emergence of the next phase in the growth and ongoing adaption of the UK shellfish industry to wider changing social and economic factors which, in fact, have always historically influenced its prosperity.

ACKNOWLEDGEMENTS

I am grateful to Dr. Eric Edwards, Director of The Shellfish Association of Great Britain, for providing data on shellfish industry values and to Mr. Paul Sockett at the Communicable Disease Surveillance Center, Colindale for details on the incidence of shellfish-associated gastroenteritis

REFERENCES

Ayres, P.A. 1978. Shellfish purification in installations using ultraviolet light. MAFF Directorate of Fisheries Research, Lowestoff, England, Laboratory Leaflet No. 43, 20 pp

Ballentine, C. 1987. Weighing the risks of the raw bar. Dairy and Food Sanitation 7(3) 121-123

Dodgson, R.W. 1928. Report on mussel purification. Fisheries Investigations Series II 10(1)1-498.

Edwards, E. 1988 On the Shellfish Scene. Fish Farming International 15(11)18.

Galbraith, N S. 1982. Communicable disease surveillance. Recent Advances in Community Medicine 2 127-141..

Gill, O.N., Cubitt, W.D., McSwiggan, D.A., Watney, B.M. and Bartlett, C.L.R. 1983. Epidemic of gastroenteritis caused by oysters contaminated with small round structured viruses. British Medical Journal 287 1532-1534.

Millard, J., Appleton, H. and Parry, J.V. 1987. Studies on the heat inactivation of hepatitis A with a special reference to shellfish. Epidemiology and Infection 98(3) 397-414.

Pain, S. 1986. Are British shellfish safe to eat? New Scientist III 29-33

Power, U.F. and Collins, J.K. 1989 Differential depuration of poliovirus, Escherichia coli, and a coliphage by the common mussel, Mytilus edulis. Appl. and Env. Micro. 55(6) 1386-1390

Richards, G.P. 1987. Shellfish-associated enteric virus illness in the United States, 1934-1984 Estuarines 10 84-85.

Richards, G.P 1988. Microbial purification of shellfish: a review of depuration and relaying Journal of Food Protection 51 (3) 218-251.

Sockett, P.N., West, P A and Jacob, M. 1985. Shellfish and public health. Public Health Laboratory Service Microbiology Digest 2(2) 29-35.

Turnbull, P.C B , and Gilbert, R.J. 1982. Fish and shellfish poisoning in Britain. In: Jellife, I.F.P. and Jellife, D.B. (eds.) Adverse Effects of Foods. Plenum Press, London, P. 297-306.

Waldock, M.J., Thain, J.E. and Waite, M.E. 1987. The distribution and potential toxic effects of TBT in UK estuaries during 1986. Applied Organometallic Chemistry 1(3) 287-301.

West, P.A., Wood, P.C. and Jacob, M. 1985. Control of food poisoning risks associated with shellfish. Journal of the Royal Society of Health 105(1)15-21.

West, P.A. 1986 Hazard Analysis Critical Control Point (HACCP) concept: application to bivalve shellfish purification systems Journal of the Society of Health 106(4)133-140.

West, P.A and Coleman, M.R 1986. A tentative national reference procedure for isolation and enumeration of Escherichia coli from bivalve molluscan shellfish by most probable number method Journal of Applied Bacteriology 61(6)505-516.

Wood, P.C. 1961. The principles of water sterilization by ultraviolet light and their applications in the purifications of oysters. Fisheries Investigations Series II 23(6) 1-48.

Wood, P.C. and Ayres, P.A. 1977. Artifical seawater for shellfish tanks. MAFF Directorate of Fisheries Research, Lowestoff, England, Laboratory Leaflet No. 39, 11 pp.

PLATE 1. Typical tray culture of oysters: New South Wales

THE STATUS OF SHELLFISH DEPURATION IN AUSTRALIA AND SOUTH-EAST ASIA

Peter A. Ayres

INTRODUCTION

The brief for this chapter is extensive in that it endeavours to present a report on the status of depuration of molluscan bivalve shellfish in Australia and the ASEAN countries of South East Asia. Apart from obvious geographical differences between the two areas, depuration in Australia is now a well established commercial practice with over three hundred plants in New South Wales. In contrast, depuration in ASEAN is still largely experimental and at the pilot plant levels of development   However there are similarities in that ASEAN has adopted some of the basic technology and plant design from Australian experience and a major problem facing both areas is a lack of water quality data on which estuaries may be classified   For ease of reference this paper deals separately with Australia and South East Asia.

AUSTRALIA - BACKGROUND

Until the late 1970's commercial oyster production in Australia was largely confined to the eastern seaboard over the natural geographic distribution of the Sydney rock oyster <u>Saccostrea commercialis</u> (Plate 2) ie. from Moreton Bay in southern Queensland, throughout New South Wales and south to Wingan Inlet in eastern Victoria (Figure 1).   Carbon dating of shell remains in Aboriginal kitchen middens suggests that the rock oyster was widely used for food by indigenous peoples as far back as 6000 B.C. The colonisation of New South Wales by early white settlers saw the harvesting of oysters not only for food but also for building purposes. Vast quantities of oysters were gathered for their shells which were burnt to provide lime for mortar. Demand increased as the colony expanded and by 1868 the depletion of natural stocks was so severe that legislation was introduced prohibiting the destruction of live oysters for this purpose and attempts made to initiate cultivation. Thomas Holt constructed a series of shallow ponds and canals near Sydney to emulate the French "claire" system where oysters are fattened for market. Siltation and high temperatures combined to cause oyster mortality and the project was abandoned.

Oyster farmers developed a variety of other more successful techniques including the erection of stone slabs on the foreshores to collect natural spatfall, construction of raised shell beds on which young oysters would be seeded and the collection of oyster spat on sticks cut from local mangrove stands. Although some examples of these types of cultivation still remain the industry catches its oyster spat on frames of tarred hardwood sticks placed on intertidal racks   When the young oysters have attached to the sticks they are transferred further up the estuary to a depot area where they continue to develop protected from fish and other predators. The bundles of sticks are then separated and nailed out in single layers on intertidal racks. They may remain on sticks until they reach market size or be removed and grown out on wire mesh trays. In recent years alternative catching surfaces such as plastic strips which can be flexed to remove the spat, have been developed and the single oysters grown out on trays normally produce a better grade and shape of oyster than those produced by the more traditional areas. Climate in the growing areas ranges from subtropical in the north to temperate in the south of New South Wales and it may take from $2^1/_2$ to 4 years to produce a market grade oyster.

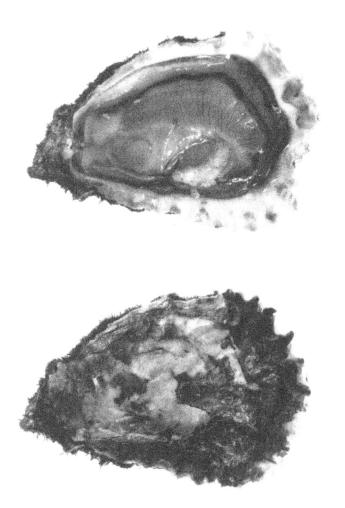

PLATE 2.   The Syndey Rock Oyster (*Saccostrea commercialis*)
(Market grade - natural size)

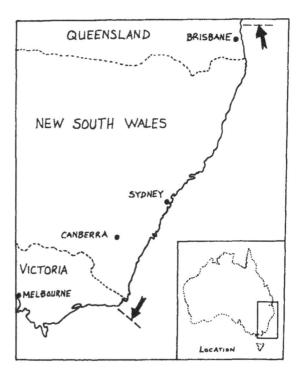

FIGURE 1. Distribution of the
Sydney rock oyster
(Saccastrea commercialis).

Oysters are marketed in two basic grades, bottles and plates. Bottle oysters as the name suggests have been traditionally sold as shucked meats in jars and are normally the smaller and often misshapen stick oysters. Plate oysters vary in size but average 50-55 g in total weight depending on the area where they are grown. Although some of the more progressive farmers are now marketing their product in cartons the standard unit for sale is a hessian bag that may contain between 85 and 120 dozen oysters, average 100 dozen or 1200 oysters. Some of the larger oyster farming businesses are vertically integrated and grow, process and market their own product, but the vast majority sell direct to processors who open the oysters on the half shell and market them as fresh or frozen product   Production increased dramatically from the 1930's to reach a peak of almost 147,000 bags (approx. 150 million oysters) in 1975 (Figure 2)   It now fluctuates just above 100,000 bags a year with a farm gate value of about A$35 million. Farmers receive A$3.40 a dozen for plate grade oysters which may retail for between A$12 and A$24 a dozen in clubs, restaurants and hotels. All in all the total value of the industry in New South Wales alone is estimated at A$200 million a year.

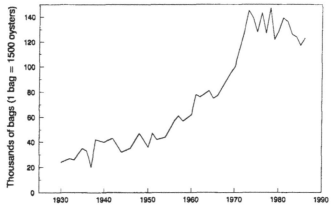

FIGURE 2. Total Oyster Production in N.S.W. 1931-1985.

Although some use has been made of hatchery reared spat in New South Wales, consistent natural spatfalls have generally meant that using hatchery spat is not cost effective. In contrast Tasmania has no indigenous rock oyster and although attempts were made to introduce the Pacific oyster (Crassostrea gigas) at intervals from the mid 1940's the species failed to establish itself except in one location in northern Tasmania. Repeated poor natural spatfalls and heavy metal pollution from mineral workings hindered commercial development until the first hatchery was established in 1977. Whereas "oysters" in Australia had been synonymous with the Sydney rock oysters for over a century the rapid development of the Tasmanian industry and imports of Pacific oysters from New Zealand means that the market is now shared by the two species Currently Sydney rock oysters comprise about 85% of Australia's oyster production with the remainder primarily Pacific oysters from Tasmania and a few mud oysters (Ostrea angasi) from Tasmania, New South Wales and Victoria. Market surveys suggest that demand still exceeds supply for much of the year but that the consumer is becoming more discerning about quality and presentation. In this area both New Zealand and Tasmania have been very successful in promoting their oysters but the New South Wales industry has tended to rest on its past monopolistic status and has undoubtedly lost ground to these, its major competitors.

THE INTRODUCTION OF DEPURATION

Although Tasmania is developing export markets for its oysters the rock oyster has never featured as a major export commodity. Prior to 1978 oysters frozen on the half shell were exported to Europe and to a lesser extent to Hong Kong and Singapore which were the major export markets for live oysters Strong domestic demand and escalating government inspection and certification fees have combined to effectively halt export markets from New South Wales although the Tasmanian industry has recently been signatory to a memorandum of understanding (MOU) for export of their product to the United States. Similarly the development of mud oyster (Ostrea angasi) culture is focused on the European market where the native oyster (Ostrea edulis) has been decimated by disease. The real significance of exports in the context of this paper has little to do with the export of a commodity but rather the control and political aspects which are implicit in certifying a product as wholesome and safe for export.

During the 1960's and 1970's there is evidence, much of it anecdotal, that consumption of oysters harvested in waters of New South Wales had been associated with outbreaks of gastroenteritis The author personally investigated such an outbreak in the United Kingdom where the consumption of Australian oysters at a formal dinner resulted in gastroenteritis with an attack rate of close to 100% amongst consumers. The oysters in question were exported as frozen half shell from New South Wales to a major seafood distributor in the United Kingdom. Earlier in 1964, such outbreaks in Australia had prompted some oyster farmers to establish depuration plants using u/v (Anon., 1964a, b) and based primarily on the information published in the United Kingdom (Wood, 1961). In the succeeding years those plants fell into disrepair and were abandoned

A major turning point came in the winter of 1978 when reports of over 2000 cases of gastroenteritis came in from both domestic and export markets. While domestic problems had clearly occurred earlier control lay primarily with State based organisations who had no established mechanism for reporting such incidents, nor it would seem, procedures for investigation and regulatory action

However exports of primary industry products come directly under the auspices of the Federal Export Inspection and Quarantine Service who issue inspection certificates for produce and licence export premises under the Export (Fish) Regulations. In 1978 the Federal authorities placed an immediate ban on further oyster exports from New South Wales pending advice from the New South Wales Government who were investigating the problem The then Health Commission of New South Wales placed a ban on all further sales from suspect areas and large quantities of oysters ready for sale were disposed of in land fill sites by oyster farmers. Epidemiological investigations confirmed oysters as the primary vehicle of infection and although the precise source of the oysters could not be determined the search was narrowed to leases in areas close to dense urban development, particularly in the Sydney region Microbiological investigations both in oyster producing areas and on oysters themselves confirmed faecal contamination and subsequently Norwalk like viral particles in fecal specimens from some of those affected. Feeding trials using human volunteers added further evidence of an oyster transmitted viral gastroenteritis with a mean incubation period of 42 hours (Grohman, et al., 1981). In addition a further outbreak some months later was traced to oysters that had been held in cold storage since harvesting and processing at the time of the initial outbreak.

The 1978 outbreak and subsequent investigations highlighted a number of factors which should be considered as they clearly not only have an important bearing on the future of the Australian, and particularly the New South Wales oyster industry but from the authors experience are important elsewhere in the world also. While possibly not an exhaustive list it is hoped that the discussion which follows illustrates the problems, the solutions and the pitfalls which face the shellfish industry and those whose role it is to regulate and control it and ensure that a safe product reaches the consumer.

1   Lack of microbiological data on oysters and shellfish growing waters even in heavily urbanised areas.

2.   Lack of dedicated analytical facilities and staff to support a shellfish sanitation programme.

3.   Lack of co-ordinated reporting, surveillance and response mechanisms within and between Government (and industry) bodies.

4   Lack of awareness by oyster farmers of the need for hygiene and care in handling a food product and their responsibility both to the community and each other to produce a safe product

5.   Poor urban planning and lack of development controls on the part of local and central Government authorities.

6.   Inadequacy of the sewage disposal system to cope with high and often seasonal storm water flows.

7.   Inadequate labelling and regulatory control of oysters to enable suspect batches to be quickly and accurately identified so that appropriate action can be initiated.

Dealing with each of the above in turn it should be emphasised that much of the comment which follows relates to lessons of the 1978 experience and in a later section dealing with the future it will be apparent that some are being or will be addressed

1.   LACK OF DATA

Apart from a few isolated samples, primarily of water, there was no historical or even current information on the sanitary status of New South Wales estuaries in 1978 These samples had been taken on an ad-hoc basis from the more urbanised areas where waters had been classified under the Clean Waters Act 1970 by the State Pollution Control Commission. This classification has no particular reference to shellfish cultivation as such and concentrates on chemical and physical parameters In general what samples were taken were taken to fulfil a minimum statutory requirement by the Commission and the Sewerage and Drainage Board From 1978 to 1981 sampling was directed towards providing information on the levels of faecal contamination in estuaries where oysters were cultivated   The frequency and extent of sampling was linked with the oyster production of particular areas so that major production areas were sampled more frequently than those of limited oyster production. Coincidentally major production areas such as the Georges and Hawkesbury Rivers in the Sydney region are also amongst the most urbanised in terms of development and boating activity. High risk areas were identified by the percentage of oyster samples which exceeded the Health Commission's limit of 230 fecal coli/100 g under the Public Health Pure Food Act (1908) Regulations and estuaries were ranked accordingly. Having made a decision to implement a depuration programme oyster farmers were required to install depuration facilities in the high risk areas first and gradually over a three year period progressively to less or unpolluted areas until all oysters produced passed through a depuration process. Because the percentage failure system was employed estuaries with a very low sampling frequency were inevitably under or over rated in terms of pollution risk and many pristine waterways were included while others where pollution risk was potentially high were probably ranked too low in order of priority.

2.   LACK OF ANALYTICAL FACILITIES

During this extensive sampling programme the Analytical Laboratories of the Health Commission were virtually dedicated to water and oyster sampling. While input from the University of New South Wales lightened the load somewhat in New South Wales had, and still has, only the one central testing facility for all foodstuffs. Once the depuration programme was implemented sampling virtually ceased and apart from a few isolated shellfish samples submitted by food inspectors no routine sampling has been conducted since. Similarly the New South Wales Department of Agriculture and Fisheries commenced a water classification programme in two of the fifteen or so major shellfish production areas but this was terminated prematurely due to lack of resources and shows no sign of being restored. Currently, as so often in the past concerned oyster farmers have to resort to Government approved private testing laboratories on a fee for service basis. The logistics of collecting, transporting and analyzing samples from over a 1000 km of coastline at one central testing laboratory or Sydney based private laboratories suggests that this problem will be an on-going one.

3.   LACK OF COORDINATION BY AUTHORITIES

The official response to industry in 1978 was to find a solution to the problem or risk being closed down permanently   Industry responded by seeking advice from overseas and commissioning the University of New South Wales to develop depuration facilities which could be implemented by oyster farmers   Although the public health is covered under the Public Health Pure Food Act and Regulations the oyster industry itself is largely administered by the New South Wales Department of Agriculture and Fisheries. Oyster leases are Crown Land and administered by the Department under the Fisheries and Oyster Farms Act 1935. This Act has no public health considerations directly but gives the relevant Minister delegated power to control the industry and the methods it employs under various regulations and sections. In effect the Fisheries and Oyster Farms Act 1935 and Regulations controls the oyster industry through the cultivation and harvesting stages to the point of sale when responsibility shifts to the Department of Health and provisions of the Pure Foods Regulations. When depuration was introduced and made a statutory requirement for all oysters sold in New South Wales advice on the design, construction and operation of depuration plants was provided by the Fisheries Department as an extension service to the oyster industry   Similarly the operator instructions for plants, registration of plants, permits to operate, log books and other administrative procedures were all under the auspices of Fisheries rather than Health. This remained in force until very recently when total control of depuration passed to the Department of Health.

Besides Fisheries and Health, the State Pollution Control Commission has a statutory role also in that it is the licensing or consent authority for waste discharges including sewage but it also sets, monitors and enforces the regulations pertaining to such discharges In some respects the Commission may find itself to be the judge, jury, prosecutor and defendant all at once. In spite of its responsibilities under the Clean Waters Act 1970, the State Pollution Control Commission does not have the resources to carry out its functions except in a token sense. At the time of the 1978 problems the oyster industry itself was represented by two separate associations who originally represented the Sydney area growers (and those who had been linked with previous food poisoning scares) and Northern area growers who wished to disassociate themselves from any possible link with food poisoning. Leaders of the industry and the more progressive individuals quickly realised that adverse publicity affected consumer confidence in all oysters not just those originating in particular areas or causing problems. As a result the oyster farmers combined to form a single association and for a few years at least sought to represent all oyster farmers equally and were recognized by government and others as a considerable force.

Overall it must be said that key members of the oyster industry were the driving force behind the introduction and development of depuration technology in New South Wales and in formulating the guidelines under which it was used. This situation prevailed but as the cohesion of the industry weakened government assumed a more dominant role. Ten years later nothing has changed for the better unfortunately and there is no overall co-operation between the agencies concerned. The Health Department control depuration by issuing the permits and approvals to operate but are grossly understaffed and ill equipped to perform other than a basic policing function Fisheries have abandoned any involvement in the issue and the State Pollution Control Commission remain at best as observers. In short there is considerable need for a single co-ordinating body to seriously address the problems and for government to commit the necessary resources to stabilising the situation and hopefully even initiating some improvements. The oyster industry in New South Wales constitutes the major aquacultural industry in Australia and is being strangled by bureaucracy on account of a problem not of its making and not within its power to resolve.

4   LACK OF AWARENESS BY OYSTER FARMERS

The Sydney rock oyster was well named as much for its natural habitat as for its appearance and strength It is quite unique in that it can withstand prolonged exposure at high ambient temperatures and will remain alive and healthy for two weeks or more out of water. The rough and irregular shape of stick cultured oysters and their general toughness seemed to have bred a certain element of contemp amongst growers who routinely shovelled the oysters into hessian sacks and treated them very harshly Premises in which oysters were handled were often simple lean to's with not even basic facilities nor concern about cleanliness or hygiene. With the introduction of depuration power was required for UV lamps and pumps, electrical installations and plant needed protection from the weather and a whole new era began. New premises, well lit with washing and toilet facilities were constructed to house depuration plants and with the emphasis on handling a food product rather than a mere commodity For the first time oyster farmers actually observed oyster activity in the depuration plants and had a conscious awareness that they were dealing with a living animal, capable of effectively filtering large volumes of water and thereby posing a hazard to consumers if they were cultivated in polluted waters The importance of salinity, temperature and dissolved oxygen were observed first hand by oyster farmers who generally had no prior awareness of these parameters in relation to the animal they cultured

Gradually the level of inspection, advice and surveillance virtually ceased Now plants were built without guidance and many existing plants were modified to the extent that they no longer conformed to the guidelines under which they were originally constructed and approved The net result has been an overall deterioration in standards and a resurgence of problems with shellfish transmitted gastroenteritis.

Clearly, the whole system is in urgent need of a thorough overhaul and oyster farmers must be re-educated about the importance of depuration, plant hygiene, and of the consequences to their own livelihood, to their industry as a whole, and the health of the consumer. The Department of Health has moved to try to correct the many deficiencies but unless adequate manpower and resources are provided for the task it seems unlikely that any lasting improvement will result. Ultimately unless control can be effected many existing areas of oyster cultivation are destined for closure and many oyster farmers will lose their livelihood.

5.   LACK OF COORDINATED PLANNING

At the turn of the century areas close to Sydney such as the Georges River were relatively pristine waterways flanked by open woodland and natural bush Commercial fishing and oyster farming were virtually the sole activities in most areas and the fishermen and oyster farmers often constituted the entire local population. Gradually the population increased, and with prosperity and leisure and development spread from the coast into estuaries and inland up and down the coastline. With this urban expansion came industry and pollution from septic and sewerage outfalls Clearing of steep riverside land led to erosion, siltation and increased storm water run-off and previously pristine areas became environmentally degraded   The growth of residential and industrial development has outstripped the capacity of sewerage systems and many are overloaded and inadequately treated. A lack of consultation and planning between the responsible local and state government authorities is clearly evident and on the receiving end is the oyster industry. Recent years have seen a huge increase in recreational boating and facilities such as marinas without attention to the wastes and pollution such activities create.   As a consequence the environment and the aquatic habitat are under pressure and the problems of the oyster industry are an inevitable symptom of a serious disease. A total catchment management approach may be an answer to slow and hopefully halt this decline

6.   INADEQUACY OF SEWERAGE DISPOSAL SYSTEMS

Much of the coastline of New South Wales is rocky terrain with shallow soils quite unsuitable for soak-away septic systems that are still common in more rural areas. Long periods of drought followed by heavy and often seasonal rainfall can produce severe flooding from storm water run-off At such times already inadequate sewerage systems are bypassed or surcharged and release of untreated effluent directly into rivers occurs Such were the circumstances which led to the 1978 food poisoning problems and almost certainly contributed both to earlier and recent problems. Development of coastal land in New South Wales particularly is proceeding at a rapid rate but the effluent disposal is low in the order of priorities and the already inadequate treatment systems become even more inadequate. Government is trying to address beach pollution problems resulting from the discharge of primary treated sewage and sewage sludge but so far has not even considered estuaries and rivers.

7.   INADEQUACY OF PRODUCT IDENTIFICATION

One of the most serious problems faced by the Health and Fisheries authorities in the 1978 outbreak was to trace quickly and accurately the source of suspect oysters from the point of sale back to their origin. Under the Fisheries and Oyster Farms Act oyster farmers are required to indicate on their bags or cartons the lease number and estuary where the oysters originate and the name of the producer. However as mentioned in the background to this paper most oyster farmers sell live oysters to processors who open them and sell on to the retail distribution system. Because oysters are sold by the bag and priced by the bag at grower level and because the oyster count per bag may vary from 85 dozen to perhaps 120 dozen processors have tended to purchase bags of different count (or grade) and mix them up before distribution. This means that at retail level oysters can only be traced back to a particular processor or supplier who in turn may or may not have records of the growers who supplied them originally. Mixing of oysters from different growers and locations further complicates the issue. At grower level other complications arise since some growers have leases in more than one estuary and all growers have a number of leases in any one estuary. During normal cultivation practice oysters are moved from one lease to another and even one estuary to another so invariably the lease number placed on the consignment for market may not necessarily have been its most recent origin A further complication arose from the re-use of bags by growers other than those identified on the bag originally As a consequence of all of these difficulties the precise identification of the source of oysters implicated in the 1978 outbreak was not possible and only some careful detective work revealed the suspect estuaries largely by testing remaining oysters which had not been sold

With the introduction of depuration from 1978 onwards each plant was issued with an identification number as it was constructed, eg. PP1 and so on. Records were kept of the name and address of the plant operator, location of the plant itself and details of the plant specifications It was intended that these records would constitute a profile or fingerprint of each plant Modifications or changes to that original specification or profile were not permitted without approval of the Fisheries Department In this manner the plant could be inspected by a fisheries or health inspector for conformity with its approved design and specification at any time The permit to operate the plant could be suspended or cancelled for infringement of the conditions. Each plant operator was also issued with a personal log book and required to enter details of each batch of oysters depurated. Information included the quantity of oysters treated, the date received and despatched, records of salinity and water temperature at commencement and completion of the minimum 36 hour depuration period and destination (buyer) of the product. Many oyster farmers were sensitive about recording customer information in detail for fear of losing their customers to competitors and in these cases recording of a sales docket number which could be checked back at a later date if necessary. Small farmers who didn't have their own plant put their product through someone elses and it was conditional on the owner of the plant to ensure such batches were identifiable in the logbook Log books had carbonized duplicate sheets for return to Fisheries as a permanent record. New or replacement log books were only issued on production of a completed book or evidence of loss or damage This system as described was put into place for the first 270 or so plants commissioned and looked as though it would be successful. Typical failures identified were retrospective completion of log books ie. after a batch or batches had been treated, failure to enter all details, falsification of quantities treated such as more sold than recorded, suspicious loss or irretrievable damage to the log book. However, by and large it was an improvement and while it was supervised showed great potential All bags, bottles and cartons of oysters sold were required to have a Purification Plant number on them and processors were not to be permitted to mix more than two sources of oyster in any batch sold and to keep records of the origin of their purchases.

Regrettably supervision of the permit and log book system was not maintained and advice from the Department of Health recently reveals widespread malpractice and falsification of records by oyster farmers. Oysters have reached the market place without purification and contrary to their documentation, plant numbers have appeared on batches of oysters not associated with that particular plant and operational guidelines have not been in compliance Supposedly depurated oysters have been found to contain bacterial counts in excess of regulatory limits and this is clearly associated with the increased incidence of oyster associated gastroenteritis. The Department of Health has moved to tighten the labelling regulations and recognised a number of major deficiencies in the production, distribution and marketing of oysters in New South Wales. These deficiencies extend beyond the responsibilities of oyster farmers and processors into lack of surveillance, monitoring and control by the responsible authorities so there is much to be addressed and quickly.

DEPURATION PRACTICE

In 1964 and again in 1978 when oyster pollution became a public matter, the oyster industry, and later the Government, looked towards Europe for assistance and inspiration largely because of the long history of depuration, particularly in the United Kingdom. Dr. G Fleet of the University of New South Wales had also visited Europe to inspect depuration facilities and meet with research workers and others involved On his return to Australia, Dr. Fleet was able to begin experimentation with the Sydney Rock Oyster In 1978 and 1980 the author was seconded from the United Kingdom Government to assist the Department of Fisheries and Oyster Farmers' Association in setting up depuration facilities, formulation of operational guidelines and to advise on regulatory matters. Funding for these two visits was shared between the oyster industry and State Government Fisheries budget. Given this background it is no surprise that initial tank designs and operational parameters were based on United Kingdom experience although the rock oyster and the climate bore no resemblance to their European temperate counterparts. Apart from its long shelf life out of water, the rock oyster is tolerant of a wide range of salinities and temperatures. Climatically, seasonal very heavy rainfall and high ambient water and air temperatures were very different from conditions in the United Kingdom. An early problem was spawning of oysters in depuration tanks during the summer months; this release of material quickly fouled the water and it was necessary to return oysters to the leases to recover condition.

Experimentation by the oyster farmers, Department of Fisheries and the University of New South Wales continued and most of the problems were overcome or at least understood. It would have been ideal to construct a few centralised depuration facilities, each to serve a particular area or estuary but apart from a lack of suitable waterside sites, New South Wales oyster farmers are very parochial and individualistic. In consequence the majority of the farmers expressed a desire to have their own facilities. This posed a problem because oyster farming operations range from one or two very large businesses with many leases, often in more than one estuary down to small family or one man operations. Therefore their ability to finance construction of a plant and capacity required also varied enormously. In addition water and oyster sampling had shown that pollution risks varied from area to area; some were free of obvious pollution and others were only polluted after heavy rain The climate over the 1000 kms of coastline is such that summer temperatures in the north may reach the mid-30's or more and rarely fall below 20°C in winter In the extreme south summer temperatures may reach the mid to upper 20's but in winter air temperature may fall to almost zero. Although water temperatures vary over a much narrower range heating or cooling of tank water was introduced as standard equipment in many of the early depuration plants.

In the early stages of certification and approval of depuration plants they were tested with batches of oysters seeded with E. coli. If a plant successfully depurated three such batches it was approved. These tests were conducted by the Health Commission but new plants were being built so rapidly that already stretched manpower resources could not cope and the scheme was abandoned Later plants were approved solely on the basis of design parameters which were provided by the Fisheries Department who also approved the actual plant As mentioned earlier, over the last 3 years or so the industry has virtually built and used whatever it likes because of lack of official involvement. Few of these new plants would conform to the original criteria and many which were approved have since been so modified as to constitute different plants.

However within the constraints already mentioned the compulsory purification of oysters in New South Wales was made law in 1981 and by 1 May 1983 this requirement was extended to all oysters sold in New South Wales regardless of whether or not they were produced in the State. Three basic types of depuration system have been adopted by the New South Wales oyster industry, high density fish box type, high density tray system and pool type. In general the shellfish to water ratio is highest in the fish box units and lowest in the pool type units. For simplicity a brief explanation of each type and its design application follows:

1. Fish box type

This system was based on a similar system used in the United Kingdom and more recently in Singapore and Malaysia. In the United Kingdom it was principally intended for use by wholesalers in inland markets where space was at a premium and natural seawater was not available. In such circumstances artificial seawater was used successfully but to lower costs of preparation a limitation on volume was important. Whilst these criteria are not important in the Australian situation there was a clear need for a small pre-fabricated plant which could be erected on leasehold land when permanent structures were not generally permissible and also suitable for one man operation, small volume oyster capacity and cheap to build and operate.

To keep costs down a widely available stack/nest type container made of high density polyethylene was selected and modified by cutting a slot or two holes in one end through which water could overflow. A small (5mm) hole was drilled in the base of each box to allow drainage in the event of pump failure and on completion of the depuration cycle. The boxes were stacked on top of each other either on a common reservoir tank or in some cases the lower level of boxes was employed as a reservoir. A typical fish box plant is shown in Plate 3 fitted with a single 30W enclosed u/v unit, pump and filter system (cartridge type). PVC plastic plumbing with individual perforated T-bar deliveries (valved) to each stack of boxes allowed use of all or part of the unit as required. The system installed in Malaysia is essentially the same (plate 10; Figure 8) but has the boxes mounted in a frame to permit easy access for experiments.

PLATE 3. Fish-box type high density depuration system.

The boxes each hold approximately 150 oysters and plants were often designed in stacks of eight boxes which equals one bag, the standard unit of sale for oysters. The principle advantages of the fish box system are·

* very effective aeration without turbulence due to the gravity flow of water through each stack.

* very high turnover or exchange rate: operationally each box only holds 20 litres of water and even with a small pump a minimum of five changes per hour is achievable without overflow or spillage.

* in some respects each box acts as a self contained unit and oyster faeces remain undisturbed in the bottom of the box as water flows out from the surface.

* the unit (ie. each box) is self draining once the pump is switched off either intentionally or accidentally via a power cut. In a normal system a prolonged power failure can result in oxygen depletion and weak or dead shellfish. In practice oysters survive better in moist air than immersed in deoxygenated water.

* once drained the boxes can be removed and used to transport the depurated shellfish if required.

* boxes are easily cleaned, strong and easily replaced (A$14 each)

2. Tray type

These systems are based on rectangular fibreglass moulded trays supported one above another in a steel frame over a reservoir. Various patterns are in use including two which were marketed as a complete package to oyster farmers (Milligan and Pritchard types) inclusive of UV steriliser, pump and plumbing. The major difference between the systems is in the water circulation pattern Some have individual inlets and outlets to each tray (Figure 3) draining to a common sump while others have water pumped to the top tray which then gravity feeds the trays below rather like the fish box system (Figure 4) Gravity feed systems are normally fitted with weir type sterilizing units (UV) but those with individual feeds requiring positive pumped pressure have to use enclosed sterilisers.

Although both variations seem to be effective the pumped, individual inlet type have some disadvantages namely:

* a larger pump is needed to achieve adequate water flow to each tray.

* careful regulation of flow to each tray is essential to avoid low flows and low dissolved oxygen particularly in the bottom tray.

* individual inlets to each tank requires a valve to be fitted on each inlet to regulate flow; increased cost and maintenance.

* an enclosed UV steriliser is essential, not optional. Such units may be up to ten times the cost of a weir unit to do the same job.

The most obvious disadvantage with the gravity feed system is the potential to transfer oyster faeces (and possible contamination) to the lower trays. In practice this does not seem to have occurred because the trays are typically 3 m long and 1 m wide x 0.2 deep and very high water velocities or turbulence would be necessary.

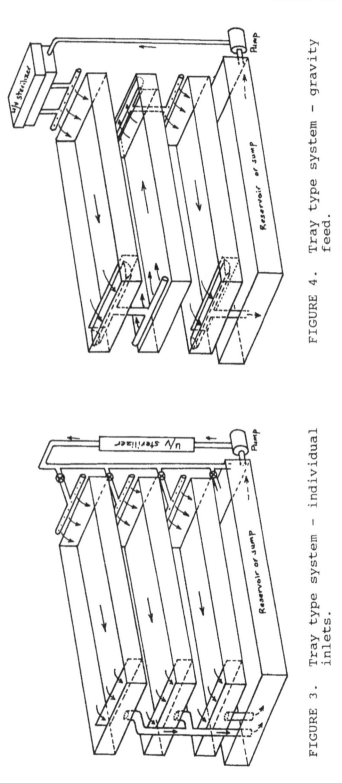

FIGURE 4. Tray type system – gravity feed.

FIGURE 3. Tray type system – individual inlets.

The major disadvantage of both types of tray systems is a practical one. Access to position or remove baskets of shellfish and to clean or flush the trays is difficult especially when the stacks of trays are high with limited space between individual trays   Plate 4 illustrates a typical tray unit and also its major drawback, access. This type of unit holds $1^{1/2}$ bags of oysters per tray (1500 approx). Fleet, Souness and colleagues at the University of New South Wales devised a very compact palletised tray unit illustrated in Plate 5 which could be positioned or removed by forklift and is ideally suited to the larger oyster farmer.

3. Pool type

Before considering the typical pool system there is an intermediate type which embraces many of the features and advantages of the conventional tray or pool system. Basically of the same rectangular shape as the trays described above the unit was designed to be used in parallel or stacked formation (supported by a frame) and deep enough to hold two layers of baskets. Designed by the author as a complete package with simple weir type u/v steriliser. The system was manufactured and sold in large numbers by a commercial GRP specialist. It proved to be ideally suited to the needs of the small oyster farmer who hoped to expand because it could be extended without major alteration in a modular fashion. The complete tank and steriliser originally retailed for under $A2000 which was less than 50% of the cost of an enclosed 8 lamp UV unit on its own. This so called "Ramsay" unit (after the manufacturer) is illustrated in Plate 6.

The "Ramsay" unit could be easily converted to double shellfish capacity by raising the overflow level, inserting a baffle wall and separating the two layers of shellfish baskets with corrugated fibreglass sheet (Figure 5). Although very effective and popular as a ready made system the weir unit was designed to accommodate two stacked or parallel tanks and when used with only one tank the recirculation system needed careful adjustment or the initial surge of water as the pump started could flood the lamps. Earlier models also had transformers mounted on the lamp housing which made it cumbersome and heavy to lift, deterring access and maintenance which is recommended and essential for all users of the system   "Ramsay" units are basically 5 bag units (5000 oysters) in the single layer format and can be increased to 10 bags capacity by the conversion described earlier.

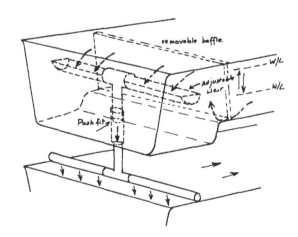

FIGURE 5. 'Ramsay' type depuration system detail
of adjustable weir system.

PLATE 5.   Modular-palletized tray
type depuration system.

PLATE 4.   Tray type depuration system with
individual inlets/outlets.

The larger swimming pool type tanks come in a variety of sizes but are basically rectangular on a minimum 2:1 length to width format with a gradient from inlet end to the outlet and drain to facilitate emptying and cleaning. Examples have been constructed in fibreglass, concrete or bricks, timber and concrete blocks painted with epoxy paint or fibreglassed. They have proved most popular with farmers who need to depurate at least 20 bags (20000 oysters) of oysters at a time. Some early examples were built to use ozone for sterilisation but the small ozone generation plants proved expensive and very troublesome and have all been replaced by u/v sterilisers. Many of the pool systems use multiple lamp (4, 6 or 8) enclosed units from various commercial sources but a Kelly Purdy type unit was designed by the author (Plate 7) in u/v resistant and corrosion proof plastic material and has been adopted for some pool systems in preference to the more expensive enclosed units. Typical examples of small pool and large pool systems are illustrated in Plates 8 and 9. Concrete pools were designed with longitudinal ridges moulded into the tank floor to act as supports for the shellfish baskets and assist in flushing and cleaning of the tank Oyster farmers commonly use loose lengths of plastic electrical conduit pipe for a similar purpose. Early pool designs only accommodated single layers of baskets but later models were designed for single or double layers as required by fitting a weir wall of adjustable height. Layers of baskets are separated by sheets of corrugated fibreglass which forms channels longitudinally down the tanks, permits a good flow around the animals and assists in collecting oyster faeces and other debris, directing it towards the sump and drain area (Figure 6) The largest pool system in New South Wales is a multiple unit holding 300 bags of oysters (300,000 approx) per cycle. This is the only plant using ozone which is economically suited to the size of the plant and its flow through design Although the majority of pool systems were originally designed to operate on a recirculation rather than flow through mode many have apparently illegally switched to flow through and this is now a common practice (Bird pers comm).

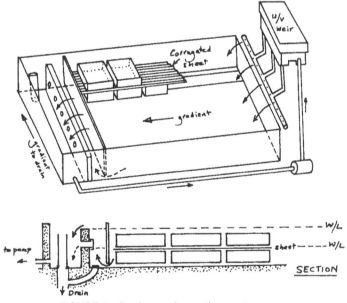

FIGURE 6. Pool type depuration system.

All depuration systems in New South Wales were designed to have a minimum water flow of at least twice the total volume per hour and most well exceed that particularly the fish box and tray type systems. A minimum dissolved oxygen level of 50% saturation at full shellfish capacity was also a design criteria. Early plants were all fitted with multiple lamp

PLATE 7. All-plastic Kelly-Purdy type u/v unit (8 x G36 lamps).

PLATE 6. 'Ramsay' type depuration system with simple weir u/v unit (2 x G36 lamps).

PLATE 9.   Partial view—large scale
pool type depuration system.

PLATE 8.   Small fiberglass pool type
depuration system.

enclosed UV sterilisers capable of achieving a 99.9% reduction in waterborne test bacteria in a single pass which in most instances amounted to a bonus for lamp suppliers and considerable financial outlay for oyster farmers. In later plants lamps and sterilisers were more discretionary and large capacity sterilisers were only recommended for large volume tanks, flow through systems and high risk locations. Elsewhere a variety of smaller enclosed systems and weir mounted units were employed on the basis that 90% of plants were designed and approved for recirculation only and all such units were capable of a 99%+ accumulative kill within one or two hours. Additionally although it is a statutory requirement to depurate all oysters in New South Wales, pollution problems are restricted to a few heavily urbanised areas and then infrequently eg. particularly after heavy rain Recent outbreaks have prompted the Department of Health to restore the requirements for a high instantaneous kill rate and has endorsed units which meet those requirements ie. 99 9% reduction of waterborne $\underline{E. \ coli}$ in a single pass.

Other basic parameters for depuration plant operation which have remained unchanged are oyster density at 500-650 oysters per square metre of tank area, and minimum flow rate of twice the total volume per hour. Water temperature between 14° and $27^{\circ}C$, salinity above $18^{\ 0}/_{00}$ and a depuration period of 36 hours minimum are also required. Full details of current operational requirements and guidelines are available in Bird (1989) Although it seems likely that future amendments will be made useful references to studies on depuration in New South Wales can be found in Souness and Fleet (1979), Son and Fleet (1980), and Eyles and Davey (1989).

THE TASMANIAN EXPERIENCE

It was mentioned earlier in the background to this paper that Tasmania has developed a very successful oyster industry over the last decade based on the Pacific oyster ($\underline{Crassostrea\ gigas}$) Tasmania as a State has relatively high unemployment and is very dependent economically on utilisation of natural resources, primary products and tourism. Besides its forests, minerals and agricultural activities, as an island State surrounded by water, fishing is an important activity The low population and abundance of sheltered bays and inlets also makes the State very suitable for mariculture and besides the oyster industry their is now a very substantial salmonid farming sector engaged in the cage culture of Atlantic salmon and rainbow trout. When the oyster and mariculture industries were planned government agencies combined to identify suitable sites to establish farms In the case of oysters it was recognised that sanitary status of the areas was an important consideration if Tasmania was to compete with the established mainland oyster industry of New South Wales and establish a viable export market for its production Identifying the United States as one such major export market public health authorities set about classifying waterways in accord with the requirements of the National Shellfish Sanitation Programme.

While it is true that some areas were highly polluted particularly after heavy rain the majority of areas were demonstrably pollution free and leases for oyster farms were granted in such areas. As a result the Tasmanian industry is able to conform the National Shellfish Sanitation Programme requirements, supported by routine monitoring and it signatory to an MOU (Memorandum of Understanding) via the federal government in Canberra It must be added that the success of the Tasmanian oyster industry is largely due to a high level of co-operation between the various government agencies and the industry itself through its association and membership Positive incentives such as subsidies, dedicated research and advisory programmes have also played a significant role

Although exports were a major target a mainland market population of some 15 million people and significant under supply situation for part of the year offered considerable local opportunities As a result Tasmanian oysters are widely marketed throughout Australia particularly in the southern States (Victoria, South Australia). Because the New South Wales Pure Food Act regulations which require all oysters sold in New South Wales to be depurated in an approved plant, one depuration plant of the flow through type is operational there and at least two others are being constructed. Under the Australian constitution relating to free trade between States many Tasmanian oyster growers oppose the need for depuration, maintaining that their oysters are grown in clean, approved waters which satisfy the most rigid criteria To date this argument has not been accepted by New South Wales authorities and is, predictably perhaps, strongly opposed by the New South Wales oyster industry. Whatever the perceived rights and wrongs of this situation the Tasmanian experience offers some important lessons namely:

1 A vital pre-requisite to the orderly development of a viable and safe bivalve industry ie. an understanding of the sanitary status of cultivation areas. Classification of waterways prior to commercial development of mariculture and continued surveillance and protection of those areas in the future ensures the long term viability of oyster culture and a safe product for the consumer The popular adage "Prevention is Better Than Cure" has much to be said for it. Putting problems right is difficult it not practically impossible unless the will and resources are available and above all the political commitment to achieve change is there.

2. The close co-operation of all responsible Government agencies and industry is vital from the outset and must be regarded as a long term and sincere commitment by all concerned.

3. Involvement of universities, Government agencies and industry in the funding and performance of relevant research and data collection on a co-operative basis.

4. Last but not least an overall pride and commitment to doing it right It may be expensive at any time but it is never going to get cheaper and easier the longer it remains undone

While Tasmania has been fortunate, through proper planning to avoid problems of sewage contamination of oyster leases other problems may have unwittingly been created. In recent years algal blooms have affected mussel and oyster growing areas in southern Tasmania and resulted in a few cases of mild gastroenteritis and temporary closure of the leases. These dinoflagellate blooms have arisen in areas where salmon farms are located and where ships from Japan dock to load wood chips. There is some evidence to suggest the dinoflagellate was introduced via the ballast water of those wood chip vessels and the blooms encouraged by localised enrichment due to waste feed and fish faeces from the salmon cages. Fortunately such problems have been recognised and a monitoring scheme with provision for season closures is now in place.

There have been no reports of toxic algal blooms affecting the New South Wales oyster industry although some of the estuaries where sewage and urban run-off are a problem have experienced diatom blooms which discolour oyster meats. Further south in Victoria the Port Phillip Bay area near the city of Melbourne has also experienced problems of algal toxicity and tainting in farmed shellfish which have led to temporary closure. In this area blooms may be symptomatic of an enrichment of the bay from urban and agricultural sources.

AUSTRALIA - THE FUTURE

It will be apparent from earlier comments in this paper that Australia is faced with two major challenges if it is to retain its existing oyster industry and progressively develop new aquaculture opportunities In Tasmania the industry is a relatively recent one and deliberately and carefully planned so that shellfish farming is located in areas free of pollution. For the future viability of these areas it is vital that this pollution free status remains which will require careful and co-ordinated planning. Development and expansion of tourism vital to the economy of the State will place inevitable pressures on currently pristine waterways and existing use activities such as shellfish farming will need particular priority in concert with such development Zoning and planning restrictions will need to be rigidly maintained and enforced. Already opposition from recreational boating and other user groups and even, in some instances conservation and environmental groups is causing conflict.

Concern for the preservation of habitat common to both environmental and shellfish culture groups suggest that co-operation between such bodies is essential and that parochial differences must be resolved. This is of course true for the whole of Australia, not just Tasmania Sunrise bivalve shellfish industries being developed in Victoria, South Australia and Western Australia must pay attention to the development in Tasmania and dedicate unpolluted waterways for this purpose unless they are to inherit the unfortunate legacy which is so evident in New South Wales. These States are less subject to pressures of industrial and urban development than New South Wales but inevitably this situation will change and careful, co-ordinated planning is essential now. Indications are that the Governments of these States recognise the problems and Victoria for example is undertaking a microbiological assessment of potential shellfish culture sites. One can only hope that the political will is there to devote the necessary resources to the task facing them Do nothing is certainly the cheapest and easiest alternative in the short term but New South Wales serves as an excellent example of the consequences.

The Department of Health in New South Wales since it assumed responsibility for the depuration programme has certainly made a promising start but more manpower and resources are clearly required as is a long term commitment to the problem The Department have a draft corporate plan which seeks to address many of the deficiencies within the industry itself; to improve education of oyster farmers, upgrade depuration plants, improve the policing and monitoring of depuration and oyster quality, provide trained staff, promote industry, self regulation and generally increase awareness of the problems and the possible solutions amongst all concerned. Frankly that is a daunting task and history would suggest, not achievable.

It is unlikely, that unless there is a very fundamental shift in Government policy, that shellfish growing waters will ever be classified because the resources simply are not available for the task. Hopefully those areas which currently enjoy a pollution free status will remain so but even that must be in doubt. The fate of oyster cultivation in urban areas is sealed unless industry can demonstrate quite categorically that they can, through the use of well designed and properly operated depuration plants, produce a safe product for the consumer The fact that they will be victims of circumstance in the end may be cold comfort. It is conceivable through a combined Government/industry approach that commercial cultivation could be concentrated and if necessary relocated to suitable sites away from heavily urbanised areas. Such areas exist but currently are in grave need of protection lest they are lost for the future.

From experiences with Australia and overseas it is the authors belief that a depuration programme can be successful if properly controlled and if regulations are made and enforced It is unlikely that bivalve shellfish can be successfully depurated within a commercially viable timeframe if they are harvested from consistently or heavily contaminated waters. Such areas are not normally very apparent even with limited sampling and should be closed to cultivation and harvesting Suitable sites and if necessary financial assistance via grants or low interest loans should be made available by Government for the establishment of centralised shellfish depuration facilities Clearly one of the major logistic problems in New South Wales is the large number of individual plants of varying standard and design located over a 1000 kms of coastline. It does not matter what regulations are made if they are not achievable nor enforceable and that simple reality seems to have escaped most control authorities. Emphasis needs to be placed on self enforcement and collective industry responsibility with appropriate penalties, well publicised and acted upon for infringements. Experience in New South Wales and in the United Kingdom suggests that close contact and liaison between industry and regulatory authorities is essential Some oyster farmers will do the right thing regardless, some will never comply but the vast majority are somewhere in the middle. Penalise the bad groups, encourage the good groups and the rest of the industry will follow. However, if there is no perceived or actual advantage in doing the right thing (and depuration is an added capital and operational cost) then the whole industry will slide towards non-compliance

In New South Wales between 1978 and 1983, the industry enjoyed a period free of incidents of gastroenteritis. It was a period of change and even enthusiasm, not perfect by any means but certainly promising for the future. Gradually the industry cohesion through its Association weakened, Government to all intents and purposes totally abandoned its responsibilities and reports of gastroenteritis began to appear with increasing frequency. Inevitably this crisis promotes renewed interest by Government and the whole cycle starts all over again. The cost to industry and Government alike is enormous and above all consumer confidence in a generally healthy and desirable product plummets. It is simply not acceptable to react by imposing unrealistic and unachievable solutions on an ad hoc basis because if that is the reality then there will be very few bivalve shellfish industries anywhere in the future. Properly designed, constructed and operated depuration plants combined with an understanding of shellfish growing area quality can provide the consumer with a safe product. Attention must be focused on systems that provide an environment for optimum shellfish activity which means fundamentally that you have to understand the animal. It is totally irrelevant and even dangerous to focus on the efficiency of u/v sterilisation or whatever in a system which is hydraulically inadequate and used to treat a product which has been handled badly and placed in a hostile or unsuitable environment. The inherent strengths and weaknesses of depuration lay primarily in the health of the animal and a happy oyster is more likely to be a healthy one. That is of course simplistic and any raw foodstuff, especially shellfish carries with it an inherent risk such that complete safety cannot be guaranteed. Society seems to accept the food poisoning risks associated with poultry even though the product is cooked and in Australia at least oyster problems pale into insignificance when compared with poultry and some other foods. Perhaps in an annual consumption of 150 million oysters we should seriously consider whether a 1000 cases of gastroenteritis a year is actually acceptable. Such an incidence would involve at most 10,000 oysters or about 1 in 15,000; how does that compare with other foods?

SOUTH-EAST ASIA - BACKGROUND

This section of the paper deals with depuration in the ASEAN countries of South-East Asia ie Malaysia, Singapore, Thailand, The Philippines and Indonesia. The author has acted as a consultant to the Fish Working Group of the ASEAN Food Handling Bureau based in Kuala Lumpur, Malaysia under the ASEAN-Australia Economic Co-operation Programme and made a number of working visits to the region. Although the views and opinions expressed are those of the author, the paper draws heavily on the activities of the many official agencies and individuals involved in shellfish sanitation in ASEAN and due acknowledgement is made to their endeavours.

Mollusc culture in Asia is an important activity and in 1985 over 2 8 million metric tonnes were produced which represented some 65.5% of total world landings at that time. The average annual growth in aquaculture for the period 1977 to 1987 was 15% (Lovatelli, personal communication, 1989) far outstripping any other food commodity group. The bivalves most widely harvested from natural fisheries or cultured belong to the families <u>Ostreidea</u>, <u>Mytilidae</u> and <u>Arcidae</u>. Amongst the oysters, various species of the genera <u>Crassostrea</u> and <u>Saccostrea</u> predominate but <u>Ostrea</u> sp. are locally important in some areas. Mussels belonging to the genera <u>Mytilus</u> and <u>Perna</u> together with various cockle species (<u>Anandara</u> and <u>Arca</u>) are also widely captured and cultured. More specific mention of locally important species will be made in the later sections dealing with activities in each of the ASEAN countries. However, the region not only has very significant existing fisheries for bivalves but also has enormous potential for continued development and expansion. The constraints are largely environmental, biological and social and are similar to those experienced in many other parts of the world. As these countries undergo rapid urban and industrial development attendant pollution problems often accentuated by poor sanitation and standards of hygiene are also increasing at a rapid rate. Biologically, major restraints to development come from a lack of seed supplies and suitable culture areas but red tides are also becoming a significant localised problem. Social and institutional restraints vary from country to country and are many and varied. Lack of trained personnel and resources are a major problem and in spite of abundant shellfish resou s limited demand for some bivalve species due to cultural and health related preferences may be important. Figure 7 taken from Lovatelli (1989) illustrates the results of a survey carried out in fifteen Asian and Pacific countries to identify perceived restraints to mollusc culture. From the authors experience there is some degree of association between poor quality, environmental pollution and limited demand and export which together add a significant constraint to development. Indeed it is these factors both singly and in combination which have been the major impetus in the development of depuration systems and shellfish sanitation programmes in ASEAN countries. Not least they have certainly heightened official awareness of the existing and potential problems and the need for urgent action. However there are important differences in the philosophies adopted by the various ASEAN countries in view of local needs and priorities and it is worthwhile looking at each country in turn to examine what the problems are and what action is being taken or proposed. Although depuration is still largely at the research and pilot stage throughout ASEAN with less than a dozen plants, remarkable progress has been made in relatively short time within the inevitable constraints of manpower and financial resources. For a fuller and more detailed summary of the situation the Proceedings of the ASEAN Consultative Workshop on Mollusc Depuration held in Penang, Malaysia in 1988 (A.F.H.B., 1989) are recommended reading and contain useful references to previous work in the region.

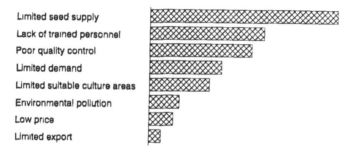

FIGURE 7. The Mollusc Industry in S.E. Asia and the Pacific:
Major restraints and their degree of importance
(after Lovatelli, 1989)

SINGAPORE

Singapore is the smallest but most highly developed of the ASEAN nations separated from mainland Malaysia by the Straits of Johore. As an important industrial, commercial and tourist centre of the region and with major air links to the rest of the world it is an important market for shellfish both to serve the local population and the many international standard hotels and restaurants. Interest in depuration and shellfish sanitation studies have been stimulated by a small but locally important mussel culture industry and also by import and export of shellfish.

A predominantly Chinese population has created a ready market for the blood cockle Anadara granosa which is commonly consumed as a lightly steamed product, purchased from the many street markets and open air food stalls and normally eaten virtually raw with a chili sauce. Malaysia, Singapores mainland neighbour is the worlds largest producer of cockles and large quantities are shipped across the straits to Singapore. Public health problems including gastroenteritis and infectious hepatitis have been associated with consumption of these cockles in Singapore and Government agencies have began to institute controls and monitoring of the product in co-operation with the Malaysian authorities.

Opportunities for aquaculture in Singapore are limited given the small land mass, relatively high population and concentration of industry such as ship building and repair However both fish and shellfish are farmed in the straits between Singapore and Malaysia. The green mussel Perna viridis is the only bivalve cultured and in 1987 approximately 1000 metric tonnes were produced from suspended raft culture, accounting for some 50% of total aquaculture production by weight.

Depuration studies by the Department of Primary Production, Marine Aquaculture Section, were initiated in the early 1980's (Davy and Graham, 1982) and have focused both on shore based and in-situ farm based systems (Cheong, 1989). Shore based depuration plants were seen as desirable and economically feasible with increased mussel production above current levels, with perhaps one large centralised facility handling the product from several farms

1. Shore based system

Experiments were conducted with a multi-tiered shallow fibreglass tray unit using UV treated recirculated seawater and based on the high density unit described by Ayres (1978) Details of the system were as follows·

Trays:	Four vertical stacks each of five trays measuring 72.7 cm x 38.2 cm x 11.5 cm depth mounted over a fibreglass reservoir or sump holding 2 cubic metres of seawater.
Pump·	0.33 kw capacity delivering 100 litres/minute or three water changes per hour.
Sterilisation	Enclosed tubular UV steriliser utilising four 30w germicidal lamps. Manufacturers performance data. 28,000 mW/sec/cm^2 at 100 litres per minute.
Shellfish Density	Mussels (Perna viridis) in clumps at 36 kg/m^2 tray area or 100 kg/m^3 of water.

After 24 hours treatment the system was drained, hosed down and refilled for a further 24 hours treatment ie. 48 hours in total. Mussels with an initial E. coli count of more than 2400 MPN/g were successfully depurated to less than 20 E. coli MPN/g within 48 hours A mortality of between 1 and 6.5% was observed and this was thought due to handling and poor initial condition of some of the mussels. Problems due to spawning usually in the first few hours led to some trials being abandoned. Mussels collected at low tide were observed to be significantly more polluted than those taken at high tide and it was suggested therefore that harvesting should be confined to high tide periods

2 In-situ farm based systems

Because of the offshore location of the mussel farms logistic difficulties in maintaining shore based facilities prompted investigation of methods which could be applied in-situ Two systems were tried, firstly the multi-tiered system mounted on a pontoon but with flow through seawater. At 100 litres/minute such a system was equally as effective as the shore based one. To reduce handling and possible damage to the mussels as they were removed from the culture ropes a flexible plastic polyvinyl bag 3m x 2m x 2m deep was made and suspended in the water from a pontoon. 16 to 18 ropes of mussels each 2m long and holding 40 to 50 kg of mussels were suspended in the bag to give a stocking density of 100 kg/m^3. Water was pumped into the bottom of the bag through a six lamp, enclosed u/v steriliser and venturi device and allowed to overflow from the top. The flow rate of 100 litres/minute gave a water exchange of 75% of the total volume per hour. E. coli and fecal coliform levels were reduced to less than 20 MPN/g within 48 hours. A high (70%) mortality of mussels was linked with a low dissolved oxygen level in the bag, probably as a result of the low exchange rate and inadequacy of the venturi device

By reducing mussel mortality the estimated cost of depuration (excluding manpower and amortisation costs) was $S0 05 per kg in the on shore system and $S0.36 per kg in the on farm system. The higher cost of the latter was due to costs associated with generating electricity for pump and steriliser units

The future of depuration in Singapore depends primarily on the development of the mussel industry and perhaps culture of other bivalves such as oysters. Clearly the technology is now available for commercial development of successful depuration systems. The mussels currently produced are sold primarily as a cooked product and so depuration is not seen as a commercial or regulatory necessity. However the sale of live mussels or those lightly steamed may make depuration mandatory given pollution conditions in the mussel culture areas

MALAYSIA

Although molluscan bivalve shellfish have been harvested from wild populations in Malaysia for centuries, aquaculture of bivalves is relatively new to the country. However it is now an important sector of the economy and receiving considerable attention from government who are anxious to see it develop to its full and undoubted potential. Production of the blood cockle Anadara granosa makes Malaysia the largest cockle producer in the world although production has fallen from a peak of 121,000 metric tonnes in 1980 to almost 41,000 tonnes in 1987. One of the major reasons for the decline is shortage of natural seed due to increases in cost of the spat and illegal transhipment for reseeding depleted beds in nearby Thailand However this is being addressed by the Department of Fisheries who are establishing new cockle beds by transporting seed from over-populated but less accessible areas. Similar attention is also being given to the establishment of mussel culture (Perna viridis) and oyster culture (Ostrea folium, Saccostrea culcullata, S echinata and Crassostrea belcheri) to expand on the production from existing natural capture fisheries in Malaysia.

The Fisheries Act of 1985 (Act 317) empowers the Minister of Agriculture to make regulations pertaining to standards and methods for the quality control of fish products however currently no such regulations have been drafted. The Environmental Act of 1974 (Act 127) contains provision for regulation of marine pollution but again no regulations are yet in place specifically for shellfish growing waters. Shellfish are covered under the Food Act of 1983 but that does not apply to collection of shellfish for sale to the public. However, the Food Regulation of 1985 has a presumption clause whereby it is assumed any food cultivated or bred for human consumption shall have been procured from a clean environment At present there are no regulatory controls regarding public health aspects of the production of shellfish which might include for example, depuration but it is anticipated these will be developed in the future.

The impetus to consider depuration came from the threat to continued exports of Malaysian cockles as a result of illness in importing countries, particularly Singapore and to a lesser extent Thailand. Examination of samples in Singapore revealed fecal contamination in cockles and epidemiological investigation demonstrated a correlation between consumption of cockles and gastroenteritis illness, and more importantly hepatitis. As a result the regional World Health Authority office recommends that ASEAN region countries initiate or expand shellfish control programmes (WHO, 1987).

Malaysia planned a National Shellfish Sanitary Control Programme which is still being co-ordinated by the Department of Fisheries in co-operation with other Government agencies and the shellfish industry. As a preliminary the author was requested to design and install a small experimental pilot scale depuration plant for treating cockles (<u>Anadara granosa</u>) under the auspices of the ASEAN Food Handling Bureau Fish Working Group and Malaysian Government. Elsewhere in the world it is the usual practice to boil cockles prior to sale which significantly reduces, though not removes, risks arising from faecal contamination. However as mentioned earlier under "Singapore" it is customary for the Chinese population in particular to eat cockles in a raw condition, sufficient heat being applied only to assist in opening and this seems to have increased the incidence of illness from that source. Although no one had attempted to depurate cockles before, the market requirement for a "safe" raw cockle dictated a consideration of depuration for Malaysian cockles.

In June 1986, the author in co-operation with scientists at the Fisheries Research Institute, Penang, constructed a pilot scale depuration system. This was based on the high density system using stacked plastic fish boxes mounted in a frame, drawer fashion over a sump or reservoir tank and with re-circulated, u/v sterilised seawater The general management of the plant is shown in Figure 8 and the actual unit in Plate 10.

An experimental programme was designed to evaluate the following factors:-

1.   Effect of stocking density on rate of depuration and efficiency of depuration.

2.   Effect of depuration on cockles mortality during and post treatment.

3   Effect of pre-depuration handling/transport and time on mortality of cockles during and after treatment.

4   Survival of cockles after processing ie during transport to market

5   Physical and chemical parameters; salinity, dissolved oxygen: temperature; NH_3 - nitrogen and pH at commencement and completion of depuration cycle.

Additionally, the optimum salinity for depuration was determined as between 25 and 32 $^0/_{00}$ and the UV sterilisation unit reduced an initial waterborne fecal coliform MPN count of 270/ml to undetectable levels in 60 minutes. Full details of this and other work is given by Ishak (1989).

In summary, the studies demonstrated that:

*   Cockles should be depurated immediately after harvest to avoid post depuration mortality. Such cockles had a shelf life of two days in live storage so treatment systems would be ideally located close to the point of sale (and harvest if possible).

*   Successful depuration from an initial faecal coliform MPN of as high as 1260/g could be achieved within 36 hours.

*   Depuration was successful at a stocking density up to 37 kg cockles per square metre of tray area.

*   Process cost excluding labour was $M0 03 per kilo of cockles. The major limitations to commercial application of the system lie with the cockles themselves which have a short shelf life out of water, a condition which is accentuated by the high ambient air temperatures characteristic of the region.

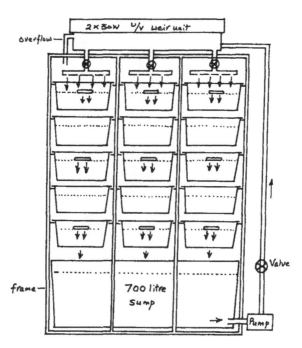

FIGURE 8. Pilot-scale high density cockle
depuration plant-malaysia.

Specifications of this pilot scale plant are as follows:-

Trays: Three stacks each of five polyethylene fish boxes ie. 15 in total. Each box has a
                maximum capacity of 45 litres but was modified to hold 20 litres before
                over-flowing.

Reservoir tank: 700 litre fibreglass tank.

Pump.   300W (0.4 hp) delivering up to 100 litres per minute.

Plumping: UV resistant PVC, 30mm diameter with valves and rubber hose connections for
                easy maintenance.

Steriliser: Simple DIY weir type unit of marine plywood and fibreglass using two 30W
                germicidal tubes (G36) (3 fitted - one as a stand-by for experimental purposes).

        Ng and Wong (1989) of the University Sains Malaysia conducted some independent cockle
depuration experiments using chlorine instead of UV. Cockles artificially polluted with faecal
coliforms were treated in filtered, chlorinated and dechlorinated seawater at various flow rates from
one exchange per hour to one exchange every four hours in a flow through system Depuration
efficiency was affected by initial fecal coliform loading and by flow rate with high counts and low
exchange rates giving the worst results. At an exchange rate of one volume per hour initial fecal
coliform levels in cockles of 55×10^3 MPN/100 g were reduced to less than 230/100 g within 24
hours By halving the exchange rate successful depuration was only achieved within 24 hours
when initial counts were in order of magnitude lower ie. $5 5 \times 10^3$ MPN/100 g. With initial fecal
coliform levels of 420×10^2 MPN/100 g successful depuration could not be achieved at even the
highest flow rate. At low flow rates and high fecal coliform levels some evidence of
re-contamination was noted.

PLATE 11.   Cockle unloading site-malaysia.
(Note latrine on jetty at the
left).

PLATE 10.   Pilot scale high density
cockle depuration system-
malaysia.

In the future it is proposed to construct a large semi-commercial scale plant holding a minimum of 1000 kgs of cockles As part of the National Shellfish Sanitary Control Programme growing waters are being evaluated and this will be combined with relaying trials of cockles from polluted to clean areas. One of the major problems in Malaysia and indeed in other ASEAN countries is clearly illustrated in Plate 11. Although many cockle beds are seriously affected by faecal contamination poor standards of hygiene and sanitation in the landing areas and during handling can seriously prejudice the sanitary quality of shellfish harvested from clean areas and render them unsafe. If depuration and growing area classification is to be adopted then to be ultimately successful it must be in concert with an educational programme to heighten awareness of the problems and dangers inherent in poor post-harvest handling. Without such an approach, depuration or indeed any other course of action will be not only useless but may be positively hazardous in inferring an unwarranted level of safety.

INDONESIA

In spite of large natural bivalve shellfish resources, commercial exploitation of the resource is very much secondary to fisheries for tuna and shrimp, primarily because shellfish are categorised as a product of low commercial value by fishermen and traders. Fish farming of fresh water fish is also a major aquacultural activity but apart from small scale farming of the green mussel Perna viridis very little culture of bivalves is undertaken Clams are exported as a canned or frozen product but otherwise most bivalves are consumed domestically, particularly by coastal communities (Wiryanti, 1989) An expansion of export markets and development of mariculture are thought likely to necessitate attention to quality both of the product and the water in which it is grown At such a time commercial application of depuration technology seems probable also.

Some preliminary trials with depuration of the green mussel have been conducted by the Agency for Agricultural Research and Development (Heruwati, 1989) using a pilot scale plant modified from the design of Souness and Fleet (1979) and a small laboratory scale plant. Results from the laboratory scale plant demonstrate that a water re-circulation rate of three changes per hour and a shellfish density of one mussel per litre gave the best performance over a 48 hour depuration period The pilot scale plant consisted of ten fibreglass tanks each holding 40 litres of water mounted over a reservoir tank from which water was recirculated through a weir type steriliser fitted with four 30W u/v lamps. Total shellfish capacity of the plant was 50 kg at 4 animals/litre and the recirculation rate, three changes per hour. Trials with blood cockles, ark shells (Arca inflata) and green mussels E coli levels of 300/g over a two and a half to 3 day period. Little effect on total plate counts or Vibrio parahaemolyticus was observed

No economic evaluation has been reported and given the current low value of bivalves in Indonesia such an evaluation would be critical to adoption of the depuration approach.

THAILAND

The production of bivalve molluscs in Thailand in 1985 totalled 183,500 tons of which some 93,000 tons was from captive fisheries and the remainder from aquaculture Comparison with earlier years demonstrates a marked increase in aquacultural production which reflects active promotion of the industry by Government to provide product for local consumption and to gain foreign exchange earnings via export. The major species cultured are green mussels (Perna viridis), blood cockle (Anadara granosa), horse mussel (Modiolous senshausenii) and oysters (Crassostrea belcheri) and (C. lugrubris).

The Thai Government has since 1973, had a committee established for research on quality of water and living resources and this currently comprises representatives from eleven government agencies. With increasing concern about pollution the sub-committee has been collecting baseline environmental data to develop management and regulatory responses. In addition the National Environment Board has established a classification system for coastal waters which includes a class for aquaculture and shellfish (Tridech et al., 1987) 70% of coastal aquacultural production is from the inner Gulf of Thailand which is subject to a variety of pollutants and has experienced toxic algal blooms. As yet there are no specific regulations concerning the sanitary quality of shellfish or shellfish growing waters but the need for urgent action has been noted (Virulhakul, 1989). However a preliminary study on the depuration of economically important bivalves has been conducted (Sangrungruang et al , 1989).

Two types of depuration system have been examined, flow through and re-circulation and tested with mussels, cockles and oysters.

1.   Flow through system

TANK: 600 litre fibreglass tank measuring 2.1m x 1.0m x 0.7m deep, containing 12 trays of shellfish 17 cm above tank bottom.

WATER: Pumped from the sea, stored and sand filtered supplied continuously at 7 litres per minute for cockles and oysters and 10 litres per minute for mussels through an overhead sprinkler system.

SHELLFISH: Oysters (Crassostrea belcheri) arranged 12 per day (32 x 24 x 8 cm) and never exceeded 500/m^2.

                Cockles (Anadara granosa) 100 per tray averaging at 25 kg/m^2 tray area.

                Mussels (Perna viridis) clusters separated and cleaned and laid directly on floor of tanks at 20 kg per tank.

2.   RECIRCULATED SYSTEM

Tank: 2 fibreglass coated marine plywood trays 1.6 x 0.5 x 0.16 m above a reservoir tank. Whole system contained 93 litres.

Water: Sand filtered seawater sterilised through an enclosed single 38W UV unit. Flow rate less than 30 litres/minute.

Shellfish: As for flow through system. Mussel density 13 kg per tank

The following results were obtained from treatment of the various shellfish in each of the two systems (counts expressed as MPN/100 g).

Oysters	FLOW THROUGH		RECIRCULATION	
Initial total coliform		14,000		14,000
Initial fecal coliform		11,000		11,000
24 hr total coliform		330		68
24 hr fecal coliform	78		45	
Mussels				
Initial total coliform		1,300		1,300
Initial fecal coliform		790		790
24 hr total coli form		220 (36 hrs)		110
24 hr fecal coliform	78		45	
48 hr fecal coliform		<20		<20
Cockles				
Initial total coliform		2,100		2,100
Initial fecal coliform		260		260
24 hr total coli form		220		20
24 hr fecal coliform	68		20	

A slight reduction was observed in TPC with all species and in both systems. Overall depuration rate was observed to be more rapid in the recirculated system and all species depurated to below 230/100g in 24 hours or less compared with up to 48 hours in the flow through system. High mortality of mussels was attributed to damage incurred when the clumps were separated and by low final oxygen levels in the recirculation system

THE PHILIPPINES

The Philippines is unique amongst ASEAN countries in terms of work on bivalve depuration having had a number of separate projects underway in various parts of the country during the 1980's and even one small privately owned commercial plant. In common with other countries of the region the Philippines Government is actively promoting aquaculture as a means of boosting fish and shellfish production. As a result culture of oysters (Crassostrea iredalei) and mussels (Perna viridis) now far exceeds the production from natural capture fisheries. There is however growing awareness of pollution and studies have confirmed that many areas are subject to faecal contamination particularly after heavy seasonal rainfall (Palpal-Latoc, 1986).

The current situation has been summarised concisely in a paper by Guevara (1989) but it is worthwhile considering here some of the depuration work conducted primarily because much of it has been done quite independently by different groups of researchers.

The South-East Asian Fisheries Development Centre (SEAFDEC) established an experimental system at Iloilo in 1983 which basically consisted of 6 rectangular marine plywood tanks coated with fibreglass and each measuring 1.9m x 0 96 x 0 42) Each tank held 15 plastic trays of shellfish, 2 5 to 3 5 kg per tray and then u/v sterilised to achieve a 95 to 99% instantaneous kill. At optimum salinity, temperature and a flow rate of 7 to 10 litres per minute SEAFDEC researchers depurated grossly contaminated shellfish (up to 2.0×10^6/100g) within 48 hours More detail of the SEAFDEC studies is given by Gacutan et al., (1986).

In 1982, the Philippine Human Resources Development Centre (PHRDC) was established in Northern Luzon with assistance from the Japan International Co-operation Agencu (JICA) The Seafarming Research and Development Centre (SRDC) or Program II of the PHRDC is responsible for the development of the oyster industry in the Philippines and carried out some research on oyster depuration as part of this responsibility. The system applied was based on Japanese principles and consisted of tanks each 1.90 x 0.9 x 0.59 constructed of glass reinforced plastic. The volume of each tank was maintained at 580 litres by an overflow system and water was continuously flowed through an overhead perforated pipe at 7 litres per minute after u/v sterilisation.

Oysters (C. iredalei) were conditioned by holding in tanks or natural waters for a minimum of one day to allow detection of weak oysters and assist recovery. After being cleaned and culled the oysters were placed in suspended mesh baskets in the tanks so that they were covered by 3 to 4 cm of water. Sixteen trials were performed between 1986 and 1987, twelve resulted in successful depuration down to below 230 faecal coli/100 g in 36 hours or less and the remaining four trials failed to reach 230/100 g and were accompanied by oyster mortality. Full details of the results have been reported by Caoile (1989) but it was concluded that problems with the health of the oysters accentuated by a low circulation rate (one exchange every 1.4 hours) in the system were responsible for the 25% failure rate in the experiments.

Using the PHRDC facilities a group from the Government Bureau of Fisheries and Aquatic Resources also looked at oyster depuration (Guevara et al., 1989). Seawater was filtered through sand and cartridge filters and either used directly or passed through an 8 lamp enclosed UV steriliser. Oysters in all trials were laid out in baskets at 2 kg per basket, a total of 16 kg per tank (tanks as described earlier) and subject to flow rates of 20 litres/minute or about 2 volume changes per hour. Parameters such as pH, salinity and dissolved oxygen were measured at intervals throughout the experiments. Successful depuration was achieved in both systems being faster (less than 24 hours) in the filtered seawater system. The researchers concluded that effective depuration was influenced by the measured parameters and care exercised in handling the animals Organoleptic assessment failed to reveal any difference between treated and untreated oysters

A team at the UPV College of Fisheries, Quezon City, working under the German Agency for Technical Co-operation Ltd (GTZ) has completed what is probably the most detailed multi-disciplinary study so far conducted in ASEAN. This study took into account the cultural methods employed for green mussels, the pollution regime of the culture areas and the various socio-economic factors involved both in the industry and in relation to the adoption of shellfish depuration technology (Barile, 1989).

The depuration plant constructed was remarkably similar to the fibreglass tray system routinely employed in New South Wales, Australia, and indeed was subsequently modified in the light of Australian experiences   Brief details of the plant are as follows.

Tank: Five tanks each 2.4 m x 0.76 x 0.14 deep stacked in a metal frame. Each tank fitted with a baffled inlet and weir type overflow.

Sterilisation.   Enclosed UV unit. Recirculated water.

Shellfish density·   Each tank contained 3 baskets each holding 10 kgs of shellfish ie 30 kg per tank or 150 kg in the system.

Fecal coli levels were successfully reduced to less than 3 MPN/g in as little as 24 hours but Vibrio counts were unaffected and actually showed some evidence of multiplication when depuration was extended to 48 hours. Detailed costings of construction and operation of the plant in relation to final meat yield of the depurated product were also made

A survey of potential users revealed a number of problems notably that the mussel industry was reluctant to adopt the technology without some organisational support and an assurance that the necessary investment in time and money would be warranted and achievable by increased sale value. People were also lacking in an understanding of what depuration actually involved and why it was necessary. Oyster farmers were more positive, perceiving depuration as a means of achieving export markets. This factor also attracted interest from investors and businessmen who had a better perception of the economic advantages of export over domestic sales.

The only commercial plant in the Philippines was a small system depurating oysters and mussels This plant was situated at an island site requiring water to be transported to it. Although it had all the right components including a small UV steriliser from Australia it was poorly constructed and operated. The operator had endeavoured to copy what he had seen in Australia without an understanding of the principles involved. However it was interesting in that his motivation to build the plant came from a desire to export a product that was safe and overcome what he saw as a market resistance to Philippine shellfish.

SUMMARY

It is difficult to present a concise summary about the status of depuration in ASEAN countries or about the future because of the enormous differences between the countries However some general comments can be made about those things which are common throughout ASEAN -

* There is an overall desire on the part of government to increase bivalve shellfish production primarily through aquaculture both for domestic and export consumption.

* There is a growing awareness by government and to a lesser extent, industry of pollution problems and the risks posed by polluted shellfish.

* Governments generally perceive depuration as the answer to these problems particularly in respect of export potential and foreign exchange earnings.

* Pressure from rapid, uncontrolled industrial development increases the problems and is outstripping regulatory controls.

* Although statutory powers exist in principle to regulate the sanitary control of shellfish and shellfish growing waters virtually no actual regulations have been put in place.

* There is a shortage of trained personnel to support research and regulatory programmes.

* Many bivalves species are currently of a low domestic value and if depuration is to be adopted the system should be simple and cheap (and effective!).

* Education programmes to increase industry, consumer and official awareness of the problems and solutions are urgently needed.

* With current resources, identification and maintenance of clean areas for shellfish cultivation is the most logical and achievable alternative.

REFERENCES

AFHB. 1989. Proceedings of the ASEAN Consultative Workshop on Mollusc Depuration (ed P A Ayres) Penang, Malaysia, 4-7 October 1988. ASEAN Food Handling Bureau, Kuala Lumpur

Anon. 1964(a). Oyster Cleansing in New South Wales Pioneered - The Fisherman, Sydney 1(11);14-15.

Anon 1964(b). Building an Oyster Cleansing Plant - Aust Fish 23(7):5

Ayres, P.A 1978 Shellfish Purification in Installations Using Ultra Violet Light - Ministry of Agriculture, Fisheries and Food. Fish Research Laboratory Leaflet No. 43 Lowestoft United Kingdom. 20p.

Barile, L E. 1989. Bacterial Depuration of Oysters and Mussels Harvested from Bacoor Bay. In: Proceedings of the ASEAN Consultative Workshop on Mollusc Depuration (ed. P A Ayres) Penang, Malaysia, 4-7 October 1988. AFHB Kuala Lumpur.

Bird, P. 1989. Purification Technology for the Sydney Rock Oyster. Department of Health, Sydney NSW Australia.

Caoile, S.J.S. 1989. Depuration - SRDC Experience In: Proceedings of the ASEAN Consultative Workshop on Mollusc Depuration (ed. P A Ayres) Penang, Malaysia, 4-7 October 1988. AFHB Kuala Lumpur.

Cheong, L 1989. Status of Depuration Work in Singapore In: Proceedings of the ASEAN Consultative Workshop on Mollusc Depuration (ed. P A Ayres) Penang, Malaysia, 4-7 October 1988, AFHB Kuala Lumpur.

Davy, F.B. and Graham, M. (Ed) 1982. Bivalve Culture in Asia and the Pacific: Proceedings of a Workshop Held in Singapore, 16-19 February 1982, Ottawa, Ont., IRDC, 1982, 90 p·111.

Eyles, M.J. and Davey, G.R. 1984. Microbiology of Commercial Depuration of the Sydney Rock Oyster (Crassostrea commercialis). J. Food Prot. 47:703-706

Gacutan, R.Q., Bulalacoa, M.L. and Baranda, H.L (JNR) 1983. Bacterial Depuration of Grossly Contaminated Oyster (Crassostrea iredalei) In: Maclean, J.L., Dizon, L B., and L.B. Hosillos (eds) - The First ASEAN Fisheries Forum. Asian Fisheries Society, Manila, Philippines. pp 429-431.

Grohmann, G.S., Murphy, A.M., Christopher, P.J., Auty, G. and Greenberg, H.B. 1981. Norwalk virus gastroenteritis in volunteers consuming depurated oysters. Aust. J. Exp. Biol. Med. Sci. 59:219-228.

Guevara, G. 1989. Status of Mollusc Depuration in the Philippines In: Proceedings of the ASEAN Consultative Workshop on Mollusc Depuration (ed. P A Ayres) Penang, Malaysia, 4-7 October 1988. AFHB Kuala Lumpur.

Guevara, G., Abella, F F. and Repito, N. 1989 Study of the Depuration of Philippine Oyster (Crassostrea iredalei) In: Proceedings of the ASEAN Consultative Workshop on Mollusc Depuration (ed. P A Ayres) Penang, Malaysia, 4-7 October 1988. AFHB Kuala Lumpur.

Heruwati, E.S. 1989. Shellfish Depuration Using Laboratory and Pilot Scale Plant in Indonesia. In· Proceedings of the ASEAN Consultative Workshop on Mollusc Depuration (ed. P A Ayres) Penang, Malaysia, 4-7 October 1988. AFHB Kuala Lumpur.

Ishak, I.B. 1989. Cockle Depuration in Malaysia. In: Proceedings of the ASEAN Consultative Workshop on Mollusc Depuration (ed. P A Ayres) Penang, Malaysia, 4-7 October 1988 AFHB Kuala Lumpur.

Lovatelli, A. 1989 Status of Mollusc Culture in Selected Asian Countries In: Proceedings of the ASEAN Consultative Workshop on Mollusc Depuration (ed P A Ayres) Penang, Malaysia, 4-7 October 1988. AFHB Kuala Lumpur.

Ng, W.K. and Wong, T.M. 1989. Depuration of Cockles (Anadara granosa. L.) Using Chlorine Sterilised Seawater. In: Proceedings of the ASEAN Consultative Workshop on Mollusc Depuration (ed. P A Ayres) Penang, Malaysia, 4-7 October 1988   AFHB Kuala Lumpur.

Palpal-Latoc, E.Q., Caoile, S.J.S and Cariaga, A.M. 1986.   Bacterial Depuration of Oyster (Crassostrea iredalei) Faustino, in the Philippines. In: Maclean, J.L., Dizon, L.B., and Hosillos, L.B. (eds). The First ASEAN Fisheries Forum. Asian Fisheries Society, Manilla, Philippines, pp 293-295.

Sangrungruong, K., Sahavacharin, S. and Ramanudom, J. 1989   Preliminary Study on Depuration of Some Economic Bivalves in Thailand. In: Proceedings of the ASEAN Consultative Workshop on Mollusc Depuration (ed. P A Ayres) Penang, Malaysia, 4-7 October 1988. AFHB Kuala Lumpur.

Son, N.T., and Fleet, G.H. 1980.   Behaviour of Pathogenic Bacteria in the Oyster, (Crassostrea commercialis), During Depuration, Relaying and Storage. Appl. Environ. Microbiol. 40:994-1002.

Souness, R.A. and Fleet, G.H. 1979   Depuration of the Sydney Rock Oyster (Crassostrea commercialis). Food Technol Aust. 3/:397-404.

Tridech, S., Unkulvasapaul, M , and Simachaya, W. 1987. Coastal Water Quality and Establishment of Water Quality Standards for Thailand. The Fourth Seminar on the Water Quality and the Quality of Living Resources in Thai Waters. pp 394-404 National Research Council of Thailand, 1987.

Virulhakul, P. 1989. Environmental Effects on Molluscs in Thailand. In: Proceedings of the ASEAN Consultative Workshop on Mollusc Depuration (ed. P A Ayres) Penang, Malaysia, 4-7 October 1988   AFHB Kuala Lumpur.

WHO. 1987. Report of the Second Meeting of the WHO Regional Working Group on Food Safety Kuala Lumpur, Malaysia, 17-21 August 1987.

Wiryanti, J. 1989.   Shellfish/Mollusc Depuration in Indonesia. In: Proceedings of the ASEAN Consultative Workshop on Mollusc Depuration (ed. P A Ayres) Penang, Malaysia, 4-7 October 1988. AFHB Kuala Lumpur

Wood, P C 1961. The Principles of Water Sterilisation by Ultra Violet Light and Their Application in the Purification of Oysters.   Ministry of Agriculture, Fisheries and Food   Fishery Investigate Series II, 23:1-47.

OVERVIEW OF THE MOLLUSCAN SHELLFISH INDUSTRY IN NEW ZEALAND RELATIVE TO DEPURATION

Graham C. Fletcher and B.J. Hayden

The shellfish industry in New Zealand has grown dramatically over the last few years Fifteen thousand tons of farmed shellfish were produced in 1986, making New Zealand the sixth of the OECD countries (NZTDB, 1989). The growth of the industry is illustrated by the export figures shown in Table 1.

TABLE 1. Shellfish Industry export figures (tons of product)

SHELLFISH	YEAR					
	1983	1984	1985	1986	1987	1988
Mussels	998	1910	1756	2707	4127	6141
Oysters	688	778	764	972	877	731
Scallops	375	319	281	443	249	405
Tuatua	18	-	-	-	-	14
Cockles	-	-	-	-	-	129
Other	344	220	428	364	547	355
Total	2423	3227	3229	4486	5800	7775

- data unavailable
Source· NZFIB (1983-1988)

Mussels

The shellfish industry is now dominated by the production of greenshell mussels, Perna canaliculus. Although previously dredged, this species is now almost exclusively a farmed product Seventy-seven percent of the 404 licenses are in the Marlborough Sounds and Golden Bay while most of the remaining leases are in Auckland, Coromandel, Hauraki Gulf and Mahurangi areas (Fig 1). Mussel farming began in 1975 and is an adaption of Japanese long-line technology with mussels grown on ropes suspended between floats. They are harvested 12-22 months after spat settlement.

Mussels make up 3 4% of New Zealand's fishing export industry which had a total value of $721 million in 1988. Over the last six years the total production of greenshell mussels has increased from 7000 to 17000 tons green weight and it is estimated that full development of existing mussel farm licenses could yield 60,000 tons annually (NZTDB, 1989)

The industry was initially based around the production of freeze-dried mussel powder which is claimed to have therapeutic benefits for sufferers of arthritis. The market for this product suffered a major setback in 1981 and other markets were sought for mussels. Food products now make up 95% by value of the mussel market (Table 2) In 1988, the major markets for New Zealand greenshell mussels were Japan, U.S.A , Australia, and the United Kingdom (Table 3)

FIGURE 1. Major shellfish areas of
New Zealand
and the United Kingdom
(Table 3).

OYSTERS

The second most important mollusk is the oyster. The dredged oyster, <u>Tiostrea</u> <u>lutaria</u>, supports an industry which produces about 9000 tons (Table 4) from the deep, unpolluted waters of Foveaux Strait (Fig. 1). These are mostly consumed by the domestic market

Rock oysters are farmed in the North of the North Island, particularly in the Bay of Islands, Mahurangi, Hauraki Gulf and Coromandel Peninsula areas (Fig 1). A small oyster farming industry based on the native rock oyster (<u>Saccostrea</u> <u>glomerata</u>) began in the 1960's In the 1970's an exotic species, the Pacific oyster (<u>Crassostrea gigas</u>), arrived in New Zealand. This faster growing species has quickly come to dominate the local oyster farming industry although a few farms still produce significant numbers of the native species. The oysters are grown on racks or trays in the intertidal zone of sheltered estuaries. They are harvested 12-24 months after spat settlement. Annual production is estimated at 4000 tons whole shell weight (Curtin, 1988) and export figures have been fairly constant over the last six years (Table 1). There are 146 leases covering 430 ha of suitable ground. Of this 300 ha are currently fully developed.

Most of the industry is oriented towards supplying half-shell oysters for the restaurant market. Approximately 50% of the oysters are exported with the Australian market accounting for $26 million (56%) of these (Table 3). Other major markets are New Caledonia, French Polynesia, Hong Kong, USA and Singapore.

SCALLOPS

Scallops (<u>Pecten novaezelandiae</u>) are dredged from the Marlborough Sounds, Tasman Bay and the north eastern coast of the North Island (Fig. 1). These are not generally a problem in shellfish sanitation, particularly as only the roe and abductor muscle is eaten.

CLAMS

There are small industries based around the tuatua (Paphies subtriangulata and P. donacina), the cockle (Chione (Austrovenus) stutchburyi) and the pipi (Paphies australis). The cockle is finding a ready demand in the U.S.A. where it has been called the southern littleneck clam. These intertidal species are taken commercially for canning and pate as well as for the fresh trade. There is a significant recreational fishery and the Maori gather them for traditional cooking.

TABLE 2. Mussel exports by product form

PRODUCT FORM	WEIGHT (ton)	FIB VALUE ($NZ000's)
Crumbed, battered	8	42
Powder in capsule	14	892
Prepared can, jar	1	34
Prepared other	26	140
Smoked	8	95
Chilled meat	210	834
Freeze-dried powder	5	230
Frozen meat	1,181	5,473
Frozen whole	365	1,234
Half-shell chill	221	569
Half-shell frozen	2,771	10,990
Other, frozen	394	880
Live	938	2,950
TOTAL	6,141	24,367

Source: NZFIB (1988)

TABLE 3. Major markets for New Zealand Bivalves ($NZ,000)

COUNTRY	SHELLFISH				
	Mussels	Oysters	Scallops	Cockles	Total
AUSTRALIA	4137	2649	5206	15	12007
JAPAN	9313	92	36	1	9442
U.S.A.	5911	274	347	559	7091
U.K.	1489	56	0	0	1545
FRANCE	227	21	793	0	1041
OTHER	3290	1438	233	25	4986

Source: NZFIB (1988)

Table 4. Production figures for bivalves excluding aquaculture (tons).

SHELLFISH	YEAR			
	1983	1984	1985	1986
Mussels	600	600	100	100
Oysters (dredge)	9700	9400	8800	6700
Scallops	4000	4700	3200	4600

Source: Dept. of Statistics (1988-1989)

SHELLFISH SANITATION

With an area of 27 million hectares, a population of only 3.4 million and an industry based largely on pastoral agriculture, New Zealand does not have any serious water pollution problem. The shellfish sanitation is similar to the U.S. National Shellfish Sanitation Program with growing waters classified as approved, conditionally approved, restricted, and prohibited. The New Zealand and American governments signed a memorandum of understanding in 1981 allowing New Zealand to be treated in the same way as U.S. states with respect to shellfish sanitation.

Most of the shellfish farming areas in New Zealand are conditionally approved. An increase in fecal coliform levels is often observed after rainfall as a result of run-off from the surrounding pasture lands. Because the risk of contamination with human sewage or virus is low, the main control of harvesting from conditionally approved areas can be based on rainfall criteria. Farmers are forbidden to harvest after a certain level of rainfall e.g. for the 3 days following a 48 hour period in which rainfall exceeds 40mm. The exact period of prohibition and the levels of rainfall are determined for each area. The Health Department is the agency responsible for surveying growing areas. It issues harvesting criteria indicating the conditions under which shellfish from a conditionally approved area are required to be depurated.

DEPURATION PROCEDURES

To export any shellfish, including those from a depuration plant, an export license is required. This license must have the approval from the Ministry of Agriculture and Fisheries (MAF) that the product is fit for human consumption. Depurated product is accepted if the shellfish have been depurated in an accepted plant.

Although there are no regulations on depuration, guidelines have been written by MAF in consultation with the Health Department and the shellfish industry (Hayden, 1987). A document, entitled "Quality Assurance Procedures for Depuration Plants", which reports these guidelines has formed the basis for approving shellfish from depuration plants to date.

DEPURATION CRITERIA

Shellfish from approved and conditionally approved areas may be depurated while shellfish from restricted areas must be depurated. The harvesting criteria issued by the Health Department indicate the conditions under which shellfish from conditionally approved areas must be depurated if they are harvested. Shellfish from prohibited areas must not be depurated.

Because shellfish needing depuration are more likely to harbor pathogens than shellfish grown in approved waters, stricter guidelines are used. In 10 depuration runs in which the 0 h fecal coliform (FC) levels exceed 230/100g, 80% must reach 50FC/100g in the approved depuration time and none may exceed 230FC/100g. The rationale for expecting these low levels is that treated shellfish should reflect the process seawater quality rather than that of local approved waters.

Guidelines are given for various depuration plant design considerations. Outdoor tanks must be protected against pests and sunshine. The area surrounding a tank must be self-draining. Suitable provision should be made for clean-up and sanitizing of the whole plant including the ultraviolet lamp and the water intake pipe. Plant layout should allow good flow of shellfish and adequate separation of purified product and water from unpurified. The tanks, trays, and plumbing must be easily cleaned and constructed of non-toxic and corrosion resistant materials. Tanks should allow at least 100L of water per bushel of shellfish and the pumps should be capable of delivering 3L per minute to each bushel of shellfish. Water flows must be balanced among all treatment tanks. Trays must hold shellfish off the bottom of tanks without inhibiting water flow. Seawater storage tanks should be designed to prevent sediment entering the plant. They should be covered and lined.

Most importantly, the source of seawater should be of low fecal coliform content with minimal salinity fluctuations and must not contain toxic chemicals. In tidal areas, water is to be taken just before high tide. The water inlet is to be screened and as deep as possible while keeping clear of the bottom

PLANT CERTIFICATION

A division of MAF, MAFQual is the agency responsible for certifying depuration plants. There are two stages in certifying depuration plants.

The first is an initial commissioning stage whereby approval is given to the design of the plant and the quality of the seawater to be used in it. Before issuing this initial certificate commissioning trials are carried out. Tests are carried out without shellfish to determine that the water circulation in the tanks does not leave any dead spots, that the turbidity of the raw process water is not too high, and that the water sterilization unit is capable of ensuring that 88 fecal coliforms per 100mL are reduced to nil in a single pass. A test with oysters seeded with 1,000-10,000 fecal coliforms/100g is carried out to ensure that these are reduced to below 50/100g within 48 h.

In the second stage, process verification trials are carried out for the first 10 consecutive processing batches. These are monitored at 0 and 48 h for fecal coliforms to determine the maximum level of fecal coliforms that can be depurated by the plant. Any alterations to the plant must be submitted for re-evaluation.

OPERATING PROCEDURES

Temperature, salinity, turbidity, flow rates, and fecal coliform levels in oyster samples have to monitored regularly. The actual level of monitoring for each plant is determined on the basis of the results of the certification procedures. The required level of testing for fecal coliforms also depends on the lease from which the shellfish are harvested. At least one sample is required before and after depuration for each batch of shellfish harvested from restricted areas or from conditionally approved areas which have exceeded their rainfall criteria.

Shellfish must arrive at the depuration plant in perfect biological condition. All shellfish must be clearly labelled as to their lease number and date of harvest and must be identifiable throughout the process. Accurate and complete records of all data from each day must be kept for at least 12 months.

Unless a plant is proved to be capable of consistently reaching the required level of depuration in a shorter time, the approved depuration time for all species is 48 h. The exception is the New Zealand cockle (Chione stutchburyi) and other species found to retain their tidal rhythm after harvest. Because these species actively pump for only 4-5 h in every 12 h tidal cycle, they require 120 h depuration to ensure 48 h of active pumping.

No shellfish may be added to the tanks part way through a depuration cycle. Also, wet storage of shellfish in depuration tanks before and after depuration is not recommended and should not be for more than 48 h.

The process water should have salinity levels within 20% of the normal salinity of the harvest areas and should not drop below 2%. Dissolved oxygen should remain below saturation but not drop below 2 mg/L. Water temperatures should not exceed 20°C and should be above 18°C for the native rock oyster, above 12°C for cockles and above 10°C for mussels and Pacific oysters. Process water is to be replaced after each run. Tanks are to be cleaned and sanitized before each run.

DEPURATION IN NEW ZEALAND AT PRESENT

Currently there are only two licensed depuration plants in New Zealand: the Noma Oysters plant in Whangarei and the Tide farm plant in Mahurangi (see Fig. 1).

NOMA OYSTERS PLANT

This plant is associated with an oyster processing facility in the city and receives oysters from all over Northland. It processes Pacific oysters and some New Zealand rock oysters. The main reason for depuration is to meet the export requirements for the New South Wales market, where all oysters are required to be depurated regardless of source. Many of the shellfish which are processed do not technically need depurating. During long wet periods, the plant is also used to process shellfish from harvest areas that have exceeded their rainfall criteria, but this is by far the minority use of the plant Currently two batches per week are depurated during the growing season although this can increase in times of high rainfall. The plant consists of four sets of four trays held in a vertical stack and has a capacity of 8000 large oysters per batch. Seawater, delivered in a truck from a clean water source 30 miles away, is sterilized by ultraviolet light and is gravity fed from one tank to the next. Sampling and testing for fecal coliforms is carried out by a local independent laboratory.

TIDE FARM PLANT

This facility is located at the Mahurangi Heads and is attached to a small processing factory in a major oyster growing area. The company carries out a farming operation that lands oysters directly to the processing plant. Pacific oysters are also brought in from all over Northland. The plant consists of four tanks and has a maximum capacity of 36,000 oysters. The plant is run continuously and all oysters entering the processing plant are depurated. The plant is also used for wet storage to ensure a constant supply of fresh oysters to the factory. The main reason for depuration is to bypass the rainfall criteria in conditionally approved leases. The company considers that there is a marketing benefit in being able to say that all oysters have been depurated. Most of the oysters are exported to Asian markets. Water is pumped in at high tide and sterilized with ultraviolet light. Testing for fecal coliforms is carried out by an independent laboratory in Auckland.

CONCLUSION

Because clean growing water is the main aim of the New Zealand Shellfish Sanitation Programme, New Zealand generally does not need to use depuration to guarantee the safety of mollusks harvested by its shellfish industry. Where depuration is carried out for either technical or marketing reasons, controls exist to ensure that it is carried out safely.

REFERENCES

Curtin, L. 1988. The New Zealand oyster farming industry. In "AQUANZ'88", New Zealand Fisheries Occasional Publication No. 4, MAFFish, Wellington.

Dept. of Statistics. 1988-1989. New Zealand Official Yearbook, 93 edition, New Zealand Department of Statistics, Wellington.

Hayden, B.J. 1987. Quality assurance procedures for depuration plants (Draft Discussion Paper), MAFFish Fisheries Research Centre, P.O. Box 297, Wellington.

NZFIB, 1983-1988. Fish Export Statistics. New Zealand Fishing Industry Board (NZFIB), Wellington.

NZFIB, 1988. New Zealand Fish Export Statistics. New Zealand Fishing Industry Board (NZFIB), Wellington.

NZTDB, 1989. The New Zealand aquaculture industry. New Zealand Trade Development Board, Wellington.

CURRENT PRODUCTION AND REGULATORY PRACTICE IN MOLLUSCAN DEPURATION IN SPAIN

Fernando López Monroy, Baldomero Puerta Henehe and Mariá José Prol Bao

INTRODUCTION

Due to historical, geographical and political factors, Spain has one of the highest "per capita" consumptions of seafood in Europe. In spite of one of the world's most important fishing fleets and an important bivalve production, Spain is a net importer of shellfish, and many other seafood as well. Depuration in Spain, which has been legally recognized since 1964, has had an important role in the shellfish industry, in both commercial and regulatory aspects. However, due to E.E.C. regulations presently being drafted, the future of depuration "per se" is not clear.

In referencing Spain's resources and commerce, the terms "molluscan shellfish" and "bivalves" shall be used as synonyms realize that 600 - 800 metric tons (Tm) of periwinkles (Littornia, sp.) and purple dry murex (Murex brandaris, Linnaeus) are harvested annually. Although in many cases these marine Gasteropoda are sold through the same commercial channels, they do not influence the depurated shellfish market.

BIVALVE PRODUCTION IN SPAIN

Although over twenty species of bivalves are harvested in Spain, many of them are produced in very limited amounts, and in most cases, their geographic distribution is also very restricted (Figure 1). Only six or seven species are important commercially. two of them, the Mediterranean mussel (Mytilus galloprovincialis, Lamarck) and the European, or flat oyster (Ostrea edulis, Linnaeus), are produced by suspended (raft) culture. Other commercially significant bivalves are the autoctonous clams; grooved, pullet and banded carpet shells (Venerupis decussata, Lin., V. pullastra, Montagu, and V. rhomboideus, Pennant), striped venus (Venus gallina, Lin.), and the common cockle (Cerastoderma edule, Lin.). These species, along with another 15 to 20 bivalves of lesser importance, are harvested by dredging from natural beds that are controlled by local associations of shellfish gatherers ("cofradias de mariscadores"), or from privately managed concessions (over 80 km² the former, and over 3.5 Km² the latter). While the private concessions can sell their shellfish production throughout the year, natural beds can only be harvested from October to February. The major part of the captures from the natural banks (over 75% of the tonnage from the intertidal banks, and 25% of that of the sublittoral waters) are landed in the first 15 days of the gathering season.

Over 95% of the 225,000 to 275,000 Tm total shellfish harvest is produced in the five Galician Rias (estuaries) in northwestern Spain; another 2.5% is produced in the Ebro river delta on the Mediterranean coast, and most of the remaining production comes from the southern Atlantic coast, between the Portuguese frontier and Algeciras/Gibraltar. The production is completely dominated, in quantitative terms, by the blue mussel, with 220,000 - 250,000 Tm of which the Galician Rias produces 97%, and the Ebro delta the rest. The Rias also produce 99% of the 2,500 - 3,000 Tm of flat oysters, over 80% of the 3,000 - 4,000 Tm of the cockles (with a peak production in 1976 of 16,500 Tm), 60 - 70% of the 7,000 - 9,000 Tm of clams, and 50 - 70% of the rest of the bivalves harvested (5,000 - 10,000 Tm).

FIGURE 1. Shellfish production and depuration in Spain.

Spain imports large amounts of bivalves from other countries. For example, 10,000 -
20,000 Tm of striped venus from Italy and Turkey; over 3,000 Tm of flat oyster (both commercial-
sized and seed) from Italy, France, Greece, Yugoslavia, Turkey, the U.K., and Scandinavia;
increasing amounts of Japanese little-neck (Venerupis semiducussta) from Italy, France, U.S.A., and
Canada; and scallops (Pecten Maximus, Lin.) from the U.K. Aquaculture is a growing industry in
Spain. Many areas along the northern and southern Atlantic coasts, as well as the Mediterranean,
are trying to recover exhausted natural banks, or are developing small, local beds by improving the
substrate, seeding, predator control and carefully controlled harvests. Attention is focused on the
valuable autoctonous species such as Venerupis descussata, V. pullastra, Ostrea edulis, but the
lower priced, hardy, foreign species V. semiducussata, and the Pacific oyster, Crassostrea gigas,
which have the added advantage of hatchery-produced seed, are becoming increasingly important.
The number of hatcheries has been steadily growing over the last several years, and hatchery seed
is becoming more and more popular with private growers.

DEPURATION IN SPAIN

COMMERCIAL ASPECTS

After the publication in 1961 of the "Ley para la Renovacion de la Flota Pesquera" ("Law for the
Renewal of the Fishing Fleet"), the shellfish industry (both gathering from natural banks, and raft
culture of mussels and oysters) steadily increased its production due to a governmental promotion
of the productive sector, and by establishing refrigerated storage in central markets of the major
population centers. Many species of bivalve, previously harvested only for local fresh consumption
or canning, were beginning to reach the markets of large population centers in the interior of the
country. This distribution elevated harvest and began to pose a serious public health problem.

All the shellfish producing areas were evaluated to assess both the productive capacity that could
be attained and the sanitary conditions of the growing waters. By 1964, the harvesting areas had
been classified into three categories, according to microbiological criteria: "clean (the majority),

hazardous, and contaminated". Only the shellfish from areas certified "clean" could be sold for direct consumption, while those harvested in the limited number of zones of remaining categories had to be depurated before commerce. By the end of the decade, over 30 depurating plants, with a total annual capacity of 260,000 Tm, had been authorize, while bivalve production had reached a total of 50,000 - 70,000 Tm.

Several years later, in 1973, the criterium used to determine which bivalves should be depurated had to be changed to suit stricter microbiological standards for foods, and to account for increased localized contamination. The new regulations were based on the possibility of the shellfish being eaten raw (or being steamed open) such that depuration became compulsory for several species, regardless of harvest origin, (e.g. Mytilus edulis, M. galloprovincialis, Ostrea edulis, Crassostrea angulata, Cerastoderma edule, Venerupis pullastra and V. decussata). During the period between 1975 and 1981, the striped venus clams (Venus gallina) was imported from Italy in large amounts (in 1976, 31,400 Tm - 37% of the total tonnage depurated that year), to complement the diminishing national harvest. This clam was included in the group for compulsory depuration. In 1981, a commercial agreement was reached between both countries to remove striped venus clams from the list.

No new depurating plants were authorized until 1983. At the end of 1988 there were 63 authorized plants in Spain of which 59 were operating with a total operating capacity (t.o.c) of 345,000 Tm (87% of total authorized capacity). Of these, 43 are in Galicia (NW Atlantic - Rias: 74.2% of the t.o.c.), 6 are in Cataluna (4 in the Ebro River delta; a total of 13.3% of the t.o.c. in the N. Mediterranean), 6 more on the NE Atlantic coast (6.3% t.o.c.), 3 are on the southern Atlantic coast (6.1% t.o.c.), and, finally, 1 is located on the central Mediterranean coast (0.1% of the total operating capacity) (Figure 1).

The initial 32 depurating plants have authorized annual capacities varying between 26,000 Tm and 3,315 Tm "per annum". Of the 31 plants authorized after 1983, only 8 have capacities above 3,500 Tm (average 6,500 Tm), while the other 23 average 1,560 Tm p.a., and 13 of these are below 1,000 Tm p.a. (thus their Spanish name - "Minis"). The trend has been towards smaller plants which process a lower proportion of low value shellfish, such as mussels (F.O.B. price at plant of $0.65 - $1.0 per kg), and a higher amount of clams and oysters ($6.50 -$30 per kg F.O.B.): while mussels average 93% of the total tonnage depurated by the larger plants, they only account for 50% of the tonnage depurated by the small plants, in which clams represent over 45% of the total. Because of this, and lower installation and running costs, and more aggressive and flexible commercial strategy, the shellfish processed by the smaller plants usually cover 40 to 45% of their authorized annual capacity, while larger plants are lucky if they reach 20% of the same.

Surprisingly, the majority of the newer plants do not represent any technological improvement over "first generation" plants, although innovation which is certainly evident in the abundant recently installed aquaculture ventures. The design and the equipment used in the newer depurating establishments are basically the same as those of their older counterparts, but adapted to the lower quantities of shellfish processed.

The depurated tonnage has risen from 46,350 Tm in 1973 (94% mussels, 3.5% oysters and 2.5% clams), to over 105,000 Tm in 1988 (93% mussel, 2% oysters, and 5% clams) (Figure 2). While the raft-cultured shellfish (mussels and oysters) are harvested all through the year, the natural growth cycle and consumer preferences produce a peak in sales over in the winter months, with a maximum in the Christmas period. Between November and January, 35% of the mussels, and 57% of the oysters are sold (Figures 3 and 4). The prices paid to the producers are, in the case of mussels, fixed yearly by the producers for each of the three commercial grades (small, 6 - 9 cm; medium, 9 - 11 cm; and large - over 11 cm). In the case of flat oysters, there are several commercial grades, depending on size (the smallest, a diameter of 6 - 7 cm) but the prices vary according to the abundance of the harvest, and generally rise sharply in the Christmas period, in spite of the usually abundant offer.

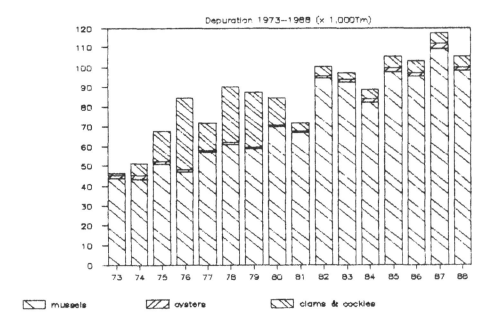

FIGURE 2. Tonnage of depurated shellfish in Spain, 1973-1988

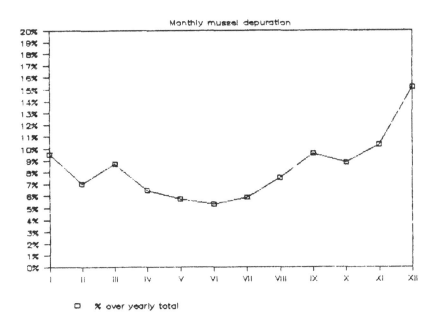

FIGURE 3. Annual depurated production of mussels in Spain.

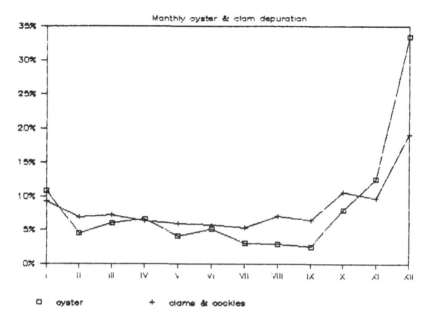

FIGURE 4. Annual depurated production of oysters, clams
and cockles in Spain.

In the case of the other bivalves (clams and cockles) the official gathering period begins in October and ends in February. Although there is a clear Christmas peak in sales (38% of the total yearly tonnage), it is not as high as could be expected. The dampening effect of the private concessions absorbs an important part of the clams gathered in October and November and sells them during the Christmas peak and between March and September when harvesting from natural banks is prohibited. As the production of Japanese little-necks increases, the peak will probably become even less pronounced. The prices depend basically on the species size and abundance (e.g. the grooved carpet shell is the most highly priced, over 3,000 pestas/$25 per Kg for the larger size). Moreover, the price of a given commercial size of any of the species may vary from one harvesting area to another by as much as 25%. In some areas, the prices are fixed by auction for the total harvest of a species during the following week, while in other areas, the auctions are held daily.

All depurating plants in Spain are private companies. In some cases, depurating tanks are leased to shellfish producers who commercialize their products once depurated. This is frequently the case of producers that run one of the private shellfish outgrowing/stocking concessions mentioned before. They are charged a fixed amount per kg for the depuration according to the species, and a smaller, also fixed, amount for holding the shellfish in the tanks after the second day

Generally, the depuration plants are owned by stockholders, or by associations of shellfish producers, who operate the plant as an independent business. These companies buy the bivalves from the producers, and have their own commercial channels, through wholesale dealers in central markets, retail outlets such as small fishmongers and restaurants. The F.O.B. price of most plants is a fixed percentage (between 15 - 20%) over the initial cost of the product, plus packing and transport.

The harvested shellfish, once weighed and graded by the producer, are loaded at the abundant, small wharves, or at the "lonjas" (the local fish/shellfish auction houses). The distance covered to the depuration plant is usually less than 20 Km. Here, they are re-washed, selected, and placed in the depurating tanks, in plastic grid boxes with a 10 or 15 kg capacity (in a layer 10 cm deep).

These boxes are laid on 15 -30 cm high, parallel ridges built into the tank floor. Sterilized sea water is supplied continuously for a minimum of 42h, after which the shellfish are re-washed, re-selected, and packed.

The type of packing depends on the shellfish, and the commercial grade. For the lowest commercial grades of mussels, plastic mesh bags of 15, 10, 5, and 2 kg are used. For better grades of mussels, plastic grid boxes or wooden boxes are used. Clams are packed in plastic mesh bags holding 2, 1, or 0.5 kg, and wooden boxes for the better grades. Oysters are usually sold in tightly closed, wooden boxes, packed with fresh fern, clean straw, or hardwood shavings, containing l2, 25, 50 or 100 oysters. The depurated shellfish are shipped to market by refrigerated trucks maintaining temperatures between 4 to 8°C. Rail and air freight are seldom used.

The major markets for depurated shellfish in Spain are Madrid, Barcelona, and Valencia. These three cities have 20% of the country' s population, and absorb around 25% of the tonnage depurated. Another 20% is exported (in 1988, 14% to Italy, 5% to France, and 1% to Germany), all of which is mussel, the only bivalve for which Spain is a net exporter. The rest (55%) is sold in the central, southern, and northeastern areas of the country.

Most of the major population centers have a central fish and shellfish market, controlled by the municipal health authorities, in which the wholesale dealers have stalls and refrigerated storage. These wholesale dealers are a major figure in the commercial channels. Possibly 70% of the depurated shellfish are sold to these dealers, who generally have a more or less fixed price for each species, commercial grade, and customer. The price fluctuations produced at origin are usually dampened such that the only variations the final customer pays are due to seasonal demands near Christmas. In cities and towns without this centralized fish market, the depurating plants deal directly with the restaurants, fishmongers, and supermarkets, and the prices generally follow the trends at origin.

One of the factors that has had a major influence on the depurated shellfish market in the last several years has been red tides. The first known occurrence of a PSP producing red tide in Spain was an outbreak during October, 1976, caused by Gymnodinium catenatum, Graham. Undetected, contaminated mussels caused intoxications in various areas of Spain and in several European counties (a total of over 200 people were affected; no deaths were recorded). In Galicia, the major shellfish producing area, a surveillance net was set up by the health authorities and the Instituto Espanol de Oceanografia, controlling both the plankton in the Rias, and the toxin content of the shellfish before harvesting and during depuration, using A.O.A.C. mouse bioassay techniques. This control began operating in 1981, just after the first Spanish DSP outbreak Since this surveillance net has been in operation many red tides have been detected in the bivalve harvesting areas, as well as toxins in the shellfish, but there have been no intoxications. The affected areas are routinely closed and no shellfish may be harvested until the toxin content is below the authorized levels.

In spite of this rigorous control, the adverse publicity represented by the publishing in the press of the red tide warnings seriously affected the commercialization of controlled, toxin-free, depurated shellfish for several months. In the last two years, the health authorities have established a communications system by which only the shellfish industry is informed continuously of the plankton/toxins situation in the different harvesting areas. This program is protecting the general public while reducing commercial harm to the product image.

REGULATORY ASPECTS

Spain, since the beginning of the decade, is politically divided into 17 "Estados Autonomicos" (autonomous states), which do not have the degree of legislative independence that other federated states enjoy. Each autonomous state has the same capacity to dictate laws and regulations in certain administrative areas. Their efforts are funded by the national government administer these laws.

The shellfish industry is regulated by two different authorities; the Fisheries authorities, which have a relatively high degree of regulatory independence, and the public health authorities, which have a lesser degree of authority in the different autonomous states.

Insofar as production is concerned, each autonomous state's fisheries authority can regulate what species can be harvested, the dates for the beginning and ending of the harvesting period, minimum capture sizes, species of shellfish which can be produced by artificial culture, and the species which can be "imported" live to ingrowing areas. Currently, the only appreciable differences in the various fisheries regulations relate to the minimum capture sizes, dates of permitted capture, and the species which can be cultivated. For example, while other E.E.C. based states permit, and even fund hatcheries and ongoing ventures for the Pacific oyster and the Japanese little-neck, Galicia does not permit the ingrowing of these two species, considering that they could eventually eradicate the autoctonous, high-priced, clams and oysters. Although, as far as we know to date, the situation has not yet occurred, it is theoretically possible that shellfish legally caught in one of the states, cannot be imported to another for depuration as, in the second state, it is below the legal minimum size.

From the public health point of view, the criteria which regulate depuration have been set by the national government and are currently in effect in all of the autonomous states. These criteria are published in two laws: Real Decreto 263/l985, published in the National Official Bulletin dated Feb. 20th., titled "Reglamento de Salubridad de Moluscos", and the "Order" of May 31st , 1985, "Norma de Calidad para los Moluscos Bivalvos Depurados", in the Official Bulletin dated June 8th., 1985.

The former is a condensed form of a law published two years previously, referring to the sanitary conditions of all products derived from fishing, shellfish gathering, and aquaculture in general, for human consumption, as well as setting the minimum hygienic conditions to be met by boats, buildings, equipment, packing, etc. to be used in the manipulation of fish and shellfish, or related products.

This Real Decreto establishes that the species that must be depurated for human consumption are those previously mentioned (Ostrea edulis, Crassostrea sp., Venerupis (Tapes) decussatus, Venerupis (Tapes) pullastra, Cerastoderma (Cardium) edule, and Mytilus edulis and M galloprovincialis). It also establishes:

    - the general minimum conditions to be met by buildings and equipment used in depuration, waste disposal criteria, weights and measures that can be employed, hygiene standards for tanks, equipment, workers, ..., washing techniques for the bivalves, etc.;
    - that the water flow used in the depurating tanks must be, at least, of 15 m^3/Tm of shellfish per hour;
    - that the separation between the water intake and outlet of the plant must be of 250 linear meters in a line parallel to the shore, and the minimum depth of the intake must be 2.5 m below the maximum spring ebb tide level;
    - that the effluents of the plant must be treated to eliminate any possible pathogenic bacterial content;
    - that the maximum stocking density in the depurating tanks is of 30 kg/m^3 (which can be augmented to 45 kg if a series of tests run by the public health authorities prove adequate depuration rates at this density.
    - that the minimum continuous duration of the depuration process is 42 hours unless the origin of the bivalves warrants a longer period.
    - that the procedure for sterilizing the sea water used in the depurating process must approved by the public health authorities. Most depurating plants are using gaseous chlorine, and a few smaller plants use U.V. lights or ozone;

- the type of analyses that must be performed in the laboratory of the plant, and the minimum qualifications of the person in charge of the laboratory, and the models of the registers that must be carried of the analytical work, and input and output of shellfish;
- minimum hygiene conditions to be met by the plant workers;
- types of material, and capacities of the packages containing the depurated products (yellow plastic mesh bags, with a 0.3 - 1.5 cm mesh size, wooden or wicker boxes, all to be used only once, and with net capacities of 1, 2, 5, 10 and 15 kg for mussels, 0.5, 1, and 2 kg for clams and cockles, and 12, 25, 50 and 100 units for oysters);
- the official description of the depuration tag, also yellow in color, must either be fixed to the exterior of the package or clearly visible from the exterior. Each tag must be indelibly marked with: date of packing (6 numbers DD-MM-YY), the legend "valid for 5 days", net content, official registration number of the depurating plant, and if appropriate, clear instructions for storage and use of the product, and;
- the temperatures admitted for storage and transport of depurated, packed products must be between 3 - 7°C.

The Order of May 31st., 1985 set the biological criteria for depurated products. Water standards relative to original source of product are:

- total plate count (20°C, 5 days): max. 1×10^5/ml;
- E. coli: max. 500/liter;
- Salmonella: none in 25 ml;
- Lancefield's D group streptococci: max. 100/g;
- Vibrio parahaemoliticus: max. 100/g;
- water soluble biotoxins: max. $80 \mu g$/100g of product (mouse bioassay);
- lipid-soluble biotoxins: none in 100g of product;
- the maximum contents of heavy metals: Cd - 1ppm; Cu - 20ppm; Sn - 250ppm; Hg - lppm; and
  Pb - 3ppm.

However, if the product is to be exported, it must comply with the importing country's standards (e.g. Hg, in many European countries is only accepted up to 0.5ppm).

This Order also specifies commercial categories for the various depurated products:

- oysters: giant, over 9 cm diameter; "flor", 8 - 9 cm; medium, 7 - 8 cm; small, 6 - 7 cm.
- Portuguese oysters: only one category - over 6 cm diameter.
- grooved carpet shell: large, up to 25u/kg, medium, 26 - 40u/kg; and small, 41 - 60 u/kg
- pullet carpet shell: large, up to 25u/kg; medium, 31 - 45 u/kg; small, 46 - 55u/kg.
- mussels: large, over 11cm (long axis); medium, 9 - 11cm; small, 5 - 9cm.
- cockles: large, over 26mm (short axis); normal, 25 - 26mm.

Other legislation also affects the depurating plants; for example, the Decreto 328/1986, of October 30th., in the Galician Official Bulletin, establishes the A.O.A.C mouse bioassay as the official method for PSP toxin assay, and a published technique for DSP bioassay in the Galician state. Other autonomous states are copying Galicia's legislation in this field, and are established their own surveillance network based on Galicia's experience. There is also a long series of E E.C. regulations and directives coming into effect, on quality criteria for industrial and urban effluents into shellfish growing areas, and public health controls on imports from non-E.E.C. nations.

We mentioned in the introduction that the future of depuration "per se" in Spain is not too clear. This is due to a regulation for intra-E.E.C. shellfish commerce that is presently being drafted to come into effect in 1990. It affects various aspects of shellfish production, public health control systems, and biological quality criteria, amongst which is the bacteriological standard for considering a bivalve apt for direct human consumption. This standard is based on coliform counts (less than 300 coliforms, or less than 230 E. coli/100g of flesh plus liquor), and absence of Salmonella in 25g of product. With very few exceptions, and many of these being seasonal, the majority of the shellfish produced in Spain will be considered adequate for direct human consumption after washing, grading and packing, without a previous depuration or relaying process. Obviously, the Spanish depurating industry will be defending the theory that "no coli is more healthy that very few coli", but the evolving regulation is expected to pass as introduced. Thus, the majority of the depurating plants will become packing plants, and those that will be selling their higher sanitary quality due to depuration will have to be very careful in how they present their publicity.

Acknowledgement

We wish to thank Miss Carmen Roca, of the Asociacion Nacional Empresarial de Depuradores de Moluscos, and Mr. Juan Hidalgo, of the Direction Comisionada del Servicio de Sanidad Exterior de Galicia for the information provided.

FRENCH SHELLFISH INDUSTRY
REGULATORY STATUS AND DEPURATION TECHNIQUES

J. Le Pauloue, B. Langlais, R. Poggi and J.Y. Perrot

SHELLFISH INDUSTRY IN FRANCE: HISTORY AND CURRENT STATUS

Molluscan shellfish production is the main marine fishing and aquacultural activity in France. The economic and social importance of this activity has for a longtime been one of the basic economic forces in several regions such as Normandy, Britanny, Poitou-Charente and Languedoc-Roussillon. The overall French production in terms of shellfish sold from harvest and farming is approaching 200,000 tons per year, corresponding to a total sales value of 2,300 MF (million Francs) per annum.

Oyster farming covers about 54,000 coastal acres. The Asian species, Crassostrea gigas, is by far the most widely farmed at the present time and is marketed under the name of "huitre creuse" or "hollow oyster". The flat "Belon" type oyster (Ostrea edulis) currently occupies only a very small amount of commerce with a yearly production of around 2,000 tons compared to 25,000 tons/yr recorded in the sixties. Obviously, oyster farming in France has been subject to dramatic fluctuations, especially during the last twenty years, due to several devastating epizootic diseases (Figure 1). Since 1980, oyster production in France has shown a small growth factor (Table 1).

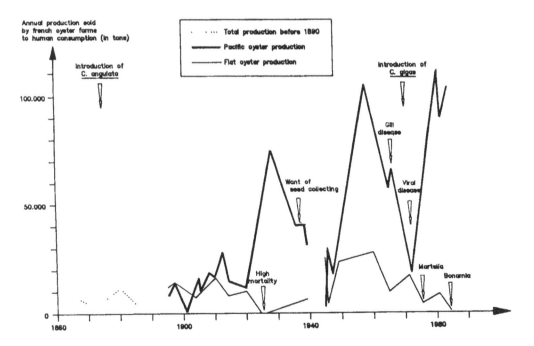

FIGURE 1. Evolution of oyster production in France from 1865 to 1983 (statistics from French Ministry of the Sea). Source· Barnabe 1989. "Aquaculture" 2 éme edition Vol. 1 and 2. Lavoiser Technique at Documentation, Paris, France.

TABLE 1. Evolution of French shellfish production (metric tons). Note, a distinction must be made between wild and cultivated mussels at the beginning of the 1980's cultivation increased during 1980-1988. Source: Central French Sea Fishing Committee (Comite Central des Pêches Maritimes) Rapport sur la Production de l'Industrie des Pêches Maritimes 1980a, 1987, Paris France.

	1980	1981	1982	1983	1984	1985	1986	1987	1988
Pacific oysters (*C. gigas*)	95 200	82 700	94 800	107 800	110 700	106 500	112 800	130 300	
Flat oysters (*O. edulis*)	4 200	2 500	1 700	1 300	2 300	1 600	2 400	1 900	
Mussels	72 700	80 400	70 900	60 500	52 900	50 900	60 600	44 500	
Manilla clams (cultivated)	2	4	10	100	200	400	420	560	450
Other shellfish (from natural beds)							26 000	29 500	
Total value (million) FF	940	1 018	1 022	1 432	1 644	1 463	1 815	1 830	
US$	144.6	156.6	157.2	220.3	252.9	225.1	279.2	281.5	

Mussel farming is performed mainly on artificial supports, the oldest method of which employs wooden sticks sunk into the sea bed. Lines of wooden poles are called "bouchots". Linear settings (1,500 km) are currently used for mussel culture on some 1720 acres of sea bed. Mussel culture in France has been stable now for several years, averaging between 35 and 50 thousand tons per year, however, production is experiencing a period of zootechnical innovation with the development of on-line culture placed at sea (ropes hung between floating rafts). On the whole, production is variable, depending on random fishing from natural sources. Two species are raised or culled in France; Mytilus edulis, the largest, on the Atlantic-Channel sea fronts, and Mytilus galloprovincialis on the coasts of the Mediterranean and North Brittany in the Saint-Brieux Bay (Figure 2).

The culture of various Veneridae or clams (e.g., Ruditapes philippinarum) is a very recent activity in France, due to the development of spawn hatcheries. The European species (Ruditapes decussatus) exists in small quantities on several areas of the French coast where it is culled on a small pedestrian scale. It is impossible to establish reliable statistics on clam fishing, but estimated cultured production of the Asian clam species results in the sale of some 500 tons/year. On the Mediterranean coast, the local species called Venerupis aurea or carpet shell represents an appreciable source of income for the local fishing population. On the Atlantic seaboard, clams are currently affected by a disease probably of bacterial etiology. This disease affects both wild and farmed specimens and explains the fall in production during 1988 and 1989 as compared to the growth recorded in 1987.

No other types of shellfish are at present raised in France. Attempts have, however, been made to cultivate edible sea urchins (Paracentrotus lividus and P. sammechinus miliaris) and scallops (Pecten maximus). The other shellfish (Venus verrucosa, Chlamys varia, Glycymeris glycymeris, Venerupis rhomboides, Donax trunculus, Tellina sp., Spisula solida and others) are continually exploited by professional fishermen (Table 2).

SPECIES	WEIGHT (t)	VALUE in million US $	MIDDLE PRICE US $ / Kg
a) Bivalves			
European bitter sweet shell (Glycimeris glycimeris)	886	0.3	0.35
Scallop (Pecten maximus)	8 003	22	2.75
Cockle (Cerastoderma edule)	6 141	4	0.65
Short neck clam (Ruditapes decussatus)	1 444	9.1	6.30
Variegated scallop (Chlamys varia)	1 157	1.7	1.50
Little neck clam (Venus verrucosa)	1 768	5.6	3.20
b) Gastéropods	4 614	4.4	0.95
c) Other shellfish	1 766	1.1	0.60
TOTAL SHELLFISH	25 779	48.2	1.9

TABLE 2. Production and values for captured shellfish, 1986 (except oysters and mussels)
Source: IREMER.

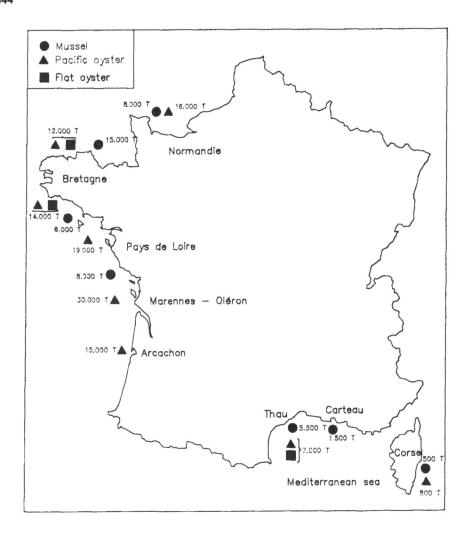

FIGURE 2. Main shellfish culture sites in France.

Compared with the total shellfish landings of the main producing countries in the world (Table 3) and specifically Europe (Figure 3 and Table 4), France ranks somewhere in the middle. France is, however, the biggest oyster producing country in Europe. The unique feature in France is consumer preference for raw shellfish. All the oysters sold are consumed raw and alive. The same applies to a large share of Mediterranean mussels and clams.

TABLE 3.   Assessment of national Mollusc Shellfish Production in 1983.
Source: Economic Status of Molluscs - Shellfish Culture.
International Seminar (J.P. Troadec) in LA Rochelle, France,
4 and 7 Mars, 1985.

COUNTRY		CANADA	FRANCE	GERMANY	ITALY	JAPAN	UK	USA	TOTAL
Production (tonnes of live weight)									
oysters	captured	-	-	-	5 000	-	-	121 600	121 600
	cultured	3 400	110 000	600	400	250 000	600	125 000	490 000
mussels	captured	-	10 000	16 000	-	-	5 200	14 500	45 700
	cultured	900	42 000	16 000	100 000	-	800	1 500	161 200
scallop	captured	-	12 200	-	-	100 000	7 800	185 000	305 000
	cultured	-	spat	-	-	77 000	-	-	77 000
clams	captured	-	-	-	-	238 000	-	66 000	305 000
	cultured	-	300	-	200	300	-	4 300	5 100
Value million US $		2 8	127	2 8	53	612	1	273	1091 6
Employment	full time and part time	2 300	40 000	600	5 500	167 000	700	4 000	220 300
potential (estimations)	oysters	175 000	50 000	15 000	30 000	30 000	15 000	-	315 000
	mussels	12 000	100 000	15 000	40 000	30 000	15 000	10 000	222 000

The domestic oyster production is sufficient to satisfy the demand of French consumers. Consumption in France varies through seasons, whereby 60% of the oysters are consumed during the Christmas and New Year festivities. A small part of the production (around 4,200 t) is exported each year, either for oyster farming (1,700 in 1988) or for direct consumption (2,500 t in 1988).

Mussel consumption is much more evenly spread through seasons with a higher proportion consumed in June through December, but the production of mussels is far below demand (about 50% of the domestic demand). The deficit is imported from Holland, the United Kingdom, Eire, and Spain. These products are either directly imported to the food markets for immediate distribution to consumers (Dutch and Spanish mussels), or kept in wet tanks in France for a short time in order to regulate distribution to the markets (mainly in the case of imports from the U.K. and Eire). Mussels for the canning and processing industry are mainly shucked, deep-frozen imports from S.E. Asia, or preserves from the North of Europe.

The French wild shellfish production, except oysters and mussels, is currently aimed at South European markets, especially Spain and Italy. The French foreign trade balance, as far as shellfish are concerned, remains in the red (Table 4).

FITNESS OF PRODUCTION AREAS

The habit of eating raw shellfish in France created serious problems due to waterborne microbial pathogens until the end of World War II. This included a form of typhoid. To counteract such a scourge, in 1918 France set up a system of controlling the wholesomeness of its oysters (all eaten raw) and extended this control to all shellfish in 1939 (Decree of 20 August 1939). This legislation provides that all coastal waters in which shellfish are fished or farmed must be classified. The contents of the technical concept (sampling method, admissible contamination thresholds, statistical reading of measurements) became mandatory as a result of the Decree of 12 October 1976. This classification allows for area designations relative to contamination along French coasts as issued by French Ministry of the Sea.

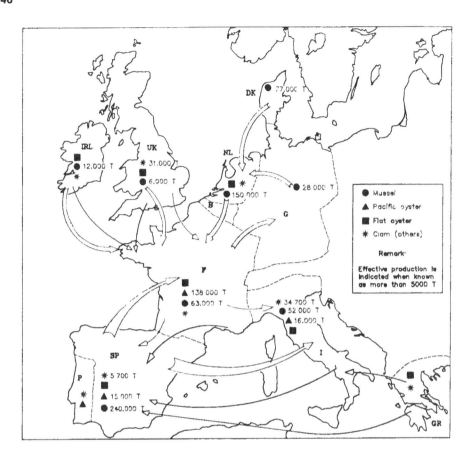

FIGURE 3. Main shellfish production sites and trade-ways in Europe, 1985.

TABLE 4. Trade Exchange for shellfish in France (Source: French Customs statistics from the Treasury).

SPECIES / FROM/TO	FLAT OYSTERS		PACIFIC OYSTERS		MUSSELS		SCALLOPS		OTHER SHELLFISH	
	IMPORT	EXPORT	IMPORT	EXPORT	IMPORT	EXPORT	IMPORT	EXPORT	IMPORT	EXPORT
EEC COUNTRIES										
BENELUX	124.73	30.82	36.77	425.27	14,863.68	30.40	177.60	55.23	443.19	291.52
DANMARK	0.30	35.26	0.31	5.40	506.42	20.68	2.06	-	0.93	2.11
GERMANY	-	172.43	-	215.74	100.80	160.35	-	28.57	-	98.36
GREECE	-	-	-	0.39	-	-	-	-	-	-
IRELAND	236.48	-	10.00	-	5,881.99	2.30	220.51	-	2,023.52	0.38
ITALY	21.12	33.65	-	2,482.67	-	197.10	-	39.16	156.92	424.90
PORTUGAL	-	-	-	-	-	-	-	3.83	16.20	16.07
SPAIN	14.69	366.91	5.00	72.40	9,679.85	165.79	0.99	229.37	63.20	1,730.77
UNITED KINGDOM	390.37	23.40	28.22	43.92	3,361.99	87.14	1,341.33	0.95	2,095.57	43.59
EUROPE except EEC										
ANDORRA	-	1.64	-	9.19	-	0.15	-	0.43	-	3.72
AUSTRIA	-	6.60	-	3.00	-	0.09	-	-	-	2.17
FINLAND	-	-	-	0.13	-	-	-	9.07	-	0.86
ICELAND	-	-	-	-	-	-	0.10	-	4.25	-
NORWAY	-	0.29	-	0.19	-	5.05	2.25	0.68	-	3.51
POLAND	-	-	-	0.32	-	-	-	-	-	-
SWEDEN	-	-	-	0.80	0.78	-	-	0.04	-	0.36
SWITZERLAND	-	68.92	-	208.99	-	325.81	-	16.60	-	176.96
AFRICA										
REPUBLIC OF MAURITANIA	-	-	-	-	-	-	-	-	2.52	0.38
SENEGAL	-	0.85	-	2.32	-	0.18	-	-	0.15	0.40
TUNISIA	-	-	-	1.50	-	-	-	-	202.96	-
Other countries	-	11.37	-	42.96	-	20.93	-	2.50	-	13.68
U S S R	-	-	-	0.37	-	-	-	-	-	-
MIDDLE EAST	-	0.90	-	5.88	-	0.60	-	1.75	-	3.37
ASIA	-	3.68	-	28.23	-	-	0.12	0.64	-	12.36
AMERICA										
CANADA	-	0.48	-	0.40	-	-	12.09	-	-	-
CHILE	-	-	-	-	-	-	6.71	-	-	-
CENTRAL AMERICA	-	0.15	-	-	-	-	-	-	-	0.43
U.S.A	-	-	-	0.15	-	-	-	-	180.29	-
TOTAL WEIGHT (t)	797.69	757.35	80.30	3,550.22	34,395.51	1,017.56	1,763.76	379.82	4,189.70	2,825.88
TOTAL VALUES (in thousand US $)	3,580	2,588	0,327	7,466	22,509	1,434	15,932	2,083	7,063	14,886
BALANCING OF FRENCH TRADE EXCHANGE (in thousand US $) * Per specie	-992		+7,139		-21,075		-13,849		+7,823	
* TOTAL	-20,954									

For the protection of the public health, French regulations admit the existence of three categories for farming or fishing shellfish

1. Approved Zones

Growing areas are waters where fishing is unrestricted although it may be controlled for the purpose of protecting the resource, and where farming is possible except for certain restrictions on the use of Public Marine Property. To be entitled to use the "APPROVED ZONE" qualification, an area of production must be subjected to in-depth investigations including:

-    an inventory of actual or potential pollutant sources (waterways, built-up areas, factories, ports, etc.).

-    a survey of the currents, that must be particularly thorough when they bring several sizeable inputs of fresh water or waste water.

-    bacteriological analyses equal in number to at least the regulatory amount established in the Decree of 12 October 1976, i.e., a minimum 26 series of samplings carried out in 12 consecutive months. These surveys are increasingly scheduled to last several years in order to allow for the climatic variations observed in the course of time (Table 5).

Table 5. Criteria for "Approved Zone" based on Decree of 12 October 1976 (Official French Gazette, Nov. 23, 1976)

	Fecal coliforms per 100 mg shellfish meat	aggregate frequency	frequency per class
Unacceptable result	3,000	0%	
Highly doubtful result	1,000	8%	8%
Doubtful result	300	19%	11%
Acceptable or Satisfactory result		100%	81%

These criteria are to be supplemented by analyses of micropollutants (hydrocarbons, PCB, detergents, heavy metals, etc.) when the National Observation Network installed on the French coasts in 1975 is complete. Since 1983, area surveys monitor for plankton cells responsible for toxic discoloration ("red tides") in the growing waters. These potentially toxic plankton blooms are a handicap to the development of mussel culture on a large part of the French coasts.

2.   Non-approved but Non-prohibited Zones

These zones are waters where fishing or farming may be authorized providing a number of prescriptions are observed, among which the obligation to give marketable sized specimens depuration treatment. Currently, these zones are primarily used for the cultivation of immature specimens intended for later culture in clean zones.

3.   Prohibited Non-approved Zones

Zones of water where any form of exploitation is prohibited.

It is to be noted that the intermediate category (Non-approved but Non-prohibited Zones), are only plausible in cases where the lack of fitness is due to pollution that can be remedied by a purification process. Such remedial possibilities exist either by soaking the shellfish for a short time in a trough full of bactaria-free seawater (depuration), or by an extended stay in an open environment in a zone acknowledged to be a clean relocation (relaying). Contamination in excess of 10,000 fecal coliforms for 100 mg of shellfish meat is assumed to present a health hazard that can hardly be eliminated by a short stay (48 to 72 hours). Moreover, when contamination is of chemical or biological origin, only a long stay (1 to 6 months) in a clean area is thought to make contaminated or polluted shellfish acceptable for consumption.

In all cases, the healthy shellfish must be given clean handling in preparation for sale. Washing, sorting or calibrating, storage and packaging must be performed in suitable vessels or firms, approved by the relevant authority. The above operations almost invariably require putting the mollusks (oysters, mussels, clams, cockles, venus clams, etc.) in finishing troughs fed with clean water of a quality controlled by the operators and the health authorities. These troughs may even perfect the health quality of the shellfish by ridding them of any small quantities of microorganisms they may contain.

REGULATION OF IMPORTS: MARKET SIZE AND SEED STOCK

The various channels for foreign shellfish imported into France are shown in Figure 4. The objectives of regulatory measures in this field are to protect the health of the livestock, and so concern exclusively the protection of the animal's health. In enactment of the Executive-Order of 9 January 1852, amended by Law 86-2 of 9 January 1986, "all measures and precautions such as to ensure the conservation of resources...can be taken as well as sanctions against any person immersing marine species in irregular conditions". These provisions, in the case of shellfish, were defined in an executive order dated 21 November 1969, setting the principle that it was forbidden to immerse foreign shellfish in French waters. The law does, however, allow certain waivers. It is due to a succession of such waivers that imported shellfish were the undoubted carriers of the latest epizootic diseases mentioned above. In this respect, many countries, particularly in the Northern hemisphere, systematically prohibit immersion of foreign shellfish in their water. The cautioning recommendations of the I.C.E.S (International Council on Exploitation of the Sea) are clear on this point (Council Resolution, 1970; 4:5/4:6/4:7).

Notwithstanding, the French government, in order to maintain the existing trade currents, has not seen fit to prohibit all immersions of foreign shellfish. Since 1982, this procedure is simply limited to animals intended for storage or depuration in isolated open-air troughs isolated from the open sea.   Adult foreign shellfish can be received for storage prior to dispatch for human consumption, in so-called storage structures. Imported shellfish from clean sources and classified as healthy are sent to storage and forwarding establishments, while imported shellfish classified unsanitary are sent to storage and depuration units.

If no livestock is available in France at a time when exceptional circumstances justify the re-seeding of a farming or fishing area, it is possible to bring in specimens or spawners of foreign origin. To avoid carrying parasites or diseases to home waters, shellfishes entering France are subject to a period of quarantine in addition to the animal health guarantees supplied by the exporting country.   The purpose of the post delivery Quarantine Unit is to isolate the foreign immature mollusks, imported to France for further culture in an open marine environment, in view of health control (or even parasite removal).

350

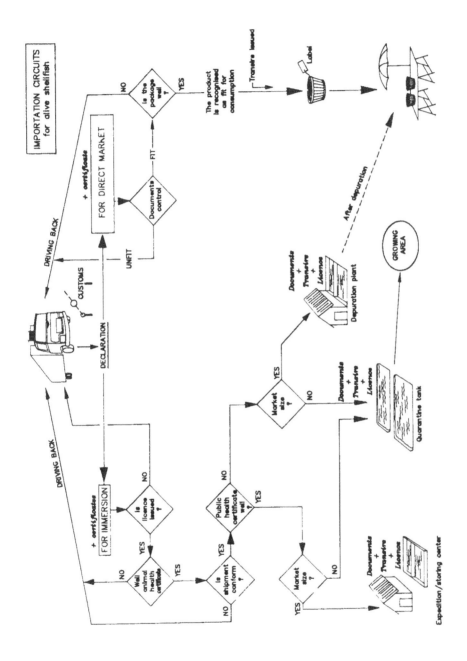

FIGURE 4. Import circuits for live shellfish in France.

Storage and quarantine facilities include a system for disinfecting the water before discharge to the natural environment. This is to avoid contamination of the latter by viruses, bacteria, parasites or competitors whose shellfish may be the carrying agent. The technical rules on the disinfection of the water aim at sterilization by a powerful oxidation process, or by discharge outside the marine environment (e.g. discharge to dunes or abandoned quarries, evaporation ponds etc.). The last of these solutions, although accepted as a principle, is rarely used for it is only feasible in very special topographical and geological conditions.

DEPURATION TECHNIQUES MEETING THE HEALTH STANDARDS DEFINED UNDER FRENCH LEGISLATION

It was remarked above that, depending on the origin of the shellfish (French or foreign produced), and the quality of the seawater in which it is sourced or stored, some form of water treatment may prove necessary. Different treatment options are available:

Shellfish finishing treatment by soaking in clean sea water without prior disinfection.

Before sale, a part of the shellfish coming from clean waters passes through finishing troughs. The main purpose of this stage is to get rid of any sand or slime that may have gathered inside the shells. These troughs are usually located outdoors and always above high-tide levels. Some may be indoor units. In regions where seawater, near the coasts or in estuaries, is muddy troughs can be fed by pre-settled or filtered water.

The surface area of the troughs must be at least 50m². They are filled with water that is changed at high tide and are cleaned each time when emptied. The minimum water height over the level of the shellfish is about 40 cm with a recommended shellfish density of 40 to 60 kg/m². This load rate can be increased in the case of running water troughs.

In cases where the oxygen content in the water is low (summer months), artificial oxygenation may be applied. It is estimated that the oxygen needs of the bivalve mollusks require a continuous dissolved oxygen saturation rate above 70%. Blogoslawski (1983) announced that the filtering activity of bivalve mollusks is elevated in proportion to the dissolved oxygen content. The oxidation techniques used are traditional methods (Todd, 1976), surface aeration using various types of turbine, air diffusion by porous diffusers, and venturi systems or waterfalls. Oxygenation must be performed after sedimentation in the case of turbid water. If aeration takes place in the treatment tanks, the device used should be designed such that the particles released by the shellfish never return to a suspended phase.

Depuration of shellfish by immersion in disinfected seawater

Whenever there is doubt as to the quality of water where shellfish are fished or farmed (microbiological pollution), product intended for sale must undergo depuration by contact with sea water of good bacteriological quality. It is highly desirable to disinfect the water prior to the immersion of the shellfish Continous disinfection makes it possible to supply permanently impeccable water quality to the troughs. Different techniques for disinfecting the sea water and immersing the shellfish are in use.

Four disinfectants can be contemplated: chlorine, bromine, ozone and ultraviolet (UV) lighting (Table 6 and 7). It is important to note that whatever the disinfectant used, except UV lighting, the final type of oxidant found in the water is thought to be hypobromus acid. Hypobromus acid, depending on the sea water pH, will decompose into hypobromite ions and H+ ions. (Carpenter and Macalady, 1978).

TABLE 6. Advantages and drawbacks of different oxidants suitable for sea water disinfection. Source: Moral et al. 1989. Purification des moules dans l'eau de ,er desomfectee aux UV IFREMER/DRV, 89-028, Nantes, France (Note, relative to ozone use - if O_3 to Br- ratio is greater than 1.8 mg/mg total oxidant residuals must be reduced before stabulization.

	CHLORINE	BROMINE	OZONE	U.V.
	BROMINE and BY-PRODUCTS	BROMINE and BY-PRODUCTS	BROMINE AND BY-PRODUCT	WAVE LENGTH 253,7 nm
ADVANTAGES	- ONLY CHLORINE GAZ - NEED MONITORING - CHEAPEST PROCESS	- IN CASE OF AMMONIAC POLLUTION, FORMATION OF INSTABLE BROMAMINES - RELIABILITY OF A SYSTEM COMPULSORILY MONITORED	- OXYGENATION OF WATER - PARTIAL DESINFECTION BY OZONE ITSELF ALLOWING REDUCED CONTACT TIME - RELIABILITY OF A COMPLETELY MONITORED SYSTEM	- NO REAGENT NO TOXICITY FOR SHELLFISH - VERY SHORT CONTACT TIME - EASY TO HANDLE
DRAWBACKS	- IN CASE OF AMMONIAC POLLUTION, POSSIBLE GENERATION OF STABLE AND TOXIC CHLORAMINES, CORROSIVE VAPORS AND DANGER TO BREATHE - LONG CONTACT TIME - TOTAL OXIDANTS RESIDUAL MUST BE REDUCED BEFORE STABULATION	- NEED TOTAL MONITORING DUE TO DANGER OF HANDLING LIQUID BROMINE - CORROSIVE VAPORS AND DANGER TO BREATHE - LONG CONTACT TIME - TOTAL OXIDANTS RESIDUAL MUST BE REDUCED BEFORE STABULATION	- RISK OF BROMATE IONS FORMATION (only > 2 5 mg/l O_3 rate) - POSSIBLE SHELLFISH BLEACHING IF OZONE OVERDOSED COSTLY PROCESS	- LAMP SHORT LIFE (CLOSED TO ONE YEAR) - LOW WATER TURBIDITY REQUIRED - UNADAPTED TO LARGE PLANTS - DIFFICULTY TO CONTROL THE TREATMENT

Disinfection with chlorine can be done with gaseous chlorine, sodium hypochlorite (javel water) or through production of the latter directly in the sea water by electrolysis. Bromine is injected in the form of bromine water, using a dosing pump. Ozone produced on site using air as a carrier gas, according to a well known drinking water treatment practice (corona discharge), is injected in the form of ozonized air. The use of these three reagents partially oxidizes the organic matter present in the sea water, increasing the consumption of oxidant, hence the dose rates required to ensure disinfection In the case of pollution by ammonia, there will be a formation of bromamines

Since shellfish are sensitive to the presence of an oxidant in the water (Galstoff, 1946), generally meaning total oxidants, the disinfection step is located upstream of the immersion troughs, and the treated water is fed to the latter only when the total oxidant residual is equal to or less than 0.10 mg per liter of the initial disinfectant used. In the case of disinfection by ultraviolet lighting, no measurable residual persists after treatment. There should be a very shallow flow of very low turbidity water to allow the UV light rays to penetrate. The main germicidal UV wavelength is 254 nm.

TABLE 7 Conditions for using seawater disinfectants (according to I.S.T.P.M., 1981).

Disinfectant	Chlorine	Bromine	Ozone	UV*
Dose rate	3 mg/l	1 - 4 mg/l	0.5-1.5 mg/l	25 milliwatts sec./cm²
Contact time required for disinfection	minimum	1 hour	6 min.	a few seconds
Retention time required for decreasing the total oxidants concentration.	from 1 to a few hours depending on the dose rate applied		15 min.	

There are basically three different methods for soaking shellfish in disinfected sea water· This can be done by batch processing, in a continuous operation, or with partially renewed seawater mixed with the seawater in the troughs (providing the recycled water does not exceed 50%). For batch processing, troughs are filled with disinfected sea water, providing the total oxidant residual is not greater than 0.10 mg/l. It is discharged after a while and replaced at varying intervals by treated water. In practice, this frequency will depend on the season and the shellfish for depuration, but usually corresponds to 2 or 3 cycles per day. Before each refill, the shellfish are hosed clean with a strong jet of water. The time they remain immersed also depends on the species and their initial level of contamination. A minimum of 48 hours is required by the regulations.

In continuous processing, troughs are pre-filled then continuously supplied with treated make-up water. This extra input changes about 10% of the volume of water every hour and offers the twofold advantage of avoiding a drop in the dissolved oxygen rate and bringing some plankton upon which the mollusks can feed.

Treatments with partial recirculation are applied only when the soaking troughs are not fitted with a running water system (e.g. when they are installed in a place where the tide characteristics do not enable them to be continuously supplied with water). In such cases, the facilities are generally fitted with a big raw water storage capacity, or a buffer trough of treated water continususly feeding the soaking troughs. In periods of exceptionally low tide, the reserve is supplied by a portion of previously used water. This system is only permitted for under 50% of the recirculation rate and should not be generalized.

TREATMENT OF SEAWATER FROM QUANRANTINE OR STORAGE TROUGHS BEFORE DISCHARGE TO THE NATURAL ENVIRONMENT

In the case of imported shellfish for immersion in French waters, our legislation requires that the waters used for storage must be disinfected. The same oxidants are employed as for disinfecting the seawater used for storage and depuration. For safety reasons, and failing sufficiently precise knowledge concerning the behavior of parasites affecting shellfish, the dose rates applied are higher (Table 8).

Table 8. Recommended treatment conditions for quanrantine storage (according to I.S.T P.M., 1981).

Disinfectant :	Chlorine	Bromine	Ozone	UV
non-filtered trough water	10 mg/l	10 mg/l mg/l	0.5-2.5 allowed	not
Filtered trough water	5 mg/l	5 mg/l		
Filter backwashing water	20 mg/l	*20 mg/l		

*In the case of bromine, the disinfected filter backwash water must be diluted before discharge to the environment. Compulsory contact times are a minimum of one hour for chlorine and bromine, and from 10 - 15 minutes for ozone

To be in a position to cope with a possible failure of the disinfection system, a retention trough must be provided with a liquid capacity equal to the sum of the usable volumes of all the storage troughs. Additionally, it must be possible to empty such storage troughs within 2 hours (in the case of high mortality rates, massive spawning, etc).

In establishments where both foreign and French shellfish are treated, the respective soaking troughs must be absolutely separate or all the waste water discharged must be treated. The schematic contained in figure 5 shows the different possibilities for treating seawater, upstream or downstream, when disinfection is necessary. There are at the present time in France about thirty facilities treating water upstream of the soaking troughs and around fifty operating downstream water that has hosted shellfish from abroad. Although more establishments use chlorination than ozonation, ozone is noted to be the most widely used reagent in terms of the annual tonnage of shellfish treated. Bromine is used very little, mainly because of the dangers associated with its handling and the emanations of corrosive fumes.

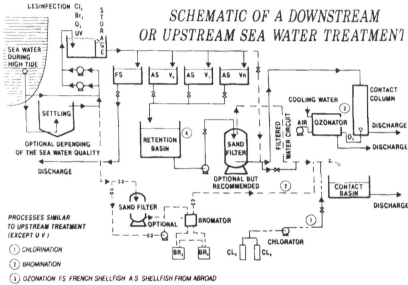

FIGURE 5. Schematic of downstream or upstream sea water treatment. Source: I.S.T.P. 1981.

DESCRIPTION OF SOME INDUSTRIAL FACILITIES

Among the plants constructed by Trailigaz, we have selectcd two different types:

1: The "Cote Bleue" shellfish depuration unit located on the edge of the Thau, open salt lake (Herault department, French South-West Mediterranean coast).
2· The Breville (Manche departement) storage and quarantine faclities, in North-West France.

1. Cote Bleue" facility

        This firm is located in the middle of a shellfish culture area in a sensitive zone, periodically subject to transient pollution   Figure 6 shows a general diagram of the water disinfection facility and distribution of the latter to the shellfish troughs. Seawater is taken from the Thau open salt lake about 50 meters from the sea front. Pumping at a rate of about 150 m^3/h, this water is disinfected with ozonized air. The ozone is produced from air with an 18-tube Monobloc type generator capable of producing 270g O_3/h, i.e., a maximum dose rate of 1.80g O_3/h/m^3 of water, at a concentration of 18g O_3/Nm^3 of air Contact between the ozonized air and the seawater takes place in a 3-compartment contact chamber. The first two compartments are fitted with porous diffusers designed to satisfy the immediate ozone demand of the seawater in the first compartment and reach a total oxidizing residual of 0.4 mg/l, expressed as ozone equivalent (contact time 2 min.). The residual is held in the second compartment for 4 minutes.

        A third compartment provides a contact time of 10 minutes reducing the total oxidant concentration from 0.4 mg/l to 0.1 - 0.2 mg/l. This final compartment is not fitted with diffusers. It is, however, equipped with a level switch controlling the pumping and seawater inlet functions, and eventually the ozonation process. The water height in the first two compartments is about 3.5m. In the third, this height is variable and the water inlets by a cascade that enables the dissolved ozone in the water to be partly stripped

        After having undergone this treatment, the water is fed to 8 soaking troughs, each with a capacity of 20m^3. They are 10 meters in length by an average 1 meter deep. The bottom slope is 2% to assist discharge of the suspended solids that settle during the soaking stage. Inlet ports spaced on the long side of the trough ensure regular distribution of the disinfected seawater. When necessary, the water can be emptied via a bottom drain located at the deepest end of each trough.

        The baskets carrying the mollusks are isolated from the bottom of the trough on longitudinal ledges. A baffle plate upstream of the bottom drain provides the possibility of emptying the trough via the surface and the bottom simultaneously. After use, the water is discharged to the open salt lake.

        As regards the types of oxidant by-products remaining in the seawater after ozonation, laboratory tests in conditions similar to those of ozonation on site, evidenced that bromates were not produced. This is not the case if a higher ozone rate is applied to bromide concentrations (Dore, 1989). This purification plant has been running since 1978 and treats about 550 metric tons of shellfish per year.

2. Breville Storage and Quarantine Facilities

        This plant established on the Channel coast is positioned to access excellent quality seawater, requiring no treatment   On the other hand, water from the troughs receiving imported shellfish has to be treated before discharge. This facility, built in 1983, with the Chamber of Commerce as sponsor of the works is described in Figure 7.

        Seawater is pumped at a depth of around 6 meters at about 900 meters offshore. Three pumps discharging 120 m^3/h supply two storage capacities (volume per unit: 500m^3). The water is then distributed as required between twenty troughs in which mussels, oysters, cockles, clams, whelks (Buccinum undatum) and winkles (Littorina littorea) are stored. Some of the troughs also serve as breeding ground for crustacean shellfish (crayfish, lobster, crab, and sea-spider).

        The troughs are rented out to seven different operators. The chamber of commerce manages the pumping, distribution of seawater and treatment required before discharging the waste water from imported shellfish storage capacities. Water from the breeding grounds and storage troughs in respect of French shellfish is discharged to the sea without treatment.

INSTALLATION DE LA COTE BLEUE

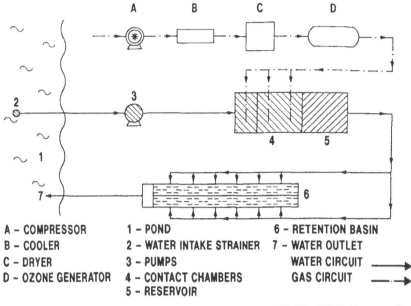

A – COMPRESSOR   1 – POND   6 – RETENTION BASIN
B – COOLER   2 – WATER INTAKE STRAINER   7 – WATER OUTLET
C – DRYER   3 – PUMPS   WATER CIRCUIT ⟶
D – OZONE GENERATOR   4 – CONTACT CHAMBERS   GAS CIRCUIT ⟶
                           5 – RESERVOIR

GENERAL DIAGRAM OF THE SEA WATER TREATMENT

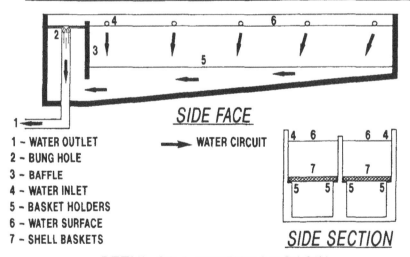

SIDE FACE

1 – WATER OUTLET   ⟶ WATER CIRCUIT
2 – BUNG HOLE
3 – BAFFLE
4 – WATER INLET
5 – BASKET HOLDERS
6 – WATER SURFACE
7 – SHELL BASKETS   SIDE SECTION

DETAIL OF A RETENTION BASIN

FIGURE 6. Installation de la Cote Bleue. (Source: Fauvel et al. 1978.)

LES VIVIERS DE BREVILLE SUR MER

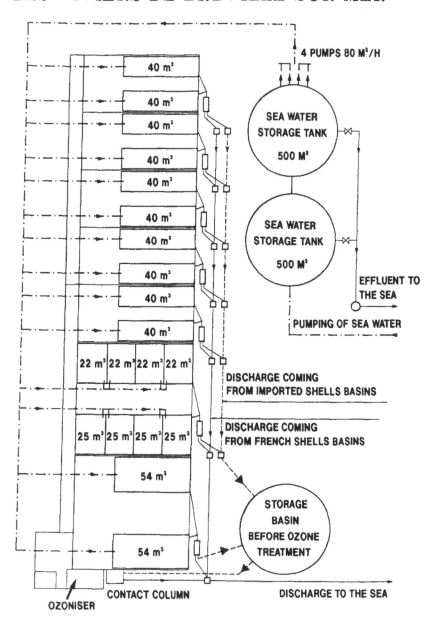

FIGURE 7.   Breville storage and quarantine facilities.

The ozonation plant consists of a 12-tube "ozobloc" type generator, with a maximum output of 216g O_3/h at a concentration of 18g O_3 /Nm^3 of air The contact column in which disinfection takes place consists of two compartments The first of these has a useful volume of 7.15 m^3 and the second 14 3 m^3 The height of the water in the column is 3 meters For a treated flow rate of 80 m^3/h, the residential time in the contact chamber is about 15 minutes with a maximum applicable dose rate of 2.7g O_3/m3 water. After treatment, the water is sent to a discharge main collecting all the waste water from the breeding grounds and other soaking troughs The pumped seawater and effluent are regularly analyzed for total and fecal coliforms, and fecal streptococci.

PERSPECTIVES OF A SINGLE EUROPEAN MARKET

The various countries of the European Economic Community have different attitudes toward shellfish. Spain, France, Holland and Italy are big producers, representing intensive farming activities or the organized management of natural resources The other countries, including Germany, Denmark, Ireland and the United Kingdom are mainly the seat of fishing activity, while culture as such plays only an accessory role. The consumption of live shellfish concerns mainly the countries of southern Europe, and above all France.

The principal producing countries adopt different strategies with regard to marketing principles and channels. For instance, Spain and Italy have opted for systematic depuration in the case of bivalve filtering mollusks, while France and Holland aim at limiting production to zones that are recognized to be salubrious. It still remains that in this field there is a strong tendency to increase outputs, including zones that are not recognized to be sanitary, giving preference to economic aspects.

In order to reach a perfectly accepted common objective of delivering shellfish fit for human consumption, concepts may differ, depending on whether the country concerned is favorable to a "Policy of Objectives" or a "Policy of Means" In this context, the commission of the European Communities must harmonize their national practices and regulations A think tank initiated in 1987, led to the proposal to draft two European regulations on shellfish One concerns the health conditions applicable to the production and marketing of bivalve mollusks and the second deals with sanitary regulations enforced by the police for the transportation of live shellfish (as an integral factor in all fishing and aquacultural products)

The objectives pursued are intended both to protect consumers and safeguard the health of the livestock. The means contemplated are:

1. To invite each member state to develop a multiparameter system for monitoring the quality of the marine waters used for shellfish raising (physical, chemical, biological and microbiological pollutions), with frequent sampling operations providing regular follow-up EEC Guideline No 79/923 of 30 October 1979, published in the EEC Journal Official (EEC Official Gazette) has already been instrumental in impoving the control of biotic parameters in recognized oyster breeding zones, in order to avoid degradation of the natural environment to the detriment of the physiology and growth of bivalve mollusks (EEC, European Economic Community)

2. To refuse a permit to market products that do not have a health level corresponding to a fecal coliform count for thermotolerant coliforms under 300/100mg of shellfish meat, and this however the result is obtained, when the products originate in zones affected by curable bacterial contaminations (<6,000 thermotolerant coliforms/100 mg of shellfish meat) It has been remarked that in the case of chemical or biological contaminants, cure requires very long-term treatments often prohibitive because of the additional cost involved.

3.   To lay down that the shellfish must pass through establishments capable of providing storage and handling facilities in optimum health conditions from which they will be directly delivered to the distribution networks

4.   To make the latest handler fully responsible for the quality of the products sold, leaving it to the latter to bring proof of his innocence and of the possible responsibility of his own supplier Thus every link in the chain is obliged to become fully aware of the problem of quality and guarantee the entire shellfish network,

5.   To regulate the conditions under which live bivalve mollusks can be transferred when they are to be immersed for culture in a zone free from a particular disease or infection. Such prospects are founded on the principle that certain diseases must be declared, as proposed by the International Office of Epizootic Diseases. This office is currently working on complete regulations on filtering bi-valve sea mollusks according to a prophylactic scheme recommended for fishes and land animals

Items 2 to 4 above are a convincing argument in favor of firms adopting structures capable of performing self-checking analyses. These principles do not diverge to any extent from those allowed in France, but it is still true that the change of scale induced by the globalization of shellfish exchange should lead to a tightening of the rules applicable in our country, especially as regards the police control to be applied   In addition, improved knowledge of the exchange of shellfish between the production media should also allow a good appraisal of production levels

Concerning prospects in the field of seawater treatment, several small firms that currently only process their own production or those in the immediate vicinity must be expected to acquire a means of disinfecting the water Seawater admitted to storage and finishing capacities must be treated if it is pumped from areas subject to slight but chronic or sudden accidental pollutions The use of disinfected water should enable the packaging and dispatching firms to work continuously with healthy shellfish whatever the quality of the water used.

RECOMMENDATIONS

It is obvious that all the very small dealers (9 4% of French authorized firms alone deliver more than 10,000 packages yearly, i e., an output of 130 metric tons per firm), should form groups in order to avoid disproportionate financial commitments.

The large existing or future firms will grow in importance and will make it their duty to treat their water upstream of the troughs to avoid the risk of shutdown due to accidental pollution in the pumping zone, thus ensuring uninterrupted operations. Moreover, these establishments with several storage capacities can if necessary be transformed into "shellfish depuration units" able to treat products from non-approved origin.

With regard to live shellfish from abroad, the French authorities, that have already tightened up their system of imported shellfish control, want to maintain the "STORAGE" and "QUARANTINE" systems, both for immature foreign specimens for culture in French waters, and fully grown specimens (of marketable size) considered as foodstuff and open to trade.

Public health protection, always the absolute priority, and the protection of the livestock both deserve careful attention as to the reliability of the treatment techniques applied and the respect of good sanitary practices. To achieve this standard it is obvious that the full automation of disinfecting processes gives optimum assurance In addition, the shellfish after depuration must be kept in a proper physiological condition, above all by a maximum dissolved oxygen content in the water. Disinfection with ozone is alone capable by itself to fulfil these two objectives.

REFERENCES

Blogoslawski W.J. 1983. Personal communication, based on presentation, Influence of water quality on shellfish culture. Presented at the International Council for the exploration of the Sea. National Marine Fisheries Service, NE Fisheries Center, Milford, CN, USA

Carpenter, J.H. and Macalady, D.L. 1978. Chemistry of halogens in seawater In "Water chlorination. Environmental Impact and Health Effects", ed. R.L Tolley, Vol. 1, Ann Arbor Science, Ann Arbor, Michigan, U.S.A.

Fauvel, Y. (other authors not provided) 1978. Seawater ozonation and shellfish depuration. Presented at the 2nd Workshop on marine and freshwater ozone applications Ozone: Its use in aquatic applications, I.O A Nov. 1-3, Orlando, FL, USA.

Galstoff, P.S. 1946. Reaction of oysters to chlorination. US Department of the Interior, Fish and Wildlife Service, Research report 11, 1-28 US Government Printing Office, Washington DC. USA

I.S.T.P.M 1981 Note Technique relative aux Equipements des etablissements ages pour l'immersion de coquillages etrangers. Service des controles sanitaires et techniques. Departement cultures marines Nantes, France.

Morel, A. (additional authors not listed) 1989. Purification des moules dans l'eau de mer des Infecteé aux UV, IFREMER, 89-028 Nantes, France.

Todd, J.J. 1978. Fundamental Characteristics of aeration systems - Water Service, 82-993, 771 and 776.

OVERVIEW OF MOLLUSK INDUSTRY AND DEPURATION IN DENMARK WITH SPECIAL REFERENCE TO OYSTERS

Grete Ellemann

HISTORICAL REVIEW

One of the few internationally known Danish words is "køkkenmødding". It cannot be translated as it is not just another word for a rubbish heap, but is unique to Denmark's early stoneage, especially the period from about 4.600 to about 4 000 Before Christ (B.C). During this time, the inhabitants were living by hunting and fishing and were settled near the coasts of the country In this period, the weather in Denmark was warmer and the sea was saltier. These conditions were advantageous for oysters, specifically the flat oyster (Ostrea edulis) As early as 1848 the "køkkenmødding's" were objects for archaeological studies that found oysters had been a fundamental nutrition in the stoneage. One of the most well known "køkkenmødding's" was found near the Limfjord in North Jutland at a place called Ertebølle and the period and the culture from these nearly 600 years is called the Ertebølleculture. The heap found at Ertebolle measures 2800 sq. meters with a depth of 1.90 meters. Estimates suggest that 20 million oysters have been consumed and deposited in this enormous heap. About 3.800 B.C. living habits changed as the population gradually became adapted to farming, and fishing became just one way amongst others to provide food.

Frederic the Second, King of Denmark, in 1587, wrote an open letter declaring "That he had been informed that a sort of fish called oysterling was found and could be fished in the North Sea off the coast of Riberhus estate and as he liked this "fish" very much ordered the vassal of the estate to let these oysters be fished and send to him. Only at the kings demand was oyster fishing allowed in this area. In this way the oyster fishing had become a royal prerogative. And it has remained so for about 300 years until State administration began in 1860.

Oyster beds have existed not only at the west coast of South Jutland, but also near Skagen and Frederikshavn. Anyone who intended to fish oysters had to pay a certain fee to the king as well as it was his duty to supply the king and the royal court with oysters. When the North Sea in 1825 broke through the tongue of land which separated the Limfjord from the sea in west, oysters settled at various places By 1840 the fishermen began selling their oyster harvests, but in 1851, referring to the order of 1587, it was stated that the Royal prerogative was still in force and shortly after this decision the oyster banks were leased to various fishermen for a certain yearly fee including the delivery of oysters to the royal family

In 1910 the Limfjord Oyster Company leased the banks in the Limfjord and some banks in the Southern part of the North Sea from the government For the first of a 5 year period 19 million oysters [units] were fished and sold, but in the 1940's production dropped to 100 thousand oysters [units] For the last 10 to 20 years the temperature has decreased so as to hamper natural oyster breeding. Oyster seed had to be imported, mostly from France, and relayed on beds for two years of growth to produce commercial size. In 1984, as the fees went up and the amount of oysters decreased due to cold winters and oyster diseases, the Limfjord Oyster Company denounced the contract with the state.

For several years, the oysters after fishing were relayed in tanks near the town Nykøbing in order to clean themselves of sand and gravel before they were packed and distributed mostly to Copenhagen and other bigger cities. But when the population in Nykøbing grew and the amount

of effluent from the city increased, new indoor facilities with depuration tanks were built at the utmost western part of the Limfjord
where the sanitation of the water was much better. Bacteriological investigations of water samples and samples of the ready to eat oysters were performed monthly during the oyster season The expenditures for relaying, bacteriological control, and other associated depuration costs increased the price for oysters year by year up to about 12 kr. ($1 50) apiece. And so it has happened that from the elder stoneage up to now the oysters have changed from being every mans daily food to a luxury which only rich people can afford. Eating 12 or 18 oysters once or twice a year has become a mere luxury in Denmark. And therefore, quotes like Sam Weller's who said "It is a very remarkable circumstance sir that poverty and oysters always seem to go together (Charles Dickens "Pickwick Papers")", can no longer refer to Danish eating habits.

DEVELOPMENT SINCE 1970

For several years the oyster trade was supplied with oysters imported from Holland and France when the Ministry of Trade in 1970 decided to release all importation of oysters. This decision caused the Fish Inspection Service to address the National Health Organization as it was well known that oysters might give rise to severe illness among which the most severe concern is for the epidemic Hepatitis-A virus. The research and health literature is profuse with scientists from many countries who have studies denoting the relation between the consumption of raw shellfish and outbreaks of various types of food poisoning. For these reasons, negotiations with the Danish Health authorities resulted in issuance of an order stating that any imports of oysters from any source have to be reported to the Fish Inspection Service, and an import license has to be issued under the conditions that each batch after import has to be relayed in indoor depuration facilities approved by the Fish Inspection Service. Such product will not be released before the result of subsequent bacteriological investigations have been approved

For several years only the Limfjord Oyster Company had access to depuration tanks and therefore this company was the only importer of oysters Since 1984 when the company denounced their contract with the State, their depuration facilities have not been in use

New depuration facilities have been built partly for relaying of imported oysters and in anticipation of oyster cultivation Since oyster diseases during the 1980's nearly eradicated the Ostrea edulis in France and Holland, mariculture of the more resistant Crassostrea gigas has resulted in Denmark. The oysters are grown in different sorts of cages and even if the Danish coastal water is too cold for the oysters to breed they have shown good conditions for growing The C gigas bred does not require as much salt in the growing waters as the Ostrea This situation direct choice for placing the cultures Because of these new oyster cultures, a new regulation was issued in 1984 after which all oysters imported or of Danish origin have to be relayed to indoor depuration facilities before they are offered for sale. These facilities are privately owned Bacteriological investigations of oyster samples and water including total counts, E. coli and Salmonella have to be performed. The samples are drawn by the fish inspector. The investigations must be performed by an approved laboratory, usually the local municipal food inspection service. The results are reported to the owner of the oysters as well as to the Danish Fish Inspection Service, who after evaluating the results will release the oysters for sale This procedure takes about one week which means that the depuration process should be long enough for the oysters to clean themselves of potentially pathogenic microorganisms. If the bacteriological results are not satisfactory, new samples are drawn before the batch can be released. The expenses for these investigations fall on the owner of the oysters.

The depuration facilities usually include two tanks with separate water supply Only one batch is allowed at a time in each tank When all oysters from the tank are sold it must be emptied of water and cleaned before a new batch can be relayed Bacteriological investigation of water samples takes place each time after refilling the tank The salt content of the water must be at least

1.2 percent. The water is pumped through a filter after which it passes through ultraviolet lamps. The water is aerated and the water supply is installed over the basin while the outlet is placed in the bottom. Following the instructions given in the publication from The English Ministry of Agriculture Fisheries and Food, by P. A. Ayers, Laboratory Leaflet no. 43, Lowestoft in 1978 recirculation of the water can be allowed. This system must always include installment of ultraviolet lighting for sterilizing the water.

In order to make a sufficient cleaning possible, the tanks must be elevated from the floor and placed at a certain distance from the wall. Of great importance for a good water hygiene is a careful cleaning of the oyster shells before relaying in trays. These trays must be constructed with holes big enough to let the water flow unrestricted. Only one or two layers of oyster are used in each tray in order to allow adequate cleansing. The temperature must be about 8-10°C which is the optimum for the activity of the oyster. Dead oysters have to be removed daily or more often if necessary. Before a depuration plant is taken into use, the building, the tanks, and the water supply must be approved by the Danish Fish Inspection Service.

Presently no case of severe diseases including epidemic Hepatitis-A virus has been reported. As long as this virus is not easily detected by routine laboratory investigations the more simple bacteriological investigations combined with a depuration treatment of 7 days length has been the insurance of a safe product.

AN OVERVIEW OF MOLLUSCAN SHELLFISH INDUSTRY IN TURKEY

Atilla Alpbaz and M.O. Balaban

INTRODUCTION

Turkey is surrounded by the Black Sea in the North, by the Aegean Sea in the West, and by the Mediterranean in the South. The length of coastline exceeds 8,000 km. Turkey produces about 625,000 tons of seafood annually. The production of molluscan shellfish has reached 25,000 to 30,000 tons annually. Shellfish, especially bivalvia is not a traditional food in Turkey. Only mussel, Mytilus galloprovincialis is consumed to any significant amount in large cities of western Turkey such as Izmir and Istanbul. Of the 55 million population in Turkey, it is estimated that no more than 1 % consume mussels regularly. Until recently, there was no interest in bivalvia in the domestic market.

A very significant rise in the shellfish demand from abroad, especially from shellfish consuming countries such as Spain and Italy changed the situation. While the toal shellfish production in 1979 was around 200 tons annually, commercial exports to Europe accelerated shellfish production. The yearly production between 1985 and 1988 is shown in Table 1 More than 99 % of this harvest is exported. Also, "rapana", which is a genus of gastropoda, is harvested from Black Sea for export. Its production reaches 10,000 tons annually It is estimated that between 2000 and 3000 families earn a living by shellfish harvesting.

Table 1 Shellfish species and their annual harvest between 1985 and 1988

Species	Harvest (tons)
Mussel (Mytilus galloprovincialis)	7,000 - 8,000
Short necked clam (Tapes decussatus)	400 - 500
Oyster (Ostrea edulis)	5,000 - 7,000
Horse mussel (Modiolus barbatus)	300 - 500
Kidonya (Venus verrucosa)	50 - 100
Lupina (Venus gallina)	8,000 - 10,000
Arca spp. and other Venus spp.	50 - 100

SHELLFISH SPECIES HARVESTED IN TURKEY

1. Black Mussel (Mytilus gallprovinialis) :

Black mussel is found from Izmir to the Black Sea, all the way to the border with the Soviet Union. It is an abundant species Its export is difficult, because mussel consuming countries have a large and established mussel fishery. For example, Spain produces 210,000 tons of mussels annually, France produces 60,000 tons, and Holland 100,000 tons. Therefore, the export price of black mussel is not high Since there is no significant domestic market, the increase in production

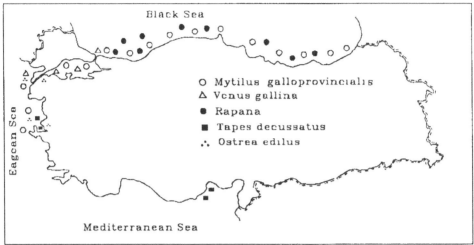

Figure 1. Shellfish production in Turkish waters.

of mussels in Turkey depends on an increase of foreign demand, with acceptable market prices
There was a mussel culture operation in Marmara Sea; however, due to the export problems and since the cost of naturally harvested mussels were 2 to 3 times lower than those of cultured mussels, the culture collapsed. Today, the potential of mussel harvest of wild stock or mussel culture is very high in Turkey, but commercialization depends completely on foreign demand

2. Horse Mussel (Modiolus barbatus L) :

The horse mussel is similar to black mussel. The shell is more convex on the ventral side, and the meat color is more pinkish. Their market value is 3 to 4 times that of the black mussel. This species is found from Izmir to Canakkale. It is especially abundant in Ayvalik bay, where the waste water of olive oil factories enhances phytoplankton activity. The mussels obtained from Ayvalik bay have a very good quality meat. There is an annual harvest of about 300 to 500 tons from Ayvalik,
and most of it is exported. About 100 fishermen make a living by harvesting mussels in Ayvalik, by diving or by simple trawling from small boats. These are boats about 6 m long, with 9 to 10 hp engines. The demand for this type of mussel is high, and there is no difficulty in exporting it

3 Short necked clam (Tapes decussatus L.) :

This is the most valuable clam harvested from Turkish waters   About 500 tons are harvested mainly from Izmir, Mersin and from other locations along the Aegean coast   The price is the highest among all the bivalvia exports from Turkey. The current harvest price is $5(U S) per kg and the market price is around $9(U.S.) per kg. This higher price is jutified by the fact that its harvest is difficult The harvest season is winter, when wheather conditions are rough and cold The clam lives in a muddy bottom It can only be harvested by a special showel Its harvest is forbidden in summer to provide a safe spawning season

4 Hard clam (Venus verrucosa L.) ·

This clam is harvested together with the short necked clam, by the same fishermen. There is no special market for this clam. It is exported together with the short necked clam. It commands a low price. Its meat is as good as the short necked clam. There is a small domestic market for restaurants catering to tourists. The rest is exported.

5. Clam (<u>Venus gallina</u> L)

Harvest of this shellfish from Turkish waters is a new practice It began 4 to 5 years ago The harvest reached nearly 10,000 tons annually. Clams are exported, either alive or after processing There is one large plant processing the clams according to customer specifications This species is very abundant in the Marmara and North Aegean regions Since it lives in a sandy bottom, its harvest is easy The boats are specially designed for this harvest After harvest, live shipments are transported by truck. Since most of the trucks are not refrigerated, this causes problems during the summer months. Most of the summer harvest is used in processing plants.

This is a small shellfish Its external color is cloudy white. The inside of the shell is bright white. The morphological data about clam are given in Table 2 The average meat of this clam is about 3 g Its total average weight is 9.8 g Therefore it can be concluded that this species is the smallest shellfish of commercial importance from Turkish waters

Table 2 Morphological data about clams (<u>Venus gallina</u> L.)

Data	N	Average	St.dev.	Min	Max
Length (mm)	40	26 8	2.6	15 5	31 3
Width (mm)	40	29.9	3.3	16.8	36 0
Thickness (mm)	40	15 7	1 3	18 3	19 0
Weight (g)	40	9 1	2 0	5 7	13 8
Meat weight (g)	40	6 2	1 4	4 1	9 4

6. Oyster (<u>Ostrea edulis</u> L.) :

This species is also found in abundance in the Aegean and in the Canakkale region There is high external demand for this species. The European customers prefer oysters harvested from the Izmir area. Those have good meat and yellowish color The oysters originating from Canakkale are more suitable for canning. They are not very durable, lose water and perish during transportation. The morphological data about oysters are given in Table 3

In one experiment in the lagoons of the Ege University, <u>Ostrea gigas</u> was grown from 0.5 cm to 8 cm in 8 months. There are great potential opportunities for the aquaculture of oysters in the Turkish waters, especially in the Aegean coast.

Table 3. Morphological data about oysters (<u>Ostrea edulis</u> L.).

Data	N	Average	St dev	Min	Max
Length (mm)	110	90.0	13.4	57 0	160 0
Width (mm)	110	72 8	9 8	53 3	97 0
Thickness (mm)	110	37 3	7 7	21 0	60 0
Weight (g)	110	148.6	57 3	46 0	361 0
Meat weight (g)	110	17.4	6 6	2 2	34 2

CONCLUSION

The production and harvest of shellfish from Turkish waters is expanding. Most of the production is exported. Therefore, the quality and safety of the shellfish is very important Proper harvesting and depuration methods, together with a clean and healthy habitat for the shellfish will ensure a growing market There is great potential for the culture of shellfish. There are 26 marine lagoons with a total area of 40,000 hectares. Especially oysters and short-necked clams are very suitable for culture in these lagoons.

Potential difficulties in exporting shellfish from Turkey can be summarized as follows :

1 - There is only one modern depuration unit for use in the export of live shellfish. This unit is in Canakkale. More depuration installations in other cities are needed to satisfy the European market.

2 - Harvesting methods are not modern Most of the fishery relies on manual labor. More research on the efficiency, cost and effect on quality of mechanical harvesting methods needs to be performed.

3 - Transport of live shellfish is a delicate operation Reliable transportation by air is a necessity

4 - There is no modern packaging operation This is an important point that needs to be addressed relative to quality and safety assurance

BIVALVE SHELLFISH DEPURATION IN ITALY BACKGROUND AND CURRENT STATUS

Walter J Canzonier

In Italy there has existed legislation referring to depuration of bivalve shellfish since 1929 (Legge 1315, 4 July l929)[1], this is reiterated in additional shellfish sanitation legislation of l934 (Legge 1265), 1962 (Legge 283) and 1963 (Legge 441) This older legislation, however, did not specifically require depuration of shellfish, rather, it permitted its use in cases where the provincial health officer felt that it might have beneficial effects as regards the public health. There is no mention in the regulation as to exactly how the plants were to be built or operated, presumably this was left to the discretion of the provincial authorities Mention is made , however, concerning the acceptability of using closed systems, including the use of artificial seawater (provided the salt and water used were clean) and the water was changed at least once every 24 hours   The reasons for this singular detailed provision are probably lost to history, but it would have favored an existing plant that operated with this type of system. Apparently, some depuration/holding facilities were operated in the 1930s as implied by Petrilli (1938, 1941) in reports on some experimental depuration performed between 1937 and 1941. As an aside, it is interesting to note the insight of this author concerning some problems of depuration and basic concepts of molluscan physiology.

In 1973 Italy experienced a minor, but widely publicized, cholera epidemic in the south central portion of the country (the regions of Campania and Puglia) (Baine et al l974). Mussels were blamed and this led to the immediate embargo of all shellfish harvesting and shipping (Decreto Ministeriale of 4 Sept. and 6 Oct. 1973)[1] This was followed by introduction of updated shellfish sanitation legislation (Disegno Leggi CD No 2458 of 30 October l973)[1]. This legislation was revised in 1977 and voted into law on 2 May l977 (Legge 192). This law covers all aspects of harvest and commercialization (distribution/processing and marketing) of bivalve shellfish. Depuration was included in this new legislation as an obligatory component of shellfish marketing. Subsequently, regulations and protocols were developed by a selected commission of shellfish specialists and microbiologists. These rules,specifying the details for the application of the process, were partially published at the end of 1978, followed by addenda in several successive years up to l987. This legislation is an unusual departure from the typical health related laws in force, which placed the burden of surveillance, control and certification of the end products on the local and provincial health officials. The new concept places the major burden for both surveillance of the plant operation and end-product quality assurance on the plant operator, with rather severe criminal penalties for violations or even apparent violations (e g. shipment of a depurated product, one sample of which does not meet the requisites of the regulations) The role of the local and provincial health officers is then reduced to an oversight operation, involving periodic plant inspection and spot sampling of the finished product. Classification of the growing waters does, however, remain the responsibility of the provincial health officials. The day-to-day sampling of

---

[1]Legge Law of the Italian Republic, enacted by the national parliament and numbered consecutively within each year.

Decreto Ministeriale: Decree of a ministry of the central government, often issued as supplement to a Legge to define and amplify its technical aspects

Disegno Leggi: Interim or draft legislation, presented by members of the Chamber of Deputies (CD) or the Senate (S) and numbered consecutively within each year

All of the above can be found in the Gazetta Ufficiale Italiana, available in most large law libraries, listed by number and year.

process water and shellfish batch samples is the responsibility of the plant operator, utilizing the services of a certified in-house or contracted laboratory The new law is certainly more flexible than the previous; however, it assumes a maturity and integrity on the part of the industry that may not be justified.

This legislation requires that all "depurable" bivalve mollusks destined for retail sale in the raw state must be depurated in an approved plant/shipping center. Prior to the enactment of this legislation, a few commercial depuration plants had been constructed and operated in the northeast (as early as 1967 at Pellestrina). These mainly served as wet storage systems for the wholesale dealers and distributors. In the absence of appropriate guidelines and controls, they were probably ineffective as depuration facilities at best.

With the publication of the regulations pertaining to the design and operation of depuration plants, pressure was brought to bear on the operators to upgrade their facilities Considerable effort was exerted by the Ministero della Sanitá (Health Ministry), under the guidance of Dr Francesco D'Alessandro, to force the operators, as well as the various public health agencies responsible for supervision, to upgrade the harvesting and processing of shellfish, with special emphasis on depuration. To encourage the renovation of old plants, and the construction of new facilities, generous government subsidies were available. Funds were also made available by the government to finance research on several aspects of depuration that required clarification (Canzonier et al , 1980,a, b; Marcucci et al. 1980; Casali et al. 1981).

Earlier research on various specific microbiological and physiological topics relevant to depuration had been conducted by several workers starting as early as the 1930's (Bovo and Berzero, 1976; Cabassi and Perna, 1974; Ceredi, 1937; Crovari, 1958; Leccese 1958; Luchetti, 1966, Petrilli, 1938, 1941; Paoletti et al. 1968; Pagano et al 1970; Santopadre and Giunti, 1955) There was particular emphasis on to the blue mussel. More recently, the Ministero della Sanitá has financed work on the uptake and elimination of polio virus and a comparison of persistence of detectable loadings of Hepatitis-A and bacterial indicators in mussels (Pluchino, 1989)

There are currently more than 50 plants authorized to operate as depuration facilities; however, not all of these may be active. Most plants utilize a flow-through system, treating the water with some form of active chlorine, often followed by catalytic hydrolysis of the residual chlorine by passage through activated charcoal columns. In some plants there is an addition of iodophores prior to entry of the seawater into the distribution pipelines Large pressurized sand filters may be used to reduce the turbidity of the incoming seawater. A small number of plants utilize ozone for the primary treatment of the process water; (e g the municipally operated plant at Fano (Piccinetti, 1989).

The larger plants are usually quite sophisticated, with complex automated water treatment and distribution systems. The largest plants may have a capacity in excess of 200 tons per cycle and water intake and treatment systems that handle over 200 cubic meters per hour Some plants rely on seawater supply systems that have intakes in approved growing areas more than two kilometers from the facilities (e.g., four plants in Chioggia).

Conversation with several shellfish dealers as well as public health officials and shellfish specialists in October of 1989 has indicated some currently perceived inadequacies of the process of depuration as applied in Italy Most frequently mentioned by the producers is the added cost of applying the process for all "depurable" species as prescribed by the regulations This was reiterated by one very knowledgeable shellfish biologist (Piccinetti, 1989) as being the major stimulus for circumvention of the process by the industry. Indeed, the practice of circumvention greatly clouds the epidemiological data as regards certain shellfish associated diseases such as Hepatitis-A.

Another disconcerting aspect of the application of the process on a country-wide scale is the considerable variation in plant design and operational protocol This is compounded by local variations in enforcement that places an inequitable burden on the operator that is trying to follow the regulations Public health officials (Pluchino, 1989) feel that there was definite need for improvement as regards uniformity of design criteria and enforcement of operation protocol

One technical shortcoming that has been uncovered by recent studies financed by the Ministry of Public Health is the need for rather high residual chlorine levels (as high as 0 9 ppm) for rather long retention times (up to an hour) in order to insure acceptable inactivation of enteroviruses such as Hep-A. These studies also indicate the inadequacy of iodophores to inactivate enteroviruses under the current operating procedures, though excellent reduction of the coliform indicators is achieved with this agent . Though they do not appear to be as effective as free chlorine in the primary treatment of the process water, there is the feeling that iodophores have a role as a secondary additive to maintain water quality within the holding tanks. There has been a proposal by technical administrators within the Ministry to increase the required titre of chlorine residual and/or the retention times, but this has not been incorporated into the regulations

There seems to be considerable skepticism concerning the use of ultraviolet light (UV) for primary treatment of the process water The major objection is the perceived need for "absolutely limpid" source water This reservation concerning the limitations of UV disinfection of the process water seems to be related to the findings and conclusions of only one study (Pagano et al 1970), with little recognition of the information available from other studies that used more appropriate systems of irradiation

ACKNOWLEDGEMENTS

The author thanks the numerous public health officials and research scientists for supplying technical information on various aspects of depuration as conducted today in Italy I am also indebted to the commercial shellfish dealers for the opportunity to visit their depuration facilities and for furnishing their opinions on the application of the process I am especially appreciative of the efforts of Dr Michele Pellizzato for supplying background information and in providing assistance in contacting several plant operators in the Veneto

REFERENCES

Baine, W. B., Zampieri, A., Mazzottil, M , Angioni, G , Greco, D., DiGioa, M and Izzo, E. 1974 L'epidemiologia del cholera in Italia. Atti Sem. Intl.: Diffusione e Trattamento dell'Infezione Colerica. The epidemilogy of cholera in Italia [Proceedings of an International Seminar on the Diffusion and Treatment of Cholera Infections]. Instituo Superiore della Sanitá Roma, Apr. 1974, pp 9-27

Bovo, F. and Berzero, R 1976 Prove di depurazione dei molluschi [Mollusc depuration trials] Igiene Moderna 69 163-184

Cabassi, E. and Perna, A 1974. Indagini batteriologiche di campioni di lamellibranchi raccolti nelle acque del litorale della regione Abruzzo nei mesi di settembre e ottobre del 1973 [Bacteriological studies on samples of lamellibranchs collected from the coastal region of Abruzzo in the months Sept and Oct. of 1973]. Nuova Veterinaria 50 253-258.

Canzonier, W J , Casale, P., Marcucci, M C., and Poletti, P. 1980a Depurazione di Venus gallina Report submitted to the Min. Marina Mercantile Italiano, Jan 1980 36 pp plus 16 figs.

Canzonier, W J , DaRos, L., Marcucci, M C. and Casale, P. l980b Depurazione dei Molluschi Bivalvi ms proj report of COSPAV, Coastal Resources Applied Res Lab , Chioggia to Min Sanitá Italiano, Nov l980 14 pp plus 21 Tables

Casali, P , Marcucci, M.C , and Canzonier, W.J 1981. Depuration of the clam, Venus gallina, in a commercial plant. Igiene Moderna 75:669-692.

Ceredi, A. 1937. Molluschi eduli e infezione tifoidea [Edible molluscs and typhoid infection]. Igiene Moderna 30:210-219

Crovari, P. 1958. Sulla depurazione dei mitili nei confronti del virus poliomielitico [Depuration of mussels as regards the polio virus]. Igiene Moderna 51: 22-32.

Leccese, A. 1958. Attuali sistemi di bonifica dei molluschi eduli. [Current systems for improving the sanitary quality of edible molluscs]. Igiene Moderna 51.376-384

Luchetti, G. 1966 Un semplice impianto di lavaggio e disinfezione dei molluschi marini di allevamento [A simple plant for the washing and disinfection of cultured marine molluscs]. Riv. Ital. Igiene 26:670-674.

Marcucci, M.C., Casali, P , DaRos, L and Canzonier, W.J. 1980. Rapid and sensitive method for monitoring the depuration of bivalve mollusks Igiene Moderna 74 393-402

Pagano, A., Braga, A., and Paolucci, S 1970 Ricerche sulla depurazione dei mitili in un impianto sperimentale a raggi U.V. [Research on the depuration of mussels in an experimental plant utilizing ultraviolet radiation] Igiene moderna 63·289-307.

Paoletti, A., Ferro, V , and DeSimone, E. 1968 Problemi igienici dei mitili: Moderni concetti di ecologia e fisiologia quali fondamenti di una efficace profilassi delle malattie da essi trasmesse. [Problems of mussel hygiene Modern concepts of ecology and physiology tht are the basis of an efficacious prophylaxis for the diseases transmitted by these animals]. Ann. Sanit. Pubblica 29. 1033-1072 .

Petrilli, F.L. 1938. Ricerche sull'autodepurazione dei mitili. [Research on the self-cleansing of mussels]. Igiene Moderna 3l 309-324.

Petrilli, F.L. l941. Ulteriori ricerche sui mitili con speciale riferimento ai metodi di controllo della loro salubritá. [Further research on mussels with special reference to the methods used to check their sanitary quality] Igiene Moderna 34 275-285.

Piccinetti, C. 1989. Personal Communication, Oct. 1989, Lab Biologia Marina di Fano, Fano 61032 Italia.

Pluchino, G. 1989. Personal Communication, Oct. 1989, re. unpubl. reports of research conducted for the Health Ministry and current status of regulations. Ministero della Sanita, Roma 00144 Italia

Santopadre, G. and Giunti, G 1955 Osservazioni preliminari su di un procedimento di purificazione dei frutti di mare mediante stabulazione [Preliminary observations on the purification of marine invertebrates by the process of wet-storage]. Igiene e Sanita Pubblica 11·382-389.

Index
_____

INDEX

A

Adenoviruses, enteric, 76
Aeration requirements, 130
Aeromonas hydrophila, 47
Agency for Agricultural Research and Development (Indonesia), 315
Algae chemostat, 201, 202
Algae culture system, design of for Rangia clams, 199
Algae growth chamber, design of, 200
Allergic reactions, to shellfish, 53
American Public Health Association methods of, 8, 151
Amnesic shellfish poisoning, 51, 52—53, 239
Anadara granosa
in Malaysia, 311
market for in Singapore, 309
Analytical Laboratories of the Health Commission (Australia), 292
APHA method, 185—186, 192
ASEAN countries, 287, 308—319
Association of Official Analytical Chemists (AOAC) methods, 151, 220, 221
protocol of, 220, 221
Astroviruses, 76
Australia, shellfish depuration in, 287—290
future of, 306—308
introduction of, 290—296
practice of, 296—305
Australian oysters, microbiological history of, 59—60

B

Bacillus cereus, elimination of, 65
Backwashing operation, 199—201
Bacteria
depuration of, 115
elimination of compared with virus elimination, 102—103
inactivation by irradiation, 251—252
ionizing irradiation effects on, 255
in molluscan-borne illness, 47—49
oyster density effects on, 63
in oyster feces, 62
in oysters, 228—230
rates of elimination of from polluted shellfish, 61
in shellfish depuration, 59—67
water flow rate effects on, 62—63
water salinity and dissolved oxygen effect on, 63
water temperature effects on, 61—62
Bacterial monitoring
of commercial depuration plant in Maine, 181—186

of commercial oyster plant, 228, 230—232
Bacteriological analysis
in Denmark, 362—363
minimum shellfish sampling schedule for, 154
Bacteriological indicators, 82—83
Bed certification, 5
Belon type oyster, 341
Bench scale automated recirculating system, 195—202
Biological filter, 197, 198
Bivalve mollusk industry, 275—284
Bivalve mollusks
capacity of to sustain intense radiation exposure, 253—254
cleansing of contaminated, 7, see also Depuration
commercially important in UK, 275
depuration of, 369
in Italy, 370—371
and human enteric viruses, 71—104
detoxification of naturally contaminated with domoic acid, 239—243
exposure of to fecally contaminated waters, 71
gathering period for, 335
illness associated with, 277—279
major markets for from New Zealand, 325
production figures for in New Zealand, 326
production of in Spain, 331—332
Black mussels, harvested in Turkey, 365—366
Bluepoints Co Inc., 9
Breville storage and quarantine facilities, 355—358
Bromine disinfection, 351
Business incentives, 174

C

Calcium hypochlorite suspension, 7
Caliciviruses, 76
Campylobacter species
in molluscan-borne illnesses, 47, 49
jejuni, 47
water temperature effects on, 62
Canada, 271—274
Capital investment, 160
Carbamate derivatives, structures of, 218
Centers for Disease Control, foodborne outbreak surveillance data of, 53
Cesium-137 radionuclides, 247
Chemical tests, 153
Chemical toxins, in molluscan-borne illness, 51—53
Chlorinated water
depuration demonstration with, 7—8
holding shellfish in, 11
Chlorination, 351—352
benefits of, 8—9
compared with UV and ozone treatments, 133
effectiveness of, 147

method of, 146—147
Cholera
    in Italy, 369
    molluscan-borne outbreaks of, 49
"Claire" system, 287
Clam depuration plants, 9, 159—162
Clams
    annual depurated production of in Spain, 335
    consumption of, 3—4
    depuration of, 241, 156—162
    design of bench scale automated recirculating
        system for, 195—202
    early depuration work on in New Jersey,
        9—10
    exports of, 261—262
    global production of, 261
    harvested in Turkey, 366—367
    imports of, 262
    leading exporters of, 263
    leading importers of, 263
    in New Zealand, 325
    organoleptic properties of live irradiated,
        249—251
    post-irradiation survival of, 249
    produced in France, 343
    studies on microbiology, physiology, and epi-
        demiology in, 10—11
    U.S. imports of by MOU status per country,
        267, 269
    Vibrio uptake and retention by, 119—120
Clean Waters Act (Australia), 291, 293
Cleansing operations, economic success of, 213
Clostridium perfringens, elimination of, 65
Cobalt-60 radionuclides, 247
Cockles
    annual depurated production of in Spain, 335
    in Malaysia, 311
    pilot-scale depuration plant for, 312—313,
        314
    unloading site for in Malaysia, 314
Coliform bacteria, see also Fecal coliform
    and enteric viruses in shellfish, 83
    montitoring for, 4
Coliform most probable number (MPN) test,
        153, 155, 274
Coliphages, 83
Communicable Disease Surveillance Centre
        (UK), 277
Computer control system, 201
Conference for Food Protection, 20
Containerization, 213
Container racks, 210
Container-relaying
    cost considerations in, 210, 213
    low cost longline device for, 211
    of oysters, 205—213
    post-treatment harvesting in, 209
    process description of, 206
    process evaluation of, 209—213
    raft used for, 207
    site requirements for, 209

vessels for, 212
Containers, 208—210
Controlled purification, see Depuration
Coronaviruses, enteric, 76
Cote Bleue facility, 355, 356
Coxasackie viruses, 73—74
Crassostrea gigas
    ascending mid-gut of, 234
    depuration of, 227—236
    virus in cells lining artery of, 236
    virus in epithelium of digestive diverticula tu-
        bules of, 234
    virus in mid-gut mucus of, 235, 236
Cricket paralysis virus
    depuration rates of, 233
    in oysters, 228
    samples of during depuration, 231

D

Decisions of Courts in Cases Under the Federal
        Food and Drugs Act, 35
Denmark, 361—363
Depuration
    bacterial efficacy of, 65—67
    in Canada, 11, 271—274
    in clean seawater, 278—279
    criteria for in New Zealand, 326—327
    current U.S. commercial practices of, 25—28
    definition of, 7
    economic considerations for, 159—162
    effectiveness of, 71
    enhancing of, 145—147
    first attempt at, 8
    in Florida, 26—27
    history of commercial, 7—13
    by immersion in disinfected seawater,
        351—353
    industry perspective on, 3—5
    industry responsibilities in, 22
    in-situ farm-based systems of, 310—311
    introduction of in Australia, 290—296
    in Italy, 369—371
    laboratory support for, 22
    limits to efficacy of, 12, 145
    in Maine, 26
    major proponent of, 7
    in Massachusetts, 26
    methods of, 145—147
    in mollusk industry of New Zealand,
        323—328
    1989 status survey of, 12
    NSSP Manual of Operations for, 21
    operating procedures for in New Zealand,
        327—328
    for oysters, 163—175
    of Pacific oysters, 227—236
    practices of in Australia, 296—305
    process description of, 205—206
    purpose of, 129
    regulation of in Spain, 331—339
    regulatory perspective on, 19—23
    reliablity of process of, 134
    routine monitoring tests in, 151—156
    shore-based system of, 310

in Singapore, 310
in Spain, 332—339
state monitoring and certification for, 31—33
state responsibilities in, 22—23
in U.S., 26—27
in United Kingdom, 275—284
use of ultraviolet light in, 137—142
Vibrio species in, 115—124
virological aspects of, 103
virus elimination through, 95
Depuration plants, 129, 181—186
Canadian, 271—274
certification and approval of in Australia, 296
certification of in New Zealand, 327
cost of construction of, 170
in Denmark, 362—363
description of facilities of in France, 354—358
design of, 63—64, 101—102, 129—135
in Italy, 370, 371
layout of, 130
location and structure of in UK, 280—281
monitoring of, 33
in the Philippines, 318
reduced number of, 12
sanitation of, 67
seawater system requirements for, 130
sizing scaling of, 132—133
in Spain, 335
structural requirements for, 129
and tank requirements in, 130—132
Depuration processes
approved, 22
in France, 351—359
in New Zealand, 326—328
static versus flowing, 101
Depuration systems
balance sheets for, 173—174
capacity and rate in, 167
circulation rate in, 167—168
container design and cost in, 170
cost recovery in, 171
design of, 129—135
filling with seawater, 282
flow through system of, 316
income statement for, 173
influent treatment in, 168—169
laboratory and personnel costs in, 169—170
loading of, 282
in Malaysia, 312—313
materials for construction of, 280
monthly cash flow in, 175
pumps in, 169
recirculated system of, 316—317
recommended and experimental environmental parameters for, 118
site and system specifics in, 168
tank capacity and size in, 167
in Thailand, 316—317
in United Kingdom, 279—283
water distribution and measurement in, 169
Depuration technology, 60—65, 134—135

Depuration units, design and layout of, 133—134
Depuration water
laws regulating substances added to, 39—40
monitoring tests for, 152—153
regulatory requirements for substances added to, 35—40
substances added to, 36
Diarrheic shellfish poisoning, 51, 52
Disinfectants, 351
Disinfection experiments, 118
Disinfection methods, 351—353
Dissolved oxygen
effect of on bacteria, 63
increased seawater content of with ozone, 147
monitoring of levels of, 153
Domoic acid, 239—242

E

Echoviruses, 73, 74
Economic assessment, 159—162
Economic considerations
in container-relaying of oysters, 210—213
for hard clam depuration, 159—162
for oyster depuration, 163—175
Education, need for, 5
Effluent treatment, cost of, 169
ELISA immunological tests, PSP toxicity monitoring with, 222, 223
Enteric adenoviruses, 76
Enteric bacteria, depuration of, 115
Enteric coronaviruses, 76
Enteric viruses
chemical factors influencing, 78
contaminating shellfish habitats, 77—83
definition of, 72
and depuration of bivalve mollusks, 71—104
epidemiology of, 83—84
isolation of from shellfish, 79—81
occurrence of in shellfish and habitats, 78—82
and pathogenic bacteria in shellfish and water, 82—83
persistence of in shellfish habitats, 77—78
sources of, 77
types and transmission of, 71—74
Enteroviruses, 10, 73—74
Enzyme immunoassays, 77
Escherichia coli
depuration rates of, 233
water temperature effects on, 61—62
ETCPC method, 155
ETPC analysis method, 182—183, 185—186
European Economic Community, 284, 358—359
Export regulations, Australian, 290—291

F

Fair Packaging and Labeling Act, 19
FDA, see Food and Drug Administration
FDC, see Food, Drug, and Cosmetic Act
Fecal coliform

analysis of, 182, 183, 185—186

effect of commercial depuration on levels of, 230

and enteric viruses in shellfish, 92

final counts of resulting for depuration of *Mya arenaria*, 273

general zero hour counts for in *Mya arenaria*, 273

as indicator of virus levels, 232

levels of in Maine shellfish, 183, 184

levels of in oysters, 229

methods of detection of, 185

monitoring of in commercial oyster plant, 230

MPN of, 153, 155, 274

parameters of successful depuration of, 232

standards of and *Vibrios* presence, 115

UV light irradiation of, 121

Fecal coliform test, 155

Fecal contamination, 71

Fecal-oral transmission, 72

Federal Department of Environment, Conservation and Protection Service, Canada, 271

Federal Department of Fisheries and Oceans (Canada), 271, 272

Federal Fish Inspection Act and Regulations (Canada), 271

Federal guidelines, 20

Federal labs, revitalized depuration activity in, 10

Federal regulations, 19—20

Filter feeders, 115

Financial evaluation, of oyster flow-through system, 170—174

Fish box-type depuration system, 297—298

Fisheries Act (Malaysia), 311

Fisheries and Oyster Farms Act (Australia), 292, 294—295

Fisheries Research Institute, Penang, pilot scale depuration system of, 312—313

Florida, hard clam depuration plants of, 159—162

Florida Comprehensive Shellfish Control Code, 159

Flow-through technology, financial analysis of, 171—174

Fluorocarbon polymer encasements, 137

Food, Drug, and Cosmetic (FDC) Act, 19, 35—37

Food Act (Malaysia), 311

Food additives, 35—40

Food and Agriculture Organization data, 261

Food and Drug Administration (FDA) approval and evaluation of stage programs by, 31

approved certification criteria of, 22

controlled purification program of, 33

Interstate Certified Shellfish Shippers List of, 32

lab analysis requirements of, 152

letter from to Louisiana Department of Health and Hospitals on ozone in depuration water, 41—43

product safety standards of, 269

regulations of, 19—20, 35—40

requirements of, 22

Shellfish Memorandum of Understandings of, 266

Food processing, 247

Food Protection Unicode, 20

Food Service Sanitation Manual, 20

France

health standards in, 351—353

main shellfish culture sites in, 344

shellfish industry regulatory status and depuration techniques in, 341—359

G

Gamma rays, 247—248

Gastroenteritis

detection of viruses causing, 76—77

with diarrheic shellfish poisoning, 52

epidemiology of, 83, 84

FDA reports on, 53

molluscan-borne, 50

Norwalk, 75—76

oyster-associated in New South Wales, 290

rotavirus, 75

shellfish-associated, 85—86

small, round viruses causing, 76

viral agents of, 74—77

Generally recognized as Safe (GRAS) substances, 20, 22, 36—40

Geosmin removal unit, 199

Global production, 261—266

Good Manufacturing Practices Regulations (GMPRs), 19

GRAS, see Generally Recognized as Safe

Growing waters, clean, 213

H

HACCP, see Hazard Analysis Critical Control Point

Halifax Laboratory quarantine unit, 239

Harvest control, state responsibilities for, 32—33

HAV, see Hepatitis A Virus

Hazard Analysis Critical Control Point (HACCP) concept, 5, 283

Heat treatment, 278

Hepatitis, epidemiology of, 83

Hepatitis A

associated with shellfish consumption, 10

inactivation of by irradiation, 251—253

molluscan-borne, 50

Hepatitis A virus (HAV), 73

ionizing irradiation effects on, 255

reductions of, 71

Hepatitis E virus, 74

HEV, see Human Enteric Viruses

High energy electron beams, 247

High-performance liquid chromatography, PSP toxicity monitoring with, 219, 221—223

Horse mussels, harvested in Turkey, 366

Human enteric viruses (HEV)

and bivalve mollusk depuration, 71—104

types of and illnesses caused by, 72

Hydraulics, improper design of, 210

Hydrogen ion concentration, 153

Hypochlorite treatment, 279

I

IgE-mediated allergies, 53
Immunological techniques, PSP toxicity monitor-
ing with, 220—223
Imports, regulation of for shellfish in France,
349—351
Incentives, need for, 5
Indonesia, bivalve shellfish industry in, 315
Industry responsibilities, 22
Infectious-agents-etiology unknown, 51
Information, need for, 5
International Council on Exploitation of the Sea,
349
Interstate Certified Shellfish Shippers List, 20,
32, 35
Interstate Shellfish Sanitation Conference
(ISSC), 20—21, 31, 48
assistance of to FDA, 36
function of, 35
lab analysis methods of, 151
Ionizing irradiation, 247—248, 254—256
Irradiation, 247—256
Irradiation techniques, 248—256
Irradiator facilities, radiation sources in, 247
ISSC, see Interstate Shellfish Sanitation Confer-
ence
Italy, 369—371

K

Kelly-Purdy unit, 137, 282, 302—303

L

Laboratory analyses
costs of, 169—170
requirements for, 151—152
Laboratory scale bacterial depuration, 227—230
Laboratory support, 22
Limfjord Oyster Company, 361—362
Lipoprotein envelope, 72
Lobsters, domoic acid poisoning of, 240, 243
Long Island plants, cost of operation of, 9

M

MAFF, see Ministry of Agriculture, Fisheries and
Food
Malaysia, mollusk industry in, 311—315
Marriott Corporation, quality assurance in, 3—4
Memorandum of Agreement or Understanding
(MOU), of Canada, 271
Microbiological analyses, 153
question of appropriateness of, 12
requirements for, 22
Ministry of Agriculture, Fisheries and Food
(MAFF), U.K., 275, 278—279
Modular-palletized tray depuration system, 301
Molluscan-borne illnesses
allergic, 53

bacterial gents of, 47—49
chemical toxins in, 52—53
epidemiology of, 47—53
frequency of outbreaks of, 53
infectious agents-etiology unknown, 51
viral agents of, 49—51
Molluscan shellfish
anaerobic glycolysis in, 254
assessment of French national production of,
345
capacity of to sustain intense radiation exposure, 243—254
human enteric viruses and depuration of,
71—104
identification and certification requiremens for,
19
international trade in, 261—269
methods of analysis of paralytic shellfish toxins in, 217—223
neurological simplicity of, 254
production of in Spain, 331—332
public health problems of, 19
role of as disease vectors, 47
Mollusk industry
in Denmark, 361—363
in Indonesia, 315
in Malaysia, 311—315
in New Zealand, 323—328
in the Philippines, 317—319
in Singapore, 309—311
in South-East Asia, 308—309
in Spain, 331—339
in Thailand, 315—317
in Turkey, 365—368
Monitoring tests, 151—156
MOU, see Memorandum of Agreement or Understanding
Mouse bioassay technique, 218, 220—223
MPN procedure, 153, 155, 274
Mussels
annual depurated production of in Spain, 334
consumption of in France, 345
depuration of domoic acid from, 241—242
detoxification of, 239—243
domoic acid uptake by, 243
exports of, 262
farming of in France, 343
harvested in Turkey, 365—366
imports of, 262
leading exporters of, 264
leading importers of, 264
in New Zealand, 323, 325
production of, 262
Mya arenaria, bacterial inactivation in with irradiation, 251

N

National Fisheries Institute, survey on oyster
consumption by, 4
National Indicator Study, 12
National Marine Fisheries Service, 20

National Shellfish Sanitary Control Programme
(Malaysia), 312, 315
National Shellfish Sanitation Program (Canada),
48, 243, 271
National Shellfish Sanitation Program (NSSP),
19
  approved certification criteria of, 22
  certification system of, 25, 35
  in depuration monitoring, 151
  establishment of, 8
  guidelines of, 31, 32
    Manual of Operations of, 20, 25, 35—36
    opertional requirements of, 141
    Part II of, 22—23, 28, 129
    requirements in on substances added to
      depuration water, 40
  objectives of, 35
  quality assurance requirements of, 151
  requirements of , 305
    implementation of, 36
    for tank design, 130—132
  on shellfish sampling methods, 153—155
Nephelometric turbidity units (NTUs), 152
Neurotoxic shellfish poisoning, 51, 52
New England Fisheries Development Associa-
  tion, 5
New South Wales Pure Food Act, 305—306
New Zealand, depuration in mollusk industry in,
  323—328
Noma oysters plant, 328
Non-A, non-B hepatitis, enterically transmitted,
  74
Norwalk-like viruses, 49—50, 75—77
Norwalk virus
  gastroenteritis with, 75—76
  ionizing irradiation effects on, 255
  in molluscan-borne illnesses, 49—50
NSSP, see National Shellfish Sanitation Pro-
  gram
N-sulfocarbamoyl derivatives, structures of, 218
NTUs, see Nephelometric Tubidity Units

O

Onbottom relaying
  devices for, 211
  reduced losses with, 213
Oxidant demand free glassware, 117—118
Oxidants
  production of ozonated seawater, 122—124
  shellfish sensitivity to, 352
Oxygen availability
  as limiting factor, 145
  and virus elimination rates, 101
Oxygen enhancement effect, 253
Oyster-associated gastroenteritis, 50, 51
Oyster beds, in Denmark, 361
Oyster farmers, 287, 293—296
Oyster farming, in France, 341
Oyster fishermen, personalities of, 213
Oyster hatcheries, in New South Wales, 290

Oyster industry, 306—308, 361—363
Oyster rack, 207—208
Oysters, see also Sydney rock oyster
  in adequate product identification for in Aus-
    tralia, 294—296
  annual depurated production of in Spain, 335
  bacteria in feces of, 62
  in cholera outbreaks, 49
  consumption of, 3—4
  container-relaying of, 205—213
  cost of, 164—165, 172
    for depuration, 163—175
    for storage, 170
  depuration of, 165—170
    critical points in, 67
    of Vibrio species from, 115—124
  effect of density on bacteria, 63
  exports of, 266
  good physiology of, 213
  harvested in Turkey, 367
  imports of, 266
  lack of analytical facilities for in Australia, 292
  lack of microbiological data on, 291—292
  leading exporters of, 265
  leading improters of, 265
  market considerations for, 163—164
  marketing of in Australia, 289
  marketing success for, 165—166
  microbiological history of, 59—60
  in New Zealand, 324
  organoleptic properties of live irradiated,
    249—251
  Pacific, depuration of, 227—236
  in the Philippines, 318
  post-irradiation survival of, 249
  post-treatment harvesting of, 209
  production of, 262—266
  total production of in New South Wales, 289
  typhoid fever outbreaks associated with, 8,
    47—48
  U.S. imports of by MOU status per country,
    268, 269
  Vibrio uptake and retention by, 118—120
  Vibrio vulnificus in, 189—193
  viral location in tissue of, 234—236
  virological aspects of depuration and relaying
    of, 103
  virus in mid-gut mucus of, 235—236
Oyster vessel, 206
Ozone
  cost of, 279
  in depuration process water, 36, 41—43
  monitoring of levels of, 153
  and production of oxidants in seawater,
    122—124
  in seawater, 117
  utilization of, 4—5
Ozone disinfection, 133, 147, 351

P

Pacific oysters, depuration of, 227—236
Paralytic shellfish poisoning toxins, 51—52,
  217—223
Parvoviruses, 76

Philippine Human Resources Development Centre (PHRDC), 318
Philippines, bivalve depuration in, 317—319
Photoreactors, 140
PHRDC, see Philippine Human Resources Development
Phytoplankton, 116
Picornaviruses, 76
Plants, see Depuration plants
Plaque forming units, 252
Plesiomonas shigelloides, 47
Polioviruses, 73—74, 251—255
Pollution, need to eliminate, 5
Pool type depuration system, 300—305
Process verification, 134
Pseudomonas species, 61
Public health
    in Italy, 369—370, 371
    recommendations for in France, 359
    in Singapore, 309—310
    state responsibilities in protecting, 31—33
    in United Kingdom, 277
Public Health Pure Food Act (Australia), 292
Public Health Service, U.S., in bivalve depuration, 11

Q

Quahogs, 249—251
Quality assurance
    requirements for, 151
    state responsibilities for, 33
    in United Kingdom, 283
Quality assurance tests, 4—5
Quarantine storage, recommended treatment conditions for, 353
Quartz encasements, 137
Quartz tube UV photoreactor, 140, 142

R

Rack handling/transport vessel, 212
Radiation dosages, 247
Radiation machines, 248
Radiation sources, 247—248
Radioimmunoassays, solid-phase, 77
Radionuclides, 247
Ramsay unit, 300, 303
Rangia clams, design of bench scale automated recirculating system for, 195—202
Recirculating holding system, 130
    bench scale automated, 195—202
    computer control and monitoring of, 201
    design of for Rangia clams, 195—202
    problems in present design of, 202
Reclamation processes, 95
Regulatory practices
    in France, 345—351
    for imports, 349—351
    in Italy, 369—370, 371
    for molluscan depuration in Spain, 336—339

in United Kingdom, 277—279
Regulatory requirements
    cost of in hard clam depuration, 162
    cost of in oyster depuration, 164
    for substances added to depuration water, 35—40
Regulatory system, 19—23, 271
Relaying, 71, 145
    in clean seawater, 278
    in oyster cleansing, 163
    process description of, 205
    virological aspects of, 103
    virus elimination through, 95, 102
Relaying device, in container-relaying, 206
Reoviruses, 74
Restaurant Business, Inc., survey on oyster consumption in restaurants by, 4
Retail Food Store Sanitation Code, 20
Rock oysters, in New Zealand, 324
Rotaviruses, 75—76

S

Salinity, 63
    requirements for, 130
    and viral accumulation in shellfish, 94
    and virus elimination rates, 101
Salmonella
    charity, water temperature effects on, 62
    typhi, in molluscan-borne illness, 47
    typhosa, as indicator, 8
Saltonstall-Kennedy grants, 5
Sampling costs, 169—170
Sanitary surveys, need for, 8
Sanitation
    bacteriological standards for, 82—83
    irradiation technology in, 247—256
    need for control of, 11
    in New Zealand shellfish industry, 326
    radiation role in, 254—256
Saxitoxin, structures of, 218
Scallops, in New Zealand, 324
Scheduled controlled purification process (SCPP)
    adherence to, 155
    development of, 151
    sampling costs in, 169—170
SCOC, see Spinney Creek Oyster Company
SCPP, see Scheduled Controlled Purification Process
SEAFDEC, see South-East Asian Fisheries Development Centre
Seafish Industry Authority (SFIA), 278
Seawater, see also Waters
    chlorination of, 146—147
    disinfected, 351—353
    downstream or upstream treatment of, 354
    minimum requirements of, 130
    oxidant production in, 122—124
    ozonated, 117
    ozone concentration in, 122
    sterilization of, 282
    treatment of, 279, 353—354

Sewerage disposal systems, inadequacy of in Australia, 294
SFIA, see Seafish Industry Authority
Shellfish, see also specific types
  activity of and viral reduction, 99
  approved production zones for, 348
  capacity of to sustain intense radiation exposure, 253—254
  cost of storage of, 170
  current U.S. commercial depuration of, 25—28
  depuration of, 59—67
    enhancing of, 145—147
    FDA regulation of, 36
    in South-East Asia, 308—319
    status of in Australia, 287—308
    in Tasmania, 305—306
  eating of raw, 345
  elimination of viruses from in depuration and relaying, 94—102
  end-product standards for, 156
  enteric viruses isolated from, 79—81
  FDA regulation for production of, 35—36
  feeding and related activities of, 84
  finishing treatment for in clean seawater, 351
  fitness of production areas for in France, 345—349
  harvesting of in United Kingdom, 281
  hepatitis-A associated with, 10
  history of commercial depuration of, 7—13
  import circuits for in France, 350
  improvement in sanitary quality of, 8
  irradiation of, 248
  loading and packing of, 335—336
  main culture sites for in France, 344
  main production sites and trade-ways for in Europe, 346
  major areas of in New Zealand and United Kingdom, 324
  markets for in Spain, 336
  microbial ecology of, 66
  need for quality assurance testing on, 4—5
  non-approved but non-prohibited production zones for, 348—349
  organoleptic properties of live irradiated, 249—251
  perceptions on illness related to, 4
  potential difficulties in exporting of from Turkey, 368
  preservation of organoleptic qualities of in irradiation, 254
  producing areas for in Spain, 332
  production and values for in France, 343
  prohibited non-approved zones for, 349
  quality assurance of, 3
  rates and extent of virus elimination from, 95—99
  relationship between viruses in and bacteriological standards, 82—83
  removal of after depuration, 283
  sanitation practices for, 284, 326

species of harvested in Turkey, 365—367
studies on microbiology, physiology, and epidemiology in, 10—11
tonnage of depurated in Spain, 333—334
trade exchange for in France, 347
transportation of in United Kingdom, 281
user's perspective on, 3—5
variability among and viral accumulation in, 92
variation in viral reduction in, 99
Vibrio uptake and retention by, 118—120
Vibrio vulnificus occurrence in, 189—193
viruses accumulated by, 84—91
virus versus bacteria elimination by, 102—103
washing and culling of, 281
Shellfish-associated enteric viral disease, epidemiology of, 83—84
Shellfish habitats, 77—83
Shellfish industry
  Australian, 291—294
  evolution of in France, 342
  in France, 341—359
  New Zealand, 323
Shellfish poisoning, types of, 51—53
Shellfish research centers, establishment of, 10
Shellfish sampling, 153—155
Shellstock storage, cost of, 10
Shigella species, in molluscan-borne illness, 47
Singapore, mollusk industry in, 309—311
Siphon withdrawal reflex, 249
Snow Mountain agent, 51
Soft-shell clams
  early depuration studies of, 10
  post-irradiation survival of, 249
South-East Asia, mollusk industry in, 308—319
South-East Asian Fisheries Development Centre (SEAFDEC), 317
Spain, production and regulatory practice in molluscan depuration in, 331—339
Spinney Creek Oyster Company (SCOC), 181—186
Spud barge, 212
SSCA, see State Shellfish Control Authority
Standard Methods for the Examination of Water and Wastewater, 152
State, 20—22, 31—33
State health authorities, need for control by, 11
State Pollution Control Commission (Australia), 291—293
State Shellfish Control Authority (SSCA), 151, 154—155
STX-antibodies, 220—221
Sulfocarbamoyl derivatives, 220
Sump, design of, 192
Sydney rock oyster, 59—65, 288—290

T

Tanks, 131—132
  configuration of, 64
  design of, 195, 197—198, 210
  systems of in United Kingdom, 280
Tasmanian industry, 290
Thailand, 315—317

Tide farm plant, 328
Trading patterns, in major mollusks, 261—269
Tray type depuration system, 298—301
Tricarboxylic acid cycle, reversal of with irradiation, 254
Turbidity
    importance of, 169
    monitoring of, 152
    and viral accumulation in shellfish, 93
    and virus elimination rates, 100—101
Turkey
    mollusk industry in, 365—368
    shellfish species harvested in, 365—367
Typhoid fever infections
    demise of as serious public health threat, 10
    molluscan-borne, 47—48
    outbreaks of, 8
    reducing risk of, 11

U

Ultraviolet depuration plants, capital investment in, 160
Ultraviolet light irradiation, 61, 351
    bactericidal effects of, 115
    first use of, 137
    method of, 146
    objections to use of, 371
    in UK depuration systems, 279
    sterilizing ability of, 61
    use of in depuration, 137—142
    in *Vibrio* depuration, 116—117, 120—121
Ultraviolet reactor, 137—142, 298, 305
United Kingdom
    bivalve mollusk industry and depuration practices in, 275—284
    commercial depuration practices in, 276, 281—283
    depuration system design in, 279
        plant structure and location of, 280—281
        recent design criteria for, 279—280
        possible future trends in, 284
        public health concerns in, 277
        regulatory control procedures for shellfish consumption in, 277—279
        size and geographic distribution of mollusk industry in, 275—276
United Nations, Yearbook of Fishery Statistics of, 261
United States v. 408 Bushels of Oysters, 35
USPHS depuration demonstration, 7—8

V

Vibrio species
    cholerae, in molluscan-borne illness, 48—49
    in depuration, 115—124
    fate, uptake and retention of, 118—120
    found in UV treated seawater, 61
    ionizing irradiation effects on, 255
    in molluscan-borne illness, 47, 48—49

need to eliminate, 4—5
parahaemolyticus, 65—66, 228, 230—232
UV light irradiation of, 120—121
vulnificus
    illnesses caused by, 49, 189
    inactivation of, 122
    in New Hampshire oysters, 191, 193
    in New Hampshire waters, 192, 193
    occurrence of in water and shellfish from Maine and New Hampshire, 189—193
    ozone assisted depuration of, 122—124
    in undepurated and depurated shellfish, 191
Viral gastroenteritis, agents of, 74—77
Viruses
    accumulation of by shellfish
        factors influencing, 87—94
        feeding and related activities in, 84
        kinetics and extent of, 87
        mechanisms of, 84—87
    chemical factors influencing depuration of, 101
    composition of, 71—72
    concentration of in water, 92—93
    design factors influencing depuration of, 101—102
    elimination of compared with bacteria elimination, 102—103
    elimination of from shellfish, 94—102
    inactivation by irradiation, 251—253
    initial levels of contamination of and elimination rates, 100
    ionizing irradiation effects on, 255
    location of in oyster tissue, 228, 232
    mechanisms of elimination of from shellfish, 94
    in molluscan-borne illness, 48—51
    monitoring for in oysters, 228
    need to eliminate, 4—5
    need to understand uptake, persistence and elimination of, 104
    oyster uptake of, 228
    persistence of, 104
    physical factors influencing depuration of, 100—101
    physical inactivation of, 94
    transport and retention of, 104
    types of, 92, 99—100

W

Wastewater treatment plants, sewage discharge from, 181
Waterbottom, 209, 213
Water flow, 67, 92
    rates of, 62—63, 152—153
Water pH
    monitoring of, 153
    and viral accumulation in shellfish, 94
    and virus elimination rates, 101
Water quality
    and bacterial control, 67
    requirements for, 32
Waters, see also Salinity; Seawater; Turbidity

bacteriological studies of in Denmark,
362—363
comparison of disinfection systems for, 133
effect of salinity on bacteria, 63
pollution problem of, 8
suspended solids or turbidity in and viral accumulation in shellfish, 93
turbidity of and virus elimination rates,
100—101
Water temperature, 61
effects of, 61—62
as limiting factor, 145

monitoring of, 152
and viral accumulation in shellfish, 93
and virus elimination rates, 100
Waterways, classification of, 306
Wet storage, 27—28
vs. depuration, 25
FDA on ozone in water used for, 41—43
types of systems of, 27
water sources for, 28

X

X-rays, 247

Printed and bound by CPI Group (UK) Ltd, Croydon, CR0 4YY

22/10/2024

01777600-0010